The Story of Space Station Mir

David M Harland

The Story of Space Station Mir

Published in association with
Praxis Publishing
Chichester, UK

David M Harland
Space Historian
Kelvinbridge
Glasgow
UK

SPRINGER–PRAXIS BOOKS IN SPACE EXPLORATION
SUBJECT *ADVISORY EDITOR*: John Mason B.Sc., M.Sc., Ph.D.

ISBN 0-387-23011-4 Springer-Verlag Berlin Heidelberg New York

Springer-Verlag is a part of Springer Science + Business Media (*springeronline.com*)

Library of Congress Cataloging-in-Publication Data

Harland, David M. (David Michael), 1955–
    The story of Space Station Mir / David M. Harland.
      p. cm.
    Includes bibliographical references and index.
    ISBN 0-387-23011-4
    1. Mir (Space station)   I. Title.

TL797.H3724 2005
629.44$'$2$'$0947–dc22

2004058915

Apart from any fair dealing for the purposes of research or private study, or criticism or review, as permitted under the Copyright, Designs and Patents Act 1988, this publication may only be reproduced, stored or transmitted, in any form or by any means, with the prior permission in writing of the publishers, or in the case of reprographic reproduction in accordance with the terms of licences issued by the Copyright Licensing Agency. Enquiries concerning reproduction outside those terms should be sent to the publishers.

© Copyright, 2005 Praxis Publishing Ltd.

The use of general descriptive names, registered names, trademarks, etc. in this publication does not imply, even in the absence of a specific statement, that such names are exempt from the relevant protective laws and regulations and therefore free for general use.

Cover design: Jim Wilkie
Project Copy Editor: Alex Whyte
Typesetting: BookEns Ltd, Royston, Herts., UK

Printed in Germany on acid-free paper

Sergei Pavlovich Korolev
(1907–1966)

Other books by David M Harland

The Mir Space Station – a precursor to space colonisation
The Space Shuttle – rôles, missions and accomplishments
Exploring the Moon – the Apollo expeditions
Jupiter Odyssey – the story of NASA's Galileo mission
The Earth in Context – a guide to the Solar System
Mission to Saturn – Cassini and the Huygens probe
The Big Bang – a view from the 21st century
The Story of the Space Shuttle
How NASA learned to fly in space

with John E Catchpole
Creating the International Space Station

with Sy Liebergot
Apollo EECOM – Journey of a lifetime

with Ben Evans
NASA's Voyager Missions – exploring the outer Solar System and beyond

with Paolo Ulivi
Lunar Exploration – human pioneers and robotic surveyors

The dream of yesterday is the hope of today and the reality of tomorrow.
Robert Goddard, American rocket pioneer

The Earth is the cradle... but you cannot live in the cradle forever.
Konstantin Tsiolkovsky, Russian rocket pioneer

Mir [is] an awesome sight.
Bill Readdy, STS-79 commander

*Never in my wildest dreams
did I ever think that I would [be] on the Russian space station.*
Shannon Lucid, Mir research astronaut

*Space station Mir is an incredible laboratory and workshop...
I really enjoy it here!*
John Blaha, Mir research astronaut

*The Mir station has been on orbit for 10 years now;
that's an amazing feat in terms of duration in a harsh environment.*
Rick Searfoss, STS-76 pilot

You have to understand that initially we planned to fly [Mir] for only 3 years.
Yelena Kondakova, Mir flight engineer

Table of contents

List of illustrations xv
List of tables xxi
Author's preface xxiii
Acknowledgements xxv

1 Getting started 1
    Into the unknown 1
    Chelomei's plan 9
    The world's first space station 13
    Frustration 14
    An excellent start 15
    Abandoned station 20
    From bad to worse 21
    Filling the gap 23
    Reconnaissance from orbit 27
    More frustrations 30
    A glimpse of the future? 32
    An orbital laboratory 32
    First crew 36
    Abort! 38
    A second tour 40
    A testing time 43
    The military's turn 44
    A new camera 46
    Salyut 5 reprise – almost 47
    Another try 48

2 Routine operations 51
    Salyut 6 51

Table of contents

Frustration... 55
Second time lucky 56
Visitors!... 58
Resupply 62
A Czechoslovakian visitor 65
End of a marathon 67
A second long-duration mission................................ 68
A Polish visitor 70
A new furnace 71
External activity... 72
An East German visitor....................................... 73
Swapping ends... 74
Record breakers.. 76
Final resupply 76
Write-off?... 78
Repair crew 79
So near and yet so far 81
Working on.. 83
A radio telescope... 84
Half a year... 86
A new ferry.. 87
Yet another expedition.. 88
A Hungarian visitor... 89
Hotel Salyut!... 90
Completing a second six-month tour 93
Reprise?... 94
Gremlins 96
One final tour 96
A clear success... 99
Another Almaz?.. 102

3 **A step towards continuous occupancy** 105
Building on success 105
Salyut 7... 105
Commissioning crew 107
A Frenchman visits 108
Installing exposure cassettes................................... 111
Back to work.. 112
A woman visitor 113
The long haul 114
Trials and tribulations 116
The second expedition 118
'A slight leak'.. 120
Pad abort!... 121
Adding solar panels 121

	The third crew	122
	Engine repairs	124
	Crisis!	129
	An orbital handover at last	133

4	**A base block for modular construction**	141
	Mir	141
	An ambitious mission	143
	A new home	145
	An old home	148
	Another new ferry	150
	The wanderers return, briefly	150
	Intermission	152

5	**An astrophysics laboratory**	153
	Second start	153
	Kvant 1	154
	A new launcher	159
	The dorsal solar panel	159
	The long-delayed Syrian mission	160
	A long six months	162
	Handover	164
	A year in space?	166
	Bulgarian reflight	168
	Repairing the telescope	170
	Back to the routine	171
	An Afghan visitor	171
	Stranded in orbit?	173
	Back to repairing the telescope	174
	Going for the record	175
	A French visitor	176
	Frustrated hopes of expansion	179
	Another intermission	183

6	**A microgravity laboratory for hire**	185
	Third start	185
	Kvant 2	187
	Working outside	190
	A damaged ferry	194
	A new computer	195
	Kristall	196
	Loose insulation	198
	A maintenance mission	201
	A Japanese visitor	202
	Spacewalks and troubleshooting	203

Mir's first woman visitor 206
Construction work 207
A Kazakh and an Austrian visitor 209

7 Expansion or abandonment? 213
Upheaval 213
A visitor from unified Germany 215
A quiet tour 216
A French visitor 217
The VDU thruster 219
More advance bookings 220
End of a tour 220
Spaceport Mir! 221
The disabled crane 222
Another French visit! 223
Damage assessment 225
Poliakov tries again 227
Crisis averted 228
The European Space Agency's first visit 229

8 Shuttle–Mir 233
Close encounter 233
An American crew member 235
Spektr 237
A case of overcrowding 238
A short maintenance tour 244
Europe's long mission 245
The Docking Module 247
Back to the routine 249
NASA moves in 250
Priroda 251
Testing times 259
Crisis 264
Recovery 269

9 The final chapter 277
Mir after NASA 277
MirCorp 281
A fiery end 283
Space tourist 284

10 In retrospect 297
The incredible Soyuz 297
An evolutionary programme 299
Constructing the Mir complex 302

Crewing 305
Weightlessness 310
The frustrations 313
The work 315
The Buran shuttle 317
Mir's legacy 318

Appendices 321
Glossary 327
Reading list 411
Index 415

List of illustrations

Sergei Pavlovich Korolev	v
Vostok 1 lifts off with Yuri Gagarin	2
Preparing a Soyuz rocket	3
An early Soviet depiction of the Soyuz spacecraft	4
Andrian Nikolayev and Vitali Sevastyanov in the Soyuz trainer	5
Parachute descent	7
Soyuz 9 recovery	8
Vladimir Komarov, Boris Yegorov and Konstantin Feoktistov	8
Vitali Sevastyanov and Andrian Nikolayev after their mission	9
The Almaz reconnaissance platform	10
The TKS resupply craft	10
Alexei Yeliseyev and Yevgeni Khrunov spacewalk	12
Georgi Beregovoi	13
Nikolai Rukavishnikov, Vladimir Shatalov and Alexei Yeliseyev	14
A depiction of a Soyuz about to dock with Salyut 1	14
A cutaway model of Salyut 1	15
Viktor Patsayev, Georgi Dobrovolsky and Vladislav Volkov	16
The interior of Salyut 1	17
Funeral for Dobrovolsky, Patsayev and Volkov	19
Lost 'Salyut 2'	21
A Proton rocket with Salyut 2	23
Vasili Lazarev and Oleg Makarov	24
Pyotr Klimuk and Valentin Lebedev	25
Another lost Salyut	26
Soyuz 14 lifts off	27
Pavel Popovich and Yuri Artyukhin	28
A model of the Salyut 3 Almaz	29
The rear of Salyut 3	29
Lev Demin and Gennadi Sarafanov	31
A painting of Salyut 4	33
Alexei Gubarev and Georgi Grechko in the Salyut 4 trainer	36

List of illustrations

Soyuz 18 in the assembly building... 39
The docking collar of Salyut 4 ... 40
Pyotr Klimuk and Vitali Sevastyanov in the Salyut 4 trainer... 40
Pyotr Klimuk and Vitali Sevastyanov... 41
Vitali Zholobov and Boris Volynov... 45
Vladimir Aksyonov and Valeri Bykovsky... 47
The Soyuz descent module is designed to float... 47
Alexei Leonov with Viktor Gorbatko and Yuri Glazkov... 48
A painting of Salyut 6... 52
Valeri Ryumin and Vladimir Kovalyonok... 55
The Kaliningrad control room during the Soyuz 25 mission... 55
Georgi Grechko and Yuri Romanenko in the airlock... 56
Yuri Romanenko and Georgi Grechko... 58
Vladimir Dzhanibekov and Oleg Makarov... 59
Oleg Makarov and Vladimir Dzhanibekov... 59
Georgi Grechko, Vladimir Dzhanibekov and Oleg Makarov... 60
Georgi Grechko, Oleg Makarov, Vladimir Dzhanibekov and
  Yuri Romanenko... 61
Progress 1 approaches Salyut 6... 62
Vladimir Remek and Alexei Gubarev... 66
The Kaliningrad control room on the day of Soyuz 28's launch... 66
Vladimir Remek, Alexei Gubarev, Georgi Grechko and Yuri Romanenko... 67
The recovery of Georgi Grechko and Yuri Romanenko... 67
Georgi Grechko and Yuri Romanenko after their mission... 68
Vladimir Kovalyonok and Alexander Ivanchenkov... 69
Alexander Ivanchenkov and Vladimir Kovalyonok... 70
Miroslaw Hermaszewski and Pyotr Klimuk... 71
Vladimir Kovalyonok outside... 72
Alexander Ivanchenkov outside... 73
Valeri Bykovsky and Sigmund Jähn... 73
Salyut 6 with Soyuz 31 at the rear... 74
Sigmund Jähn signs his name on the Soyuz 29 descent capsule... 75
A view of Salyut 6... 75
Alexander Ivanchenkov taking pictures from Salyut 6... 77
Valeri Ryumin and Vladimir Lyakhov... 79
Vladimir Lyakhov and Valeri Ryumin in the Salyut 6 trainer... 80
Georgi Ivanov... 82
Valeri Ryumin releases the dish of the KRT-10 radio-telescope... 86
Valeri Ryumin and Vladimir Lyakhov in recliners... 87
Valeri Ryumin and Vladimir Lyakhov are interviewed... 87
Leonid Popov and Valeri Ryumin... 88
Valeri Kubasov and Bertalan Farkas... 90
Vladimir Aksyonov and Yuri Malyschev... 90
Viktor Gorbatko and Pham Tuan... 91

List of illustrations xvii

Valeri Ryumin, Arnaldo Tamayo Méndez, Vladimir Romanenko and
   Leonid Popov 92
Valeri Ryumin and Leonid Popov 94
Gennadi Strekalov, Oleg Makarov and Leonid Kizim 95
Vladimir Kovalyonok and Viktor Savinykh 96
Vladimir Kovalyonok and Viktor Savinykh with the shower unit 97
Jugderdemidiyin Gurragcha and Vladimir Dzhanibekov 97
A view of Salyut 6 with Soyuz-T 4 at the front port 98
Leonid Popov and Dumitru Prunariu 99
Alexander Ivanchenkov, Vladimir Dzhanibekov and Jean-Loup Chrétien 108
Salyut 7 with Soyuz-T 5 attached 109
Vladimir Dzhanibekov, Alexander Ivanchenkov and Jean-Loup Chrétien 110
Soyuz-T 6 descends by parachute 110
Valentin Lebedev outside 111
Anatoli Berezovoi outside 112
Leonid Popov, Svetlana Savitskaya and Alexander Serebrov 113
Anatoli Berezovoi, Valentin Lebedev and Svetlana Savitskaya 113
Valentin Lebedev and Svetlana Savitskaya 114
A painting of the Salyut 7–Cosmos 1443 complex 117
Alexander Serebrov, Vladimir Titov and Gennadi Strekalov 117
Salyut 7 with Cosmos 1443 118
Alexander Alexandrov in Cosmos 1443 119
Alexander Alexandrov and Vladimir Lyakhov 119
Vladimir Solovyov, Leonid Kizim and Oleg Atkov 122
Yuri Malyschev, Rakesh Sharma and Gennadi Strekalov 123
Leonid Kizim outside 124
Leonid Kizim with a spacewalker's toolkit 125
Igor Volk 126
Vladimir Solovyov, Vladimir Dzhanibekov, Svetlana Savitskaya and
   Leonid Kizim 127
Vladimir Dzhanibekov in the hatch 127
Svetlana Savitskaya tests a vacuum-welding tool 127
Soyuz-T 13 lifts off 130
Soyuz-T 13 approaches the 'dead' Salyut 7 131
Viktor Savinykh inside chilly Salyut 7 132
Viktor Savinykh and Vladimir Dzhanibekov inside chilly Salyut 7 .. 132
Vladimir Vasyutin, Georgi Grechko, Viktor Savinykh, Alexander Volkov
   and Vladimir Dzhanibekov 134
Salyut 7 with Soyuz-T 14 attached 135
The Mir base block undergoing preparation for launch 142
Vladimir Solovyov and Leonid Kizim 143
The Mir base block as seen from Soyuz-T 15 144
The Kaliningrad control room 146
Vladimir Solovyov and Leonid Kizim after landing 151
Alexander Laveikin 153

List of illustrations

Kvant 1 and its TKS tug. 155
Soyuz-TM 3 lifts off 161
Alexander Viktorenko, Alexander Alexandrov and Mohammed Faris 161
Vladimir Titov and Musa Manarov. 164
Alexander Alexandrov, Anatoli Levchenko, Vladimir Titov and
  Musa Manarov 164
Musa Manarov and Anatoli Levchenko 165
Alexander Alexandrov, Yuri Romanenko and Anatoli Levchenko. 166
Anatoli Solovyov, Viktor Savinykh and (Bulgarian) Alexander Alexandrov. . . 169
The Orlan extravehicular suit 171
Abdul Ahad Mohmand, Vladimir Lyakhov and Valeri Poliakov 172
Sergei Krikalev, Jean-Loup Chrétien and Alexander Volkov 176
Jean-Loup Chrétien. 176
How Jean-Loup Chrétien deployed the Era structure 177
Musa Manarov after 366 days in space 179
Valeri Poliakov, Alexander Volkov and Sergei Krikalev 183
Alexander Serebrov and Alexander Viktorenko. 186
Kvant 2 in place 188
Kvant 2's airlock. 189
The scan platform on the exterior of Kvant 2 189
Alexander Serebrov tests the YMK. 191
Soyuz-TM 9 lifts off 193
Anatoli Solovyov and Alexander Balandin 194
Kristall is added 197
Kvant 2's inner compartment 200
Toehiro Akiyama, Musa Manarov and Viktor Afanasayev 203
Toehiro Akiyama 204
Mir from Soyuz-TM 11. 205
Anatoli Artsebarski, Helen Sharman and Sergei Krikalev 206
Recovering Helen Sharman. 207
Takhtar Aubakirov, Alexander Volkov and Franz Viehböck. 209
Franz Viehböck and Anatoli Artsebarski. 210
Klaus-Dietrich Flade, Alexander Viktorenko and Alexander Kaleri. 215
Alexander Kaleri and Alexander Viktorenko work on a gyrodyne. 217
Michel Tognini 218
The Sofora girder with the VDU roll-control thruster block 219
The TORU system 222
Soyuz-TM 17 waits to dock 223
Jean-Pierre Haigneré. 224
Alexander Viktorenko, Yelena Kondakova and Ulf Merbold 230
Talget Musabayev, Ulf Merbold and Yelena Kondakova 230
Valeri Poliakov, Ulf Merbold, Yuri Malenchenko and Yelena Kondakova . . . 231
The APAS docking system 234
Mir as viewed by STS-63. 234
Gennadi Strekalov on the boom of Mir's crane. 236

Mir is ready to receive STS-71 239
Vladimir Dezhurov welcomes Robert Gibson 241
Norman Thagard and Gennadi Strekalov 241
Atlantis about to undock from Mir..................................... 243
Sergei Avdeyev, Yuri Gidzenko and Thomas Reiter 246
Sergei Avdeyev and Thomas Reiter outside............................ 246
The Docking Module is manoeuvred to mate with Kristall 248
Sergei Avdeyev uses the Tchibis leggings 250
Shannon Lucid in the Spektr module 251
Shannon Lucid, Yuri Usachev and Yuri Onufrienko in Priroda 253
Shannon Lucid exercising in the Mir base block 253
Yuri Onufrienko and Yuri Usachev film a Pepsi commercial.......... 254
Shannon Lucid views wheat in the Svet cultivator..................... 255
Claudie André-Deshays.. 256
John Blaha, Shannon Lucid, Valeri Korzun, Alexander Kaleri,
   Bill Readdy and Tom Akers 257
John Blaha in Priroda... 258
Mir with all four radial modules in place 259
Valeri Korzun, Brent Jett, John Blaha, Jerry Linenger and Mike Baker 260
Alexander Kaleri, Jerry Linenger, Valeri Korzun, Vasili Tsibliev,
   Reinhold Ewald and Alexander Lazutkin........................... 261
Reinhold Ewald at work 262
Vasili Tsibliev... 262
Jerry Linenger wears his Sokol suit 264
Michael Foale... 264
How Progress-M 34 collided with the Spektr module 266
Spektr's damaged solar panel 268
Pavel Vinogradov, David Wolf and Anatoli Solovyov.................. 271
Pavel Vinogradov and Anatoli Solovyov work on a Vozdukh 272
Pavel Vinogradov, David Wolf and Anatoli Solovyov celebrate christmas.... 273
Léopold Eyharts... 274
Andy Thomas... 275
Valeri Ryumin and Nikolai Budarin 275
A final view of Mir by STS-91 276
Yuri Baturin, Gennadi Padalka and Sergei Avdeyev 277
Jean-Pierre Haigneré, Viktor Afanasayev and Ivan Bella............... 279
Viktor Afanasayev and Sergei Avdeyev outside......................... 280
Alexander Kaleri and Sergei Zalyotin 282
Sergei Zalyotin.. 282
Mir burns up.. 284
Kaliningrad at the time of Mir's demise 284
Dennis Tito.. 285
Mir's legacy – the core of the International Space Station 319

List of tables

Salyut 1 docking operations 20
Salyut 1 crewing 21
Salyut 3 docking operations 32
Salyut 3 crewing 32
Salyut 4 docking operations 44
Salyut 4 crewing 44
Salyut 5 docking operations 49
Salyut 5 crewing 49
Salyut 6 docking operations 100
Salyut 6 crewing 101
Salyut 6 spacewalks 102
Salyut 7 docking operations 137
Salyut 7 crewing 138
Salyut 7 spacewalks 139
Mir docking operations 285
Mir crewing 289
Mir spacewalks 291
Space station occupancy 301
Space station launches and re-entries 305
International space endurance records 308

Author's preface

The human advance into space is akin to the first sea creatures venturing onto dry land, except that instead of adapting to the environment by genetic evolution we use technology to create a cosy life-sustaining bubble in space. A self-sustaining habitat for the large-scale exploration of the Solar System will not be feasible, however, until the basic technology has been developed; until the engineering knowledge required to sustain this has been learned; until a full regenerative atmospheric processor has been developed; until food can be grown in space; and until a full waste recycler has been developed. The Mir space station has played the key rôle in learning to sustain such environmental systems. It also facilitated a study of how the human organism adapts to the weightless state. Of course, a range of activities, including life sciences (both plants and animals), materials processing (semiconductor, metallic and biological) and observation (Earth, astronomical and space physics) were performed, but these were opportunistic exploitation of Mir's existence, rather than its *raison d'être*. Mir was often dismissed by some commentators as an 'ageing' space station, but its age indicated that it was *succeeding* in its primary mission. The first part of the complex was launched in 1986. After several false starts, it was continuously inhabited for a few days short of a decade, with one cosmonaut – a physician – spending 14 months onboard studying his own adaptation to weightlessness and, in the process, showing that a human being could live in that state for the length of time required to fly to the planet Mars. By the time Mir was de-orbited in 2001 the assembly of its successor, the International Space Station, was already underway, with modules incorporating systems developed for Mir at its core. In a sense, however, as this book will relate, Mir was the *first* international space station.

David M. Harland
Kelvinbridge, Glasgow
June 2004

Acknowledgements

I must acknowledge the sterling service provided by Neville Kidger, Craig Covault, Gordon Hooper, Peter Gualtieri, Tim Furniss and Roelof Schuiling over the years in print, and Robert Christie and Phillip Clark for their *Satellite Digest*. Most of the early pictures originate from the Soviet news agencies *Novosti* and *Tass* and more recent material derives from the Russian Space Agency, the Energiya Corporation, NASA, the European Space Agency and its affiliates, and the Tokyo Broadcasting System. I am grateful to Dietrich Haeseler for allowing me reproduce several of his pictures. In a few cases either the copyright owner was not evident, or a response could not be elicited, and I have included the picture for its historical significance. Finally, I would like to thank John Mason, Brian Harvey and Philip Harris for their helpful suggestions early on, Flo McGuire and Alex Williams for help later, Dave Woolard for a sequence of graphics, and of course Clive Horwood of Praxis for his enthusiastic support throughout.

NOTE

Unless otherwise stated, Moscow Time is used, but Moscow Summer Time (which is one hour ahead) has been used without specific mention from the final Sunday in March to the final Sunday in September. The cosmodrome and the recovery zone in Kazakhstan are two hours ahead of Moscow.

1

Getting started

INTO THE UNKNOWN

On 12 April 1961, after a 108-minute flight which took him all the way around the Earth, Yuri Gagarin landed in the Soviet Union to a hero's welcome. This historic test flight proved the Vostok spacecraft's basic systems, but how long could the human body spend in weightlessness? Several months later, Gherman Titov spent a day in space. He felt sick when he moved about in the capsule, but he suffered no lasting effects and was certainly not incapacitated, as some had predicted. In 1962 Andrian Nikolayev extended the record to four days, and in 1963 Valeri Bykovsky flew for five days. Was there no limit? In 1965, as America prepared to send men to the Moon, Frank Borman and James Lovell spent 14 days in Gemini 7 to verify that astronauts would be able to survive a trip to the Moon and back. Two weeks sitting in the cramped capsule, comparable to the front seats of a Volkswagen car, had been no fun, however, and theirs really had been an endurance mission.

The Gemini record still stood in 1969, when the Soviets switched the emphasis of their space programme from the Moon to the establishment of a space station in Earth orbit. Being able to survive a fortnight in space was sufficient for expeditions to the Moon, but it represented only the first step towards *living* in space. In 1970, with the launch of their first station still a year away, the Soviets decided to attempt a record-breaking mission in the Soyuz spacecraft to demonstrate that cosmonauts would be able to serve the planned tour of duty. Andrian Nikolayev, now head of the cosmonaut corps, assigned himself as commander and Vitali Sevastyanov, a new member of the corps who was known to television audiences as the presenter of a programme that explained scientific discoveries for a general audience, as his flight engineer.

1 June had been a blisteringly hot day at the launch site on the open desert

2 Getting started

plain, but a welcome chill had followed the setting of the Sun. It was already dark when, two hours before launch, the two men rode the elevator up the side of the service structure and settled into the capsule. The last time Nikolayev had flown, he had worn a bulky pressure suit. The Soyuz supported a 'shirt-sleeved' environment, so this time he and Sevastyanov wore lightweight track suits. The only indication that they were about to make a flight was that they wore communications headsets.

Vostok 1 lifts off with Yuri Gagarin.

The preparation of the rocket was controlled from the nearby blockhouse, so there was little for the cosmonauts to do. In fact, the R-7, or *Semyorka*, designed by Sergei Korolev in the 1950s, was the world's first intercontinental ballistic missile. It had been developed to send a nuclear warhead on a ballistic arc. It was a 'parallel-stage' rocket, due to the fact that the technology to air-start an in-tandem stage had not been available when it had been designed. Its four conical strap-on boosters were to fire together with the core for lift off. It had been used to launch Sputnik and later, with an upper stage, Vostok. The more powerful upper stage it now incorporated could place the 7.5-tonne Soyuz spacecraft into orbit. And it was the same launch pad too – a concrete pier overhanging the edge of a vast pear-shaped flame pit.

With 30 minutes to go, the twin sections of the service structure split the eight levels of wrap-around walkway and then swung down to expose the slender rocket. Ten minutes later, the topping-off of the liquid oxygen tanks ceased, the kerosene tanks were pressurised, and nitrogen gas was pumped through the propellant lines to clear them and to blow off the covers of the 20 nozzles. With only 40 seconds to go, the rocket was switched to internal power. Twenty seconds later, the umbilical arm was disconnected and swung down. A turbopump in each of the five segments began to feed fuel and oxidiser into the four reaction chambers in the segment. After a final check, pyrotechnic charges were simultaneously fired. As the engines slowly built up their thrust, the flame belched into the pit beneath. The rocket was not actually supported at its base; the core was held up by four arms just above the top of the strap-ons. As soon as the thrust overcame its 450-tonne mass, the rocket started to climb. This released the supporting arms, which immediately swung out like the petals of a flower to clear the way for the protruding strap-ons. One way or another, Nikolayev and Sevastyanov were now committed.

At the moment of launch, the spacecraft activated its flight sequencer. As the rocket pitched to a 51.6-degree inclination, the globe of the tiny moving map display on the spacecraft's control panel showed its progress along

A Soyuz rocket is transported by rail to the launch pad, and installed in the support structure over the pit excavated into the Kazakh steppe.

4 Getting started

its northeasterly ground track, and as it rose through the atmosphere it flew almost directly over the town of Baikonur.[1]

At 115 seconds (at 46 kilometres altitude) the ring of solid-rocket motors at the top of the 6-metre-tall escape tower fired to pull it free of the shroud; three seconds later the 20-metre-long strap-ons were jettisoned to fall onto the Kyzyl-Kum desert 400 kilometres downrange, which was littered with spent rocket debris. The core kept thrusting. Now that half of its propellant had been consumed, the acceleration built up rapidly. At 165 seconds (at an altitude of 80 kilometres, at the official edge of space) the aerodynamic shroud that had protected the spacecraft was jettisoned; it fell to Earth 525 kilometres downrange. At 288 seconds (at an altitude of 175 kilometres) the 30-metre-long core shut down and was jettisoned; it re-entered the atmosphere 1,500 kilometres downrange and burned up. The four-chambered engine of the 8-metre-long upper stage started immediately, and by the time it had built up to its full 35 tonnes of thrust the shallow interstage truss had been jettisoned too. At an altitude of about 175 kilometres, as the upper stage continued to accelerate to a peak of 3.5 g, it pitched over and accelerated to a horizontal speed of 7.8 kilometres per second to enter low orbit. At 530 seconds, by now far north of China, the upper stage shut down and was jettisoned. In this 175- by 225-kilometre orbit, the air drag was sufficient to cause such a large lightweight object to re-enter the atmosphere in a few hours, and so the spacecraft – now officially designated Soyuz 9 – manoeuvred to lift its perigee to avoid this fate. As soon as it was evident that the spacecraft was in good condition, a report was issued that its objective was to investigate human adaptation to weightlessness on a long-duration mission. An earlier report had said that Soyuz's life-support system could sustain a 30-day flight; this was presumed to be with a single cosmonaut, so it seemed reasonable to expect Soyuz 9 to attempt to claim the record.

Despite the late launch, and the time spent over the next few orbits checking out the systems, the cosmonauts were to operate a daily schedule synchronised with the control centre at Yevpatoria in the Crimea. The tracking network of ground stations and ships enabled them to maintain communications for lengthy periods during favourable passes, and they slept for eight hours per day during the period when the ground track was unfavourable. Though spacious in

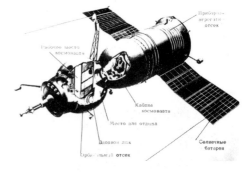

An early Soviet depiction of the Soyuz spacecraft.

---

[1] The launch site in the Soviet Central Asian Republic of Kazakhstan was built at the Tyuratam rail junction on the barren steppe. When Sputnik was launched, its place of origin was announced to be Baikonur, which, although 350 kilometres to the northeast, was the nearest large population centre on the ground track. Later, the 'closed' town of Leninsk was built at the base to house the growing rocket community.

comparison to earlier spacecraft, the Soyuz was rather cramped. Its descent module barely accommodated the couches, but the spheroidal orbital module provided room to stretch, and the crew spent most of their time in it. On the first night, in sleeping bags set across storage lockers, they had some difficulty adapting to the absence of gravity, but they slept well during the rest of the flight.

Nikolayev and Sevastyanov were to follow a strict exercise regime, perform a range of biomedical tests, carry out some scientific experiments and make Earth observations. They were to test a variety of exercise apparatus which, it was hoped, would tone up their cardiovascular system. It was also hoped to counter the muscle atrophy and avoid the significant loss of bone calcium that had been observed in dogs following 22 days in a Vostok capsule flown in the guise of Cosmos 110, a few years earlier.

The orbital module had been equipped as a gymnasium. A special 'load-suit' was fastened to the floor by a pair of expanders which created a load of 20–40 kilograms in order to simulate the force of gravity. It exercised specific muscles by transmitting skeletal loads while walking, running and jumping in place against its action. A chest expander was also used to stress the arm and chest muscles with a 10-kilogram load. If these proved effective, they were to be used onboard the orbital station. The state of health of the cosmonauts was assessed by measuring heart and respiration rates, measuring arterial pressure, testing the ability of the vestibular system in the inner ear to maintain a sense of balance in weightlessness, and testing the sensitivity of the eye to different levels of illumination and contrast. This followed up on cosmonauts' reports on earlier flights. Typical symptoms were a sensation of falling backwards, blood rushing to the head, a stuffy nose, a puffy face, decreased appetite and motion sickness. While most of these symptoms disappeared after a few days, the sensation of excess blood in the head lasted somewhat longer, and returned during periods of strenuous exercise. Visual acuity tended to degrade too, making it difficult to resolve extremely fine detail at long range, but this was believed to be reversible. The orbital module was also fitted with a small electric oven to prepare hot food and drink, and a foldaway toilet. The cosmonauts had two sets of towels (one damp and the other dry) with which to sponge themselves down, but they were restricted to ablutions once per day, and changed their underwear weekly. In addition to a traditional wet shaving kit, they evaluated an electric razor and a hand-held vacuum cleaner to clear the air of the particulates which could result in choking or blindness. Compared with Borman and Lovell's Gemini ordeal, life on the two-compartment Soyuz was sheer luxury. One of Sevastyanov's immediate recommendations was that crews assigned to long flights be issued a stock of thick socks (as the altered distribution of blood induced

Andrian Nikolayev and Vitali Sevastyanov (rear) in the Soyuz trainer.

by weightlessness caused the extremities to grow cold) and a darning needle and some wool to repair the holes made by knocking their toes against the walls as they moved about inside the spacecraft.

The primary scientific research involved taking multispectral pictures to develop a technique to interpret imagery from future missions which would permit variations in the moisture content of glaciers to be measured, the stocks of plankton in the sea to be assessed, and different rock and soil types to be determined in order to help in assessing the resources of the more remote regions of the Soviet Union.[2] Another device was an RSS hand-held spectrometer which, by measuring reflectance spectra, could distinguish soils with different moisture content.[3] Its results were to be used to map salt deposits of Kara-Bogaz-Gol Bay in Turkmenistan, adjacent to the Caspian Sea.

As the cosmonauts prepared to return to Earth, some doctors feared that they would black out from orthostatic intolerance resulting from the deceleration loading imposed by re-entry, despite an hour's strenuous exercise twice per day to maintain a high level of fitness; some even warned that they might die from the shock after *so long* in weightlessness. The real test of the mission, therefore, would come in its final phase.

On their final orbit, the cosmonauts ran through the de-orbit checklist and aligned the spacecraft with its main engine facing in the direction of motion, to serve as a brake. The de-orbit manoeuvre was performed while passing on a northeasterly track across the South Atlantic, towards Africa. This slowed it down by about 150 metres per second, which was just sufficient to dip its trajectory into the atmosphere while low over Soviet territory. (The recoveries of Soviet spacecraft were governed by several parameters: the de-orbit manoeuvre had to be in the southern hemisphere and in sunlight to enable the crew to visually confirm that the ship was properly aligned, and the landing had to be in daylight to enable the recovery team to visually find the capsule on the Kazakh steppe. If it was early morning over the Atlantic, it was late afternoon in Kazakhstan. The flight path intersected the recovery zone on only one or two orbits per day. And finally, of course, the spacecraft had to be at the proper place at the proper time to make the de-orbit manoeuvre. All of these factors led to long flights being recovered shortly before local sunset.) About half an hour after the de-orbit manoeuvre, explosive bolts were detonated to release the orbital and service modules. Four minutes later, passing 100 kilometres above the Horn of Africa, the descent module began to feel significant atmospheric drag. Although it was moving at 7 kilometres per second and entering the atmosphere at a depressed angle of about 30 degrees, the deceleration developed slowly. As soon as the cosmonauts noticed themselves being eased into their couches (at about 0.1 *g*), they settled as best they could and tightened their straps.

---

[2] Despite the fact that the Siberian wilderness supplied 10 per cent of the gas, 25 per cent of the oil and 30 per cent of the coal for the Soviet economy, it had barely been tapped.
[3] Acronyms such as RSS are explained in the glossary.

If the capsule had been spherical (as had the Vostok) it would have followed a steep ballistic trajectory that would have imposed a peak deceleration load of 10 g, but the Soyuz descent module was bell-shaped and it re-entered with its blunt base facing forward. The automated control system exploited the fact that the capsule had an offset centre of mass with a lift to drag ratio of 0.28, and selectively fired small hydrogen peroxide thrusters to control its roll, pitch and yaw in order to vary the lift vector.[4] It shallowed to fly almost horizontally along a narrow corridor in the upper atmosphere in order both to extend its trajectory by 450 kilometres and to refine its lateral motion to aim for a specific landing site. In addition to refining the trajectory, this dynamic re-entry reduced the g-load. Flying heads-up imposed a peak of only 3 g, and even when the capsule rolled into a maximum 60-degree bank to one side or the other to refine its trajectory the load rose to only 4 g. Nevertheless, after being weightless for so long, it felt much worse to the two occupants.

As the capsule dug deeper into the atmosphere, the air was unable to move out of its way rapidly enough, so it created a shock wave which stood off in front of the blunt base. This extremely hot dense plasma wrapped itself around the capsule, but did not actually come into contact with it (if it had, it would have seared through the aluminium wall of the capsule and incinerated its interior, as had occurred in early tests). The heat shield around the base was a 2-centimetre-thick titanium honeycomb with a fibrous asbestos binder. Since this was an ablative material, the outer surface blistered and peeled off as it was heated to a peak of 3,000°C, thereby protecting the cabin, whose the internal temperature did not exceed 28°C. Radio communication with Yevpatoria was blocked for several minutes by the free electrons in the ionised air surrounding the capsule, so the cosmonauts were out of contact during this most dangerous part of the descent.

Parachute descent.

An external pressure sensor detected the capsule's descent through 10 kilometres altitude, and explosively ejected a small drogue parachute. While this did not greatly diminish the 850 kilometres per hour rate of descent, it violently wrenched the capsule upright. A few seconds later, a larger drogue was ejected, which in turn drew out the main parachute, which had a catchment of 1,000 square metres.[5] The crew

---

[4] It had a pair of nozzles set 60 degrees apart near the upper rim to control the roll rate, two set between these to control the pitch rate, and two near the broad base to control the yaw rate.

[5] It was at this point during the 1967 mission of Soyuz 1 that a tangled chute condemned Vladimir Komarov to death.

8 **Getting started**

Soyuz 9 recovery.

sank deeper into their couches as the sink rate was instantly cut from 35 metres per second to 8 metres per second. A radio beacon was automatically activated to help the recovery force track its descent. Nitrogen was pumped into the shock absorbers in the couch supports, raising them by about 10 centimetres, and at an altitude of 5 kilometres a tiny valve opened to equalise the internal and external pressures. At an altitude of 3 kilometres, the base of the heat shield was jettisoned and the remaining hydrogen peroxide in the thrusters was vented. Some 52 minutes after initiating the de-orbit burn, an indicator on the control panel illuminated to warn the crew to brace for impact. When 2 metres off the ground, two pairs of solid-propellant rockets on the newly-exposed underside of the capsule fired for a fraction of a second to slow it. The shock of the 3-tonne capsule striking the ground at 2 metres per second was absorbed by the couches. Since the hatch was in the narrow neck of the capsule, if it remained upright the crew usually waited to be hauled out by the recovery team, but if it rolled over onto its side they would swing in the circular hatch and scramble out by themselves – or at least they had after flights lasting only a few days.

The trajectory had been so accurate that the heliborne recovery team watched the capsule descend near the town of Arkalyk. In this case, it settled upright. Although Nikolayev and Sevastyanov were exuberant, their muscles were so overwhelmed by 1 g that they required to be lifted out of the hatch. Evidently, fears that space-

Vladimir Komarov (left), Boris Yegorov and Konstantin Feoktistov.

farers would not survive the descent were groundless, but the men were clearly in poor shape. As their bodies recovered, the doctors confirmed that neither had suffered any irreversible effects. Dr Boris Yegorov, who had flown in space for 24 hours on Voskhod 1 in 1964, wrote in *Izvestia* the following week that, with proper

Chelomei's plan

Vitali Sevastyanov (left) and Andrian Nikolayev after their mission.

exercise, cosmonauts should be able to stay in space for at least a month. In fact, this 18-day flight had shown that the problem of extended flight was less to do with survival in weightlessness than later readaptation to Earth's gravity. As longer flights would show, although the body did change in adjusting to weightlessness, this eased off after three to four weeks, then more or less stabilised. Belatedly, it was realised that Nikolayev and Sevastyanov had flown just long enough for their bodies to adapt. Ironically, therefore, longer-duration crews would return in considerably *better* condition than had this intrepid pair. In recognition of their contribution to the effects of the space environment on the human body, Nikolayev and Sevastyanov were issued the 1970 Award of the International Academy of Astronautics. Writing in *Pravda*, Professor Vasili Pavlov said that the condition of the cosmonauts would enable the programme to pursue "flights of increasing duration". Professor Mstislav Keldysh, the president of the Academy of Sciences, added that within a few years it was intended to establish an orbital station operated "in the interests of the national economy", and that it would be manned by a succession of crews.

CHELOMEI'S PLAN

While Sergei Korolev pursued a lunar programme designed to compete with NASA, which had eagerly accepted John Kennedy's challenge to be the first to land a man on the Moon, Vladimir Chelomei, Korolev's long-time rival in Soviet rocketry, had gained support from the military to build an orbiting reconnaissance platform named Almaz (Diamond) that would be launched on his recently developed Proton rocket, which had a 20-tonne capacity. The platform was a single pressurised unit having a stepped-cylinder configuration. Its main compartment was 4.15 metres in diameter, and the 3-metre-diameter forward part was mated with a conical capsule that retained the lower part of its launch escape tower on the front in orbit, as this

10 **Getting started**

A model of Vladimir Chelomei's Almaz reconnaissance platform, with the conical crew capsule and launch escape system on its nose. Courtesy of Dietrich Haeseler.

contained the re-entry control system and parachute. The crew of three were to enter the capsule by a hatch in the side, but in orbit they were to open a hatch in the base (in the heat shield) and pass through a tunnel to another hatch into the module behind. To serve as a reconnaissance platform, Almaz was to have an ocean-surveillance radar and a high-resolution optical imaging system which developed its own film and fed it through a scanner for real-time viewing of the imagery by the crew and periodic transmission to Earth. A pair of large solar panels generated the power for this equipment and the life-support systems. The control panel divided the pressurised compartment into two 'rooms', with the small wardroom being in the narrow section. The wide section was dominated by the enormous conical bulk of the long-focal-length folded-optics camera system with its large film canisters. The platform had to be sufficiently well stocked to support a crew for at least a month and, in addition, provision was made to resupply it because such an expensive facility could not simply be used once and discarded.

In addition to the hatch to the spacecraft on its nose, the Almaz had a docking port at its rear, and Chelomei designed a cargo ferry that was nearly as large as the platform itself. Since the technology for automated rendezvous and docking had not yet been developed, this ferry was to be flown by a crew, and a succession of crews

A model of Vladimir Chelomei's TKS resupply craft, with the conical crew capsule and launch escape system on its nose. Notice the cosmonaut (arrowed) observing through a small porthole in preparation for docking. Courtesy of Dietrich Haeseler.

were to operate the platform without interruption. The ferry was referred to simply as the TKS – this being the acronym for the Russian words describing its function as a combined crew transport and logistical resupply ship. It would be launched on a Proton. In orbit, the crew would transfer to the main compartment, and because the docking system was on the far end they would fly the ship 'backwards'. During the final approach to the Almaz they would observe through a porthole adjacent to the docking unit, and once the vehicles were joined together the two docking assemblies would be disengaged and swung aside to expose a hermetically sealed tunnel. After the handover, the retiring crew would return to Earth in the ferry's capsule, leaving the main compartment of the TKS in place as a warehouse. When its cargo had been consumed, the TKS would be jettisoned. This would continue until the programme was finished, at which time the final crew would power down the Almaz, retreat to its capsule and return to Earth. The Almaz/TKS concept offered great potential,[6] but since Chelomei had no experience in designing spacecraft the development soon fell behind schedule and the date set for the test flight – 1968 – came and went.

Meanwhile, Korolev's Soyuz spacecraft suffered a series of failures culminating in the loss of its first pilot, Vladimir Komarov, in April 1967. An ambitious test involving the docking of two spacecraft (so that two cosmonauts could transfer from one vehicle to another in space) was cancelled when Soyuz 1 developed an attitude control problem and a jammed solar panel limited its power. After a successful re-entry, Komarov died when the parachute failed to deploy properly, and the capsule smashed into the ground. Korolev had not lived to witness this disaster, having died a year earlier of complications following an operation to remove a colonic tumour.

The great dream of landing a man on the Moon in 1967 to mark the fiftieth anniversary of the Revolution, was evidently impossible, so the target date was pushed back first to 1968, then to 1969. Although in January 1969 Soyuz 4 and Soyuz 5 docked and a pair of cosmonauts spacewalked from one vehicle to the other to test the procedure to be employed in transferring to and from the vehicle that was to make a lunar landing, the lunar programme suffered a setback the following month when, during the first test, the huge new rocket (the N-1) exploded a minute into its flight. The second flight, in early July, a mere two weeks before Apollo 11 was due to attempt the first manned lunar landing, fared even worse: the rocket broke up and toppled back onto the pad, obliterating the entire facility. With the lunar programme stalled, it was decided to switch the emphasis to a space station in Earth orbit. With the Proton rocket yet to demonstrate the reliability needed to trust it to carry a crew, and Chelomei's spacecraft still to fly its initial re-entry trials, the rival bureaux were told to work together to use the Soyuz launched on the trusty *Semyorka* to service the Almaz. In its new configuration, the space station would be launched unmanned, and its crew would follow a few days later in a Soyuz ferry.

Chelomei had intended to employ a pair of engines, mounted one on each side of

---

[6] Indeed, it was a brilliant design, considerably better than the Manned Orbiting Laboratory of the US Air Force, which was cancelled in 1969 shortly before its first test flight.

12 Getting started

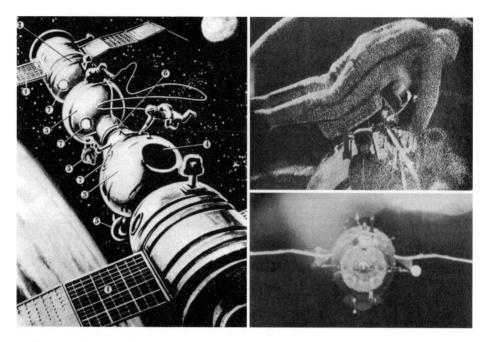

Alexei Yeliseyev and Yevgeni Khrunov spacewalked from Soyuz 5 to Soyuz 4.
Annotated graphic courtesy of *Aviatsiya I. Kosmonavitka*.

the axial docking port. The Korolev Bureau, now run by Vasili Mishin, its founder's former deputy, replaced this engine block with a Soyuz propulsion module. Because this prevented rear access, the capsule on the front of the platform was deleted, a short transfer compartment mounted at the front, and the docking port relocated to that end. This new compartment would provide access to the inner hatch which had previously connected to Chelomei's capsule. Rather than risk unproven solar panel deployment frames, a pair of Soyuz panels were mounted at each end, two on the propulsion system and two on the new docking compartment. To ensure that there was the minimum of delay in effecting the redesign, the fabrication work was done on the same assembly line at the Khrunichev factory as had been established for the Almaz, and the stepped-cylinder configuration was retained so as to be able to mate with the existing attachment ring on the Proton rocket. In this new configuration, the platform was a 14.6-metre-long craft with a 2.2-metre-diameter propulsion module at the rear of the main body (the 4.15-metre and 3-metre-diameter stepped-cylinder) with a 2.2-metre-diameter docking compartment on the front. The energy in sunlight at the Earth's distance from the Sun is about 1 kW per square metre, and each pair of Soyuz panels exposed an area of 12 square metres, but as the transducers were only 10 per cent efficient they generated a peak of 1.2 kW (falling off sinusoidally with divergence from perpendicular illumination). Unfortunately, much of this power was lost in the low-voltage (28 volt) direct-current distribution lines, and even with the contribution from the panels on the docked ferry the 3.6 kW peak would be barely sufficient to run the environmental systems required to support a crew, let alone run other apparatus.

By employing proven systems, this hybrid would become available much earlier, and the plan was to launch the first vehicle within two years. Chelomei, meanwhile, continued work on the reconnaissance platform, to be serviced by a Soyuz, and went on to complete the development of the TKS, which later played a central rôle in the development of the space station programme.

THE WORLD'S FIRST SPACE STATION

In early 1971, in expectation of launching its first orbital station, the Soviet Union began to hype the benefits. Speaking on *Radio Moscow* on 9 April, Academician Boris Petrov, the chairman of the Intercosmos Council of the Academy of Sciences, said that small orbital platforms that would be manned by several specialists would appear in the near future. He said that these would operate for periods of between one month and one year, and would later be superseded by large laboratories which would be assembled in orbit and remain in use for several years. On Cosmonaut Day, 12 April, he emphasised the importance of establishing an orbital station for sustained operations, and claimed that "all of our manned flights ... are aimed at the ultimate achievement of this goal".

The next day, a Prague news agency reported Anatoli Filipchenko, who had flown Soyuz 7, as saying that the launch of an orbital station was imminent and that it would be visited by successive crews. Andrian Nikolayev said that an orderly programme had been devised that would lead to more sophisticated orbital stations. Georgi Beregovoi, who had flown Soyuz 3, predicted that by the end of the decade the operation of an orbital station would be a matter of routine. On 19 April a Proton rocket put Salyut 1 into orbit.[7] It was put into a slow rotation for stability, with its solar panels facing the Sun so as to keep its batteries charged.

Georgi Beregovoi.

---

[7] Many years later, pictures were released showing Salyut 1 being prepared for launch. One of these showed the vehicle with the name Zarya, the original choice, but because this was the radio call sign of the Yevpatoria communications site the world's first space station was hastily renamed Salyut as a salute to Yuri Gagarin's pioneering flight a decade earlier.

FRUSTRATION

Soyuz 10, launched on 23 April 1971, carried Vladimir Shatalov, Alexei Yeliseyev and Nikolai Rukavishnikov, the latter, on his first flight, being a member of the team that had designed the station. After being checked out during its initial orbit, Soyuz 10 made the first of a series of engine firings designed to rendezvous

Nikolai Rukavishnikov (left), Vladimir Shatalov and Alexei Yeliseyev.

with Salyut 1. At each stage, radars in a network spread across the Soviet Union determined its orbit, and Yevpatoria computed the parameters for the next manoeuvre. On its 22nd orbit, the ferry's trajectory brought it within a few kilometres of its target, and Shatalov took control. Meanwhile, Salyut 1 had been commanded to orient itself to maintain the radar transponder on its nose facing the approaching ferry. Shatalov had docked Soyuz 4 with Soyuz 5 in 1969. Compared with approaching another Soyuz, he said approaching the 18-tonne station end on was "like a train entering a terminus", and more demanding of the pilot than approaching another Soyuz. The slow rendezvous had resulted in the final approach being made whilst making an optimum pass over the radar tracking network. As the bulk of the station expanded to fill the viewfinder, a television camera transmitted pictures of the docking unit on its end. Although the atmosphere was tense, the docking was achieved without a hitch. The probe on the nose of the ferry slipped into the conical drogue, and the capture latches engaged to hold it in a 'soft-docked' state. The probe was then retracted to draw the 1.5-metre-diameter annular collars together to achieve a 'hard docking'. It had been feared that unless the smaller vehicle was perfectly aligned when its probe retracted, it might be shaken so badly as to damage the connectors holding its three modular components together, but there was no such thrashing.

A depiction of a Soyuz about to dock with Salyut 1. Courtesy of *Novosti*.

When Soyuz 4 had docked with Soyuz 5 two cosmonauts had spacewalked from one vehicle to the other. A few months later Soyuz 7 and Soyuz 8 had been launched to test a docking unit incorporating an internal transfer tunnel, but they had not been able to dock, and this system was untested. Unfortunately, the electrical connections within the mated collars failed to link up, and the cosmonauts were unable to swing back the ferry's docking probe to access the 0.8-metre-diameter hermetically sealed tunnel to access the station. After only 5.5 hours together, during which, as *Tass* put it, Soyuz 10 "completed its planned experiments", it undocked

A cutaway model of Salyut 1 showing a cosmonaut at the control panel. The large conical housing of the primary instrument at the rear of the main compartment has been airbrushed out. Courtesy of *Novosti*.

and withdrew several metres. For an hour, the cosmonauts visually inspected Salyut and sent back television pictures of its docking system, and then they returned to Earth.

Over the next several days, robust support was given to the programme. *Pravda* reported Academician Anatoli Blagonrovov's view that orbital stations would soon permit space research to be elevated to a qualitatively new level in Earth sciences, biology and astronomy. He suggested that observations made from orbit would help to solve complex hydrological problems such as the assessment of rainfall intensity, the location of subsurface water and the analysis of the moisture content of soil, and would greatly assist in the study of surface and marine topography. Blagonrovov also predicted that the continuous monitoring of the Earth and its atmosphere on a global scale, when combined with observations of the Sun, the solar wind and the terrestrial magnetic field, would increase our understanding of the solar–terrestrial relationship. *Tass* reported Academician Feodor Chukhrov as saying that orbital stations would be able to produce continuous measurements of the total thermal, radiational and gravitational properties of the Earth. *Tass* also quoted Konstantin Feoktistov, who had flown on Voskhod 1 and had played a leading rôle in designing the station, as saying that the primary activities of cosmonauts on an orbital station would be determined by the economic benefit to the national economy, and would focus on the assessment of crop yields, the manufacture of extremely pure crystals and metal structures in microgravity and the collection of data for oceanographical and geological studies. He emphasised that the development of an orbital station to be used in the interest of the national economy was "top of the agenda" in the space programme.

AN EXCELLENT START

Soyuz 11 was launched on 6 June 1971. In comparison to Shatalov and Yeliseyev, the new crew were relatively inexperienced: commander Georgi Dobrovolsky was on

16 Getting started

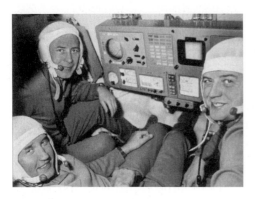

Viktor Patsayev, Georgi Dobrovolsky (centre) and Vladislav Volkov.

his first flight, as was Viktor Patsayev, and only Vladislav Volkov had flown in space before (on Soyuz 7). It was subsequently revealed that they were the backup crew, as Alexei Leonov's crew had been dropped following the pre-flight medical in which Valeri Kubasov was found to have symptoms of a lung infection.

As previously, Soyuz 11 followed a slow rendezvous and docked with Salyut 1 on its second day. This time there was no problem, the hatch was easily opened, and the crew entered the station. The downlink from the automatic television camera showed the three men performing somersaults in the enormous (by comparison with the Soyuz) main compartment. Once the station was verified to be fully operational, the ferry was powered down.

The objective of the mission was to stay in space for a month, during which a range of biomedical experiments to study the effects of prolonged exposure to weightlessness were to be performed. These included measuring bone density, studying the reaction of the vestibular system to different stimuli, measuring energy expenditure and measuring the capacity of the respiratory system in order to assess stamina, taking electrocardiographs to monitor the electrical activity of the heart, measuring the flow of blood in the major blood vessels by taking seismocardiograms, measuring arterial pressure, measuring the force and rhythm of the heart, and taking blood samples for post-flight analysis. In effect, the cosmonauts were acting as guinea pigs to prove that the human organism could survive a tour of duty aboard an orbital station and return safely to Earth.[8]

It had been decided to operate a shift cycle, whereby while one cosmonaut slept the other two would be on duty. Their staggered daily cycles involved eight hours on duty, eight hours off duty, and eight hours of sleep. In earlier, smaller, spacecraft this had not been feasible because the cosmonauts would have been in each other's way, but the ferry's orbital module was turned into a dormitory. The off-duty time included all toiletries, meals and exercises. Following the results of Soyuz 9, the exercise regime included a 40-minute session after breakfast and a further 80 minutes accumulated during the day. The gymnasium included a variety of elasticated expanders and a moving-belt treadmill (called the KTF) built into the floor across the centre of the main compartment (an elasticated harness was worn to draw the user against it to provide the necessary traction). It was hoped that a balanced exercise programme would stave off the muscular atrophy which, despite their efforts, had afflicted Nikolayev and Sevastyanov, but this imposed a significant

---

[8] Looking back, it is difficult to appreciate just how truly pioneering this mission was.

overhead, reducing the time available for work, although if that was the price of surviving in space, then so be it. Psychologists monitored the video downlink to study the crew, and quickly realised that weightlessness rendered assessment of interrelationships based on body language impractical because these were based on postures and gestures appropriate to physical orientations imposed by gravity. Despite the fact that the decor imposed a strict sense of up and down, in space the crewmen floated in random orientations, which made it difficult to assess their facial expressions; this was made worse by the pooling of blood in the head, which distorted facial features. In addition, the pitch of the voice was altered, and for a while the psychologists mistakenly believed that the cosmonauts were under a great deal of stress, whereas in fact they were delighted to be in space.

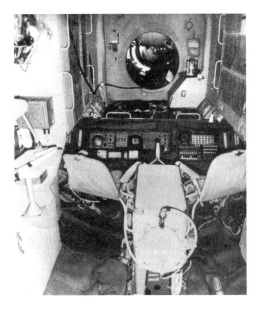

The interior of Salyut 1 looking towards the forward docking compartment, with the control panel in the floor.

As the days passed, the crew made television broadcasts of their activities, showing off the station and its apparatus, and they were featured nightly on the domestic news, enabling the audience to see that although Patsayev shaved, Dobrovolsky and Volkov had opted to grow beards. A number of scientific experiments were performed, including the growing of plants, making protein from chlorella algae, and a genetic study involving drosophilae fruit flies. As part of a study of the adaptation of the vestibular system to weightlessness, frogs eggs were developed into tadpoles, which developed receptors for sensing gravity even though they were weightless. After all of the tadpoles had developed, they were frozen for return to Earth. It was hoped that the experiment would provide data which could be applied to human beings, because frogs use similar organs for balance.

Earth observations were conducted, and a portable spectrograph was used to study aspects of the atmosphere, including airglow. The cosmonauts reported seeing noctilucent clouds, a fairly rare luminous silvery high-altitude cloud. This was the first time these had been observed from space. Whereas from the ground noctilucent clouds are almost exclusively confined between latitudes of about 60 and 80 degrees during the summer months, cosmonauts frequently detected them nearer the equator (even although they were not observable from the ground), often reporting them to cover large areas and, on rare occasions, over almost an entire hemisphere. Formed at altitudes in excess of 80 kilometres, where there is very little water vapour present, the way in which they formed was ill-understood.[9] In another experiment, coverage of water vapour cloud was assessed, and this data was correlated with that from the Meteor weather satellites in higher orbits. Whenever the station passed beneath such a satellite, a visual assessment was made so as to compare observations undertaken at different altitudes under identical lighting conditions.

Another major experiment involved using a multispectral camera to photograph the reflectance spectrum of the Earth's surface. The Caspian Sea region was selected because similar data was being collected by a pattern of aircraft flying at 300 metres and 8,000 metres to calibrate the station's sensors and to assist in the interpretation of its results. It was hoped that the characteristic reflectance spectra of different soil conditions and different crops would be discernible, and such imagery would provide a way to survey agriculture from orbit. Once the orbital platform's system had been calibrated, it could be used to scan broad swathes of the nation, measuring the extent and the rate of growth of different crops, identifying diseased crops, and enabling the likely yield to be forecasted.[10] Clearly, the station really was intended to undertake work that would make a significant contribution to the national economy.

Advantage was taken of being above the atmosphere to undertake astronomical observations. The Anna-3 telescopic spectrometer measured the gamma-ray energy spectrum of various known sources. A forward-planning experiment was conducted to determine how the space environment affected optical materials intended to be used on a later telescope. Gamma-rays and heavy ion, neutron and charged particle fluxes were measured in the immediate vicinity of the station to assess cumulative exposure to these forms of radiation. The high-frequency electron-resonance effects of the low-temperature plasma at orbital altitude on the transmission antennas were measured to establish the conditions under which resonance developed, to measure the deterioration of the signal, and to measure the distortion of the propagation characteristics of the antennas. All this produced important engineering data to assist in designing later orbital stations, as resonance seriously degrades communications. In addition, engineering tests were conducted with the station's orientation system to determine how well it would maintain its attitude when rotating at specific rates in all three axes. It was, therefore, a busy mission for the three men.

The routine was disturbed on 18 June, when a small electrical fire in a cable was discovered. The cosmonauts became so alarmed that they urged ground controllers to let them evacuate the station, but in weightlessness a fire in an oxygen-nitrogen mix is not so great a danger because in the absence of convection to draw off the

---

[9] Noctilucent clouds are probably formed by the condensation of water vapour around small nuclei, but the nature of the condensation nuclei is unknown; suggestions include concentrations of silicon and iron particles from volcanic plumes and residual particles from meteor showers. Measurements by later Salyut crews revealed that these clouds form at three distinct altitudes, each of which has a different characteristic temperature ranging from −130°C to −150°C.

[10] The Earth Resources Technology Satellite (ERTS) which NASA planned to launch the following year was intended to perform a similar function, and was later transformed into the Landat series.

resultant carbon dioxide and draw in new oxygen it suffocates. Nevertheless, they powered up their ferry just in case, a fact which did not escape the Western observers monitoring the telemetry signals. As soon as it became evident that there was indeed no danger, the cosmonauts re-entered the station and resumed work. But their pioneering spirit appeared to have been broken, and a week later it was decided to let them return to Earth early. On 29 June, therefore, they packed the film canisters and the results of their experiments into the cramped Soyuz descent module, and returned the station to its automatic flight regime. The transfer tunnel was overpressurised to check for a leaky hatch, and then vented. They undocked and, laughing and joking because they were happy to be heading home, aligned the spacecraft for the de-orbit manoeuvre. There was a premature loss of signal, but the recovery team that gathered around the capsule on the steppe did not know this, and when they opened the hatch they were appalled to find Dobrovolsky, Patsayev and Volkov dead in their couches.

An enquiry eventually determined that a tiny valve in the descent module that was designed to equalise the pressure during the final stage of the parachute descent had been shaken open by the shock from the explosive bolts that had jettisoned the orbital and service modules. Analysis of the telemetry revealed that even though the leaking air had rotated the descent module, the automatic system had acted to correct this. Although the overly violent separation would have alerted the cosmonauts to a malfunction, the status of this valve was not monitored and they would have been distracted as the air began to vent. Nevertheless, on noting that they were losing air, they had attempted to access the valve, but because this had not been intended to be accessed by the crew it was inconveniently located and they were unable to close it. Soyuz cosmonauts did not wear pressure suits for launch and re-entry, so they first succumbed to unconsciousness resulting from asphyxiation as the pressure fell, then died from embolism of the blood in vacuum. They were exposed to vacuum for fully 12 minutes before, on entering the lower atmosphere, the capsule was repressurised through the still-open valve. Fortunately, the re-entry manoeuvres and the parachute deployment were automatic.

Dobrovolsky, Patsayev and Volkov – familiar faces to the nation as a result of

After the public had paid their respects, the bodies of Dobrovolsky, Patsayev and Volkov were interred in the Kremlin wall.

their nightly television shows – were ceremoniously laid out in Moscow the next day, and their ashes subsequently entombed in the Kremlin Wall, together with their fallen colleagues Yuri Gagarin and Vladimir Komarov and former chief spacecraft designer Sergei Korolev. It took almost two weeks for the official announcement to be issued. In the interim, there was speculation that they had succumbed to the 3 g deceleration of re-entering the atmosphere. The pessimists argued that the deaths of *all three* men proved that there was indeed a limit to the time that the human body could spend in weightlessness. However, there was so little difference between their 23-day flight and Soyuz 9's 18-day flight that it was soon accepted that some other fate must have befallen them. When it was realised that they had died *prior to* re-entry it became clear that deceleration forces had played no part in their demise. The conclusion was that the station had performed well, and there was no evidence of a limit to the time a crew could spend in space; the problem with the Soyuz ferry was a secondary issue.

ABANDONED STATION

On 4 July *Tass* reported Boris Petrov's remark that, in the 1970s, orbital stations with changing crews would pursue a broadly based programme that would combine both routine research with opportunistic experiments. It also predicted that although stations like Salyut would continue to be used for some time, more complex multi-purpose stations would later be developed. At the end of that month, Petrov told a Japanese reporter in Moscow that despite the loss of the Soyuz 11 crew, spacesuits would be carried only for working outside a spacecraft, not for launch and re-entry. By October, however, it had become clear that the modifications to the Soyuz ferry to make it safe were so extensive that it would remain grounded for a year or more. This ruled out the possibility of making another visit to Salyut 1, so the station was de-orbited over the Pacific. During six months in space, its systems had performed well. On 22 October, *Izvestia* reported Konstantin Feoktistov as saying that future stations would be capable of automatic operation in order to enable crews to devote more of their time to scientific and economic activities – the Salyut 1 crew had spent much of their time on housekeeping tasks.

Table 1.1 Salyut 1 docking operations

Spacecraft	Docking		Port	Undocking		Days
Soyuz 10	24 Apr 1971	0447	front	24 Apr 1971	~1030	0.24
Soyuz 11	7 Jun 1971	0755	front	29 Jun 1971	2128	22.56

Dates are Moscow Time.

From bad to worse 21

Table 1.2 Salyut 1 crewing

Cosmonaut		Arrival	Departure	Days
Georgi Dobrovolsky	CDR	Soyuz 11	Soyuz 11	23.76
Vladislav Volkov	FE	Soyuz 11	Soyuz 11	23.76
Viktor Patsayev	FE	Soyuz 11	Soyuz 11	23.76

FROM BAD TO WORSE

Things went quiet for a while, but on 4 January 1972 Petrov said that the emphasis in space for the foreseeable future would be geodesy, agriculture and prospecting for minerals in the interests of the national economy. On 11 April, *Tass* quoted Valeri Bykovsky as saying that the primary task of the manned space programme was still to develop orbital stations. That same day, Vladimir Shatalov hinted that a manned mission would probably be launched later in the year. The next day he was quoted by *Izvestia* as predicting that cosmonauts in an orbital station would be able to track meteorological phenomena and provide timely warnings of danger, prospect for minerals, assess the state of crops and warn of disease, spot forest fires, locate plankton in the open sea and steer fishing fleets to the richest fish stocks. He also predicted that scientists would join cosmonauts on future, much larger, stations. In March, French sources had suggested that a new station was to be launched in May, but this did not occur. Significantly, these reports predicted that it would be visited by *pairs* of cosmonauts, which suggested that the modifications to the Soyuz were rather more substantial than had been admitted. Because it was Soviet practice to evaluate a new configuration in orbit prior to using it operationally, the fact that this had not been done implied the reports that another station was about to be launched were premature. In late June, however, the modifications were tested in the guise of Cosmos 496. A new station was launched in July, but the Proton's second stage malfunctioned and the payload plunged back into the atmosphere. A photograph released several years later showed a vehicle labelled 'Salyut 2' being prepared; it had the same configuration as Salyut 1. A study of the orbital data by Phillip Clark suggested that if this station had achieved orbit, the same mission sequence would have

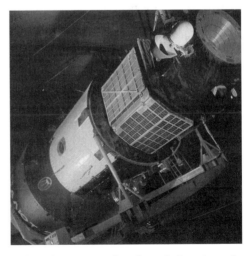

Although never assigned a designation, the space station launched in July 1972 was labelled 'Salyut 2'.

been followed as planned for its predecessor: two crews each spending a month onboard. In this case, orbital dynamics suggested the first crew would have boarded in the second week of August and the second in mid-October.

The Khrunichev factory, meanwhile, was turning out stations on a production line, and it already had another one ready. This was set up on the pad in September, and the preparations for its launch initiated; but a fault was discovered, and it was returned to the assembly building to be stripped down.

On 4 October *Pravda* marked the anniversary of the launch of Sputnik by quoting Vitali Sevastyanov as saying that cosmonauts in orbital stations would be able to carry out hydrological studies by assessing the extent and depth of snow and ice cover, and by monitoring the level of water in rivers. It was yet another argument evinced in support of a facility intended to directly benefit the national economy.[11]

By early 1973 Khrunichev had produced another hybrid and the first Almaz, and it was decided at the highest level to fly the Almaz first. It was launched on 3 April. As soon as it had been checked out, it was designated Salyut 2. The fact that there was a military reconnaissance variant was a secret, but the fact that the new station was silent on the radio frequencies used by its predecessor and had telemetry similar to that from spy satellites enabled Western observers to see through the simple ploy of calling it Salyut and talking up the benefit to the economy of operating an orbital station. Its 11.6-metre-long body comprised only the stepped-cylinder of the main compartment. It had essentially the same configuration as Chelomei had envisaged with the rear-mounted docking system set between the two engines. A pair of large solar panels were stowed tightly against the transfer tunnel for launch. In orbit, these first unfolded straight out and then released fore-and-aft flaps. The panels were able to rotate through 180 degrees in order to track the Sun to produce a combined peak of 5 kW. The capsule on the front had been deleted and the transfer hatch sealed. The major change, of course, was that the crew would fly up in a Soyuz rather than a TKS ferry.

Yevgeni Khrunov said in early April that his fellow cosmonauts were preparing for a flight in the near future. A week later, on 10 April, Boris Petrov, speaking at Helsinki University, said that a greater degree of automation had been built into Salyut 2, so it would require only a crew of two to operate it. This confirmed that the modifications to the Soyuz had reduced its crew capacity. He added that one was at that moment being prepared to pay the new station a visit. Unfortunately, at this point fate intervened, and on 14 April, just after performing a manoeuvre to refine its orbit, Salyut 2 suffered a catastrophic malfunction. Several days later, with the embarrassment of yet another failure, it was denied that Salyut 2 had been meant to host a crew, but Pavel Popovich and Yuri Artyukhin, both military cosmonauts, had been preparing for launch when news came that the station had been disabled. The loss was eventually attributed to an electrical fault in the new propulsion unit. This apparently started a fire which rapidly engulfed the internal compartment. The

---

[11] This was in marked contrast to Skylab, which was being sold by NASA as a laboratory in the sky, and whose primary instrumentation was a sophisticated solar observatory.

A Proton rocket with Salyut 2 protected by an aerodynamic shroud is about to be rolled out to the pad for launch.

sudden rise in pressure ruptured the hull and the resultant venting sent the vehicle spinning out of control. As the spin rate increased, the solar panels were torn off and the structure broke up. In light of the suddenness of the station's demise, it was fortunate that the crew had not already boarded it.

Given the clear evidence of a major systems failure, the West was astonished on 11 May when a Proton launched another station. In fact, as this hybrid did not have Chelomei's propulsion system, there was no reason not to launch it. Unfortunately, as it manoeuvred to orient itself after boosting clear of its carrier rocket stage, a fault resulted in the new station draining its attitude-control propellant, and this left it tumbling. This time the problem developed so soon after reaching orbit that it was immediately evident that the vehicle would have to be written off, and the hulk was disguised as Cosmos 557. Although it was not acknowledged, it is likely that instead of having small solar panels on the transfer compartment and the propulsion module, this configuration had three much larger panels on the narrower part of the stepped-cylinder to provide the power to operate a greater suite of instruments. It is believed that Alexei Leonov and Valeri Kubasov, who had planned to make the inaugural visit to Salyut 1, were to have occupied this new station, but following this failure they were reassigned as the prime crew on the joint Apollo–Soyuz mission being planned for 1975.

There was speculation in the West that the immediate 'replacement' of Salyut 2 by Cosmos 557 meant that it had been intended to operate them simultaneously in order to upstage the forthcoming American Skylab, which was launched on 14 May. This is unlikely, because sustained parallel operations would have been beyond the capability of the control centre at Yevpatoria, and the facility at Kaliningrad, near Moscow, was still under construction at that time. It is conceivable, however, that it had been intended to alternate missions, with only one station being occupied at any time, but there is no evidence of this. The most outrageous speculation was that it had been intended to link up the two stations. Given the state of the technology, this would have been impracticable, and, again, there is no evidence that such a union had been considered; rather the opposite, in fact, as the two programmes were distinctly divergent at that point.

FILLING THE GAP

The Khrunichev factory's production line drew to a halt while the faulty elements of each configuration were redesigned. As it would clearly be some time before another

Vasili Lazarev (left) and Oleg Makarov.

station would be completed, it was decided to make the best of what was available. This meant independent missions using the Soyuz, the revised version of which had not yet been flown by a crew. Accordingly, on 27 September Soyuz 12, with Vasili Lazarev and Oleg Makarov onboard, was placed into orbit. Its basic mission was to test the modified systems. Lightweight pressure suits were now required for launch, orbital manoeuvring and re-entry, and the one-piece suits had been designed to be easy to don in weightlessness, so that they could be put on in an emergency. In essence there was a zip from the chin to the crotch, sealed by an internal rubber bladder. The feet were inserted first, then one arm, then the head was eased through the integral metal collar, and finally the other arm was inserted. An integral hood at the rear incorporated the visor, which locked onto the collar at the front. The life support system for the spacesuits was so bulky that it replaced the third seat, so for the foreseeable future the Soyuz would be restricted to a crew of two.

Other changes had been made to the Soyuz to dedicate it to the rôle of serving as a station ferry. The solar panels had been deleted, and electrical power was provided by chemical storage batteries. Although this saved mass, it restricted the ship to just 2.3 days of independent flight in which to rendezvous with a station. Once docked, its batteries could be recharged from the station's solar panels for another two days in which to return to Earth.[12] Given this limit, it was immediately announced that this mission would last only 48 hours, so as to pre-empt criticism in the West that some failure had forced an 'early' return. The spacecraft made a series of manoeuvres to rendezvous with a hypothetical station at an altitude of 340 kilometres, which was almost 100 kilometres higher than the operating altitude of Salyut 1. This was taken to be an indication that future stations would operate in higher orbits so as to reduce the propellant required to periodically overcome the drag of the tenuous ionosphere, and, thereby, increase the station's operational life. When it was out of range of the tracking stations and communications ships, Soyuz 12 relayed through the network of Molniya satellites which spent most of each orbit thousands of kilometres above the Soviet Union, and hence were visible from low orbits virtually across the entire eastern hemisphere.

Soyuz 12 tested a more advanced multispectral camera to measure the reflectance spectrum of the Earth's surface at different wavelengths, so as to reveal information

---

[12] At the time, a rendezvous was to take place on the seventeenth orbit, at the end of the first day. As favourable recovery opportunities occurred only at 24-hour intervals, a ferry aborting a rendezvous late on would require a second day's endurance, which is why 48 hours was selected.

of economic value. Set up in the orbital module, this had nine lenses, each of which had a different filter. Any given three were chosen for each observation: two for film sensitive to visible light and the other sensitive to infrared wavelengths. The imagery had a resolution of about 100 metres. As before, data was collected by aircraft and ground teams for comparison – measurements of reflectance spectra had been made on earlier flights, and a means of interpreting the photographs devised, and this test was to determine whether the atmosphere introduced any effects that would have to be allowed for in the calibration process.

An indication of future expectations was revealed by the launch of Cosmos 613 in late November. This Soyuz was powered down for 60 days so as to demonstrate that even although it was limited to a 48-hour period of sustaining a crew, once it was docked with a station and powered down it could survive for considerably longer. In essence, the purpose of this flight was to determine the extent to which its systems degraded, because a 'service life' would set a limit to the length of a tour onboard an orbital station. When it was powered back up, it returned to Earth without incident. Meanwhile, when Soyuz 13 flew in December it became evident that there were two types of Soyuz in operation, incorporating different modifications resulting from the Soyuz 11 tragedy, and in this case the solar panels had been retained. It was the first flight to be controlled from Kaliningrad. Soyuz 13 was to test apparatus intended for future stations. Its main instrument was the Orion-2 telescope, an improved form of the apparatus used on Salyut 1. Pyotr Klimuk and Valentin Lebedev were trained in its operation at the Byurakan Observatory in Armenia. With the telescope mounted on the front of the orbital module instead of the docking unit, the spacecraft had to be oriented

Pyotr Klimuk (left) and Valentin Lebedev.

for each observation by Klimuk, in his couch in the descent module, and then the instrument was operated by Lebedev, in the orbital module. Earth-resources studies were continued using a multispectral camera similar to that on Soyuz 12, and the KSS-2 spectrograph, which was used to study the atmosphere by observing the limb at sunrise and sunset to measure the characteristic absorption by water vapour of sunlight passing horizontally through the atmosphere, and so measure its vertical distribution. Similar observations would be made routinely by future orbital stations. Throughout, the Oasis-2 closed-cycle biological cultivator manufactured protein. In a related experiment, chlorella algae absorbed carbon dioxide and liberated oxygen at a much higher rate than expected – it was found that a litre of algae produced 50 litres of oxygen per day.[13] After eight days, having caught up with

---

[13] While it would be feasible to use this process to top up the station's air supply, it would be better applied to food production. A human being cannot digest chlorella but a fish can.

26 Getting started

Although dated 1977, this painting by Alexei Leonov of a Salyut station with three solar panels was initially circulated in 1973.

crucial experiments lost on the failed stations, Klimuk and Lebedev returned to Earth.

The flight of Cosmos 573 (which had made a mysterious two-day flight in June) now made sense. Cosmos 496 evaluated the version with solar panels. This had been flown as Soyuz 13, and would later be utilised for Apollo–Soyuz. Cosmos 573 had demonstrated the battery-powered version flown as Soyuz 12, and meant to serve as a space station ferry. Something else also now became evident. Since Cosmos 496 (in June 1972) pre-dated all three of the failed stations, and Cosmos 573 had post-dated them, those stations would have been served by the variant with solar panels *rather than the stripped-down ferry*. This is deduced from the simple truth that it would have been folly to launch the station prior to proving the spacecraft type that was to deliver its crew, as that test might have revealed a design fault that would have put the station beyond reach. In the case of the one that failed to reach orbit in July 1972, it would have been necessary to use the variant with the solar panels, as the ferry was to augment the output from the small panels on the initial hybrid. Interestingly, a painting that was released in early 1973 depicted a ferry with solar panels flying alongside a station having three large rotating solar panels. With hindsight, it was evident that this was intended to be the upgraded hybrid (written off as Cosmos 557) with a ferry of the Cosmos 496 type, because we know that the station lost in the launch failure was of the initial configuration and that Salyut 2 was an Almaz. Not until Salyut 4 was this triple-panel configuration finally revealed, and by that time stations were being serviced exclusively by the stripped-down ferry.[14]

cont.
    The rapid growth of chlorella suggested that a closed-cycle fish farm might be feasible. The fish would consume the algae and the oxygen, the algae would consume the carbon dioxide exhaled by the fish. The human crew would eat the fish once they had matured. It was an interesting idea.

[14] In fact, from this point, until the more advanced Soyuz-T was introduced in 1979, the only Soyuz equipped with solar panels were those assigned to independent missions.

RECONNAISSANCE FROM ORBIT

On 25 June 1974, Salyut 3, the second Almaz, was launched. Considering the fate of its predecessors, little was announced until it was confirmed to be fully functional. It was maintained in an attitude with the instruments on its underside facing the Earth, which was feasible because its solar panels were able to rotate.[15] Its main instrument was a 6-metre-focal-length folded-optics telescopic camera. *Tass* said that the objective was to perfect a specialised automated orbital station that could be periodically visited by crews who would adjust experimental equipment, perform repairs and retrieve and replace film and other data storage media. Although scientific studies were undoubtedly intended, its function was military reconnaissance. A hatch in the side of the transfer tunnel to the docking system provided access to a small capsule for the return of film to Earth, but this fact was not announced. Compared with contemporary Soviet photographic reconnaissance satellites which used a version of the Vostok spacecraft and operated only for short periods (typically less than two weeks) a man-tended platform on an extended flight was considered to be a significant advance.[16] Salyut 3's systems were capable of sustaining a maximum of three months of occupancy over a period of six months, and it was hoped to send up two or three pairs of cosmonauts.

Pavel Popovich and Yuri Artyukhin, who were to have operated the first Almaz, were launched on 3 July in Soyuz 14. The next day, when the automatic rendezvous system closed the range to 150 metres, Popovich took over, completed the approach and docked at the first attempt. He

Soyuz 14 lifts off.

---

[15] Salyut 1, in contrast, had been oriented so that its fixed panels faced the Sun, which restricted its ability to hold other orientations.

[16] In contrast, the 'Big Bird' satellites run by the US Department of Defense, the first of which had been launched two years earlier, combined an extended mission – at that time, typically 4 months – with a Sun-synchronous polar orbit. In addition to scanning low-resolution images for transmission to Earth, they had a stock of six capsules to return important high-resolution film to Earth without interrupting their coverage. In fact, the promise of the Big Bird system was so outstanding that the decision to develop it had pre-empted the Air Force's Manned Orbiting Laboratory, which was the direct equivalent of Almaz.

28 **Getting started**

Pavel Popovich and Yuri Artyukhin.

remained at the controls while Artyukhin opened the hatches and went to check out the station's life support system, then Soyuz 14 was powered down and its batteries switched over for recharging in case they had to beat a hasty retreat. At this point in the programme, simply docking and boarding a station represented a major achievement.

The layout of Salyut 3 was different from that of Salyut 1. A control panel and storage unit formed a wall between the main compartment and the living room at the front, and a small corridor on one side provided access. The main compartment was dominated by the conical camera housing which sat in the centre and extended almost to the roof. Their living room, although compact, was unobstructed. It was arranged with lockers and two seats, and a table that incorporated recesses and storage bins for food preparation. There was a stove to heat canned food, and a spigot to supply hot water to reconstitute the dehydrated food. The psychologists had insisted that a crew would more readily adapt to weightlessness if they were provided with a distinct sense of up and down, so the floor and the ceiling were different colours and there was a carpet on the floor made of velcro so that the cosmonauts could stand in place and 'walk'. In addition, based on previous experience, all protruding apparatus was wrapped in padding to prevent the crew inadvertently injuring themselves. In the living room they had four portholes, which, in addition to providing a less claustrophobic atmosphere, facilitated navigational observations, photography and filming. Instead of sleeping in the ferry (as their predecessors on Salyut 1 had done) there were *bona fide* berths. The centrepiece of the gymnasium was a treadmill, and the daily cycle called for two hours of exercise per man. Although crews were not expected to remain aboard for more than a few weeks at a time, a deployable shower had been fitted.[17] The cosmonauts operated eight-hour shifts, but with only two men present round-the-clock operations were not possible.

Salyut 3's reconnaissance mission required its camera to be kept aimed at the Earth. One way of making a spacecraft track the Earth is to orientate it vertically, so that the force of gravity would tug slightly more on its bottom end than on its top and thereby provide 'gravity-gradient' stability. For this to be useful, however, the instruments would have had to have been mounted axially; Salyut 3's camera was in

---

[17] A thin plastic curtain device, the shower was mounted on the roof and was drawn down for use. It used a jet of air to force the water from the spray head to the suction cup at the opposite end. It was a nice idea but it proved awkward to use because it took several hours to set up, use, clean and pack away.

Reconnaissance from orbit 29

A model of the Salyut 3 Almaz showing the bulky reconnaissance camera and a cosmonaut at the control panel in the 'wall' that divides the two compartments. Courtesy of Dietrich Haeseler.

the floor of the main compartment and the station would have to be kept horizontal, which meant that a force had to be applied to make it rotate in synchrony with its progress around its orbit. It had an electrically-powered gyroscopic attitude-control system that could adjust its orientation without consuming propellant.[18] Since these 'gyrodynes' used electricity, it was also necessary to keep the solar panels facing the Sun. Because the panels rotated, they could track the Sun, but only if the Sun lay more or less perpendicular to their rotation axis. The station had therefore to be oriented with both of these considerations in mind: with its panels aimed at the Sun, and with its camera aimed at the Earth. This sometimes required it to rotate at different rates in all three axes simultaneously, and as the geometry altered, the rates required were different at different times of the year. The gyrodynes proved to be very responsive to control inputs, and could 'lock' the station with respect to the ground track. Power management was critical, however. High-capacity batteries were charged during a daylight pass, and then provided power when the station was in the Earth's shadow. Power-hungry activities had to be scheduled with this limitation in mind. Other new systems were tested, including one

A model of the rear of the Almaz version of the Salyut space station showing the still-to-be-deployed solar panels and the docking system. Courtesy of Dietrich Haeseler.

for improved thermal regulation and one for condensing water vapour from the air which could reclaim 1 litre per day per man for hygienic use and food preparation – this was a significant step towards self-sufficiency. Whereas Salyut 1 had been able to communicate only when in range of the communications network, Salyut 3 could transmit the imagery from its camera through the Molniya relay satellites. To assess the resolution of the camera, special photographic targets were displayed at the cosmodrome.

---

[18] Salyut 1 had only thrusters to alter its orientation, and manoeuvring had been limited in order not to expend its propellant.

It was not entirely a reconnaissance mission, however. Geological structures and atmospheric phenomena were studied to gain information in the interest of the national economy. Geological structures were photographed to find areas likely to contain mineral deposits, to identify land subject to salination, and to assess the condition of the soil in specific areas. Earth-resources observations concentrated on Soviet Central Asia, including the Caucasus, the Pamirs of Tajikistan, Kara-Bogaz-Gol Bay in Turkmenistan and the Ustyurt Plateau of southwest Kazakhstan. Leonid Sedov later announced that Salyut 3's imagery had indicated 67 sites in the Caspian region that were likely to yield oil and natural gas, and 84 other sites in Uzbekistan. In addition, Atlantic observations were made for the Tropex-74 project, in concert with Meteor weather satellites and science ships. The major non-imaging experiment involved the RSS-2 spectrograph, a hand-held unit to study processes affecting the Earth's thermal environment, and to gain data to assess trends in climatic changes due to aerosols resulting from industrial smoke, dust and chemical pollutants. This was important work, because it was believed that the disruption of the optical and radiation properties of the upper atmosphere would create the conditions necessary for climatic changes. The instrument could be used either to observe the atmosphere at times of orbital sunrise or sunset by measuring the selective absorption of sunlight passing through the layers of the atmosphere, or to make observations of the Earth's surface and the oceans by measuring the reflectance spectrum under high insolation. By measuring the humidity of soil, subsurface water reservoirs were identified in the parched lands by the Caspian which could be developed to support intensive oasis farming. Several purely engineering tests were performed. One (called Resonance) measured the vibration modes of the combined structure under different conditions. To carry out these measurements, the cosmonauts set up a variety of spring-loaded masses and then monitored their oscillations. These measurements were vital for two reasons: firstly to identify sources of vibrations that might blur the camera's imagery, and secondly to discover potentially dangerous vibrations that might resonate and threaten the integrity of the seals around the portholes and the docking collar. Another test evaluated the ferry's ability to control the complex's orientation.

Soyuz 14 landed on 19 July. Despite having spent a fortnight in space, Popovich and Artyukhin scrambled out of their capsule without the assistance of the recovery forces, in marked contrast to the Soyuz 9 crew. When they arrived, the doctors pronounced both men to be in good shape. This was encouraging, because it implied that the strict exercise regime had ameliorated the atrophying effects of adaptation to weightlessness.

MORE FRUSTRATIONS

A month later, Soyuz 15 set off with the second crew. As before, both cosmonauts, Gennadi Sarafanov and Lev Demin, were military officers, but neither had flown in space before. Although the automatic system made several approaches to Salyut 3, Sarafanov had to abort it with just 40 metres to go each time because the closing rate was excessive. When he finally attempted to make the docking manually, he ran low on propellant and was forced to withdraw. Now on their 23rd orbit, the ground track was far to the west of the recovery zone, and so they settled down to await the next favourable pass. The report of Soyuz 15's recovery was notable for its omission of the usual phrase, "following the completion of its planned experiment programme". Several weeks later, Vladimir Shatalov, visiting America as part of the preparations for the Apollo–Soyuz project, said that a cargo–tanker ferry was under development, that Soyuz 15 had been testing an automated docking system, and that after this had repeatedly failed to close in properly the crew demanded to dock manually, as had always been done in the past, but it had been decided to end the mission because so much propellant had already been consumed. Phillip Clark's analysis of the orbital data suggests that a three-week mission had been intended.

Lev Demin (left) and Gennadi Sarafanov.

On 23 September, a month after the aborted attempt to place a second crew on board, Salyut 3 released a capsule that had been loaded by the first crew (and would have been topped up by the second) and this was successfully retrieved. For many years the configuration of this capsule was kept secret and in the absence of specific information there was speculation that there might have been a large Vostok capsule on the nose of the station; that it was a small capsule similar to those utilised by the lunar sample-return probes; and later still (once its existence became known) that it was one of Chelomei's capsules. In fact, it was a 400-kilogram, 0.85-metre-diameter drum-shaped spin-stabilised capsule that could accommodate up to 120 kilograms of compact cargo, which was far more than could be ferried back in the Soyuz descent module.[19] No further attempts were made to send crews to Salyut 3. Although its potential had not been fully exploited, it was a success. It continued in its automated flight regime until the end of the year so that the deterioration of its systems could be evaluated, and it was then powered down and left to re-enter, which it did early in the new year.

---

[19] Several years later, a reconnaissance variant of the Soyuz was introduced which was able to carry a number of small film-return capsules, and it is likely that these were similar to those released by the military Salyut. Hence the Almaz platform proved the technology later exploited by the automated satellites.

A GLIMPSE OF THE FUTURE?

Shatalov's reference to the development of an automated cargo–tanker ferry caused considerable speculation, for it would require a station to incorporate *two* docking ports. It would be almost two years, however, before Andrian Nikolayev told the author Peter Smolders that future Salyuts would have two ports, (although he did not say where the additional port would be located) and another year before the first such station was launched. Clearly, the testing of this automatic docking system so early reflected considerable forward planning. Since the actual configuration of future stations was not announced, there was speculation that a multiple docking module was to be launched to link together two Salyuts and a pair of Soyuz ferry craft. This hypothesis was reinforced by the frequent official statements that in future orbital complexes would be assembled in space from modular components. The launching of two stations within one month during 1973 had shown that a production line was already operating. By this point, however, it had also become evident that the Salyut programme was more complex than the straightforward series of incremental steps characteristic of earlier Soviet engineering developments in space. Given the dramatic differences between the two successful stations, it was clear that these represented two types: one a scientific research station and the other a reconnaissance platform staffed by military officers. Apart from the blaze of publicity given to the scientific station, and the absence of it in the case of the military platform, the two types were readily distinguished because the scientific station's telemetry used continuous-wave pulse-duration modulation and transmitted on the traditional Soyuz frequencies and the military utilised frequency-shift-keyed pulse-duration modulation on frequencies used by spy satellites.

Table 1.3 Salyut 3 docking operations

Spacecraft	Docking		Port	Undocking		Days
Soyuz 14	4 Jul 1974	2351	rear	19 Jul 1974	1203	14.51

Dates are Moscow Time.

Table 1.4 Salyut 3 crewing

Cosmonaut		Arrival	Departure	Days
Pavel Popovich	CDR	Soyuz 14	Soyuz 14	15.73
Yuri Artyukhin	FE	Soyuz 14	Soyuz 14	15.73

AN ORBITAL LABORATORY

Placed into orbit on 26 December 1974, Salyut 4 was one of the Korolev Bureau's stations. It had the same configuration as the failed Cosmos 557, with a trio of solar panels with a total area of 60 square metres mounted on the narrower section of the main compartment which could be rotated through 340 degrees to face the Sun for a peak output of 4 kW. Although steerable panels improved the operational flexibility, the fact that the docked ferry could not make a contribution meant that the power supply barely exceeded that of Salyut 1.

Salyut 4 introduced the Delta semi-automatic navigational system. This was a significant advance over previous stations, which had required data on their path to be measured on every revolution by a network of tracking stations. The new system used sensors to determine the rising and setting of the Sun on the Earth's horizon to determine the period of each orbit. It also introduced the Argon 16 computer, which combined the information from the Delta with the readings from a radio altimeter to

A painting of Salyut 4 by A.Ryzhov. Courtesy of *Novosti*.

compute the station's orbital parameters, monitor its progress around the globe and compute the times during which it would be within communications range of the various ground stations. This could pinpoint the station's location over the surface of the Earth to within 3 kilometres, and its altitude to within a few hundred metres. The position was displayed by the Globus navigational display on the control panel. The new station also tested the Kaskad attitude control system, which used a pair of infrared sensors to determine its attitude relative to the Earth, together with an ion sensor which determined its orientation with respect to its direction of flight through the ambient ionosphere. It could align and maintain the station within 5 degrees of a specified orientation. Between them, the Delta and Kaskad systems both simplified the ground-support system and reduced the crew's workload.[20] Importantly, Kaskad resulted in a significant reduction in fuel consumption. A teleprinter (Stroka) enabled Kaliningrad to send up a lengthy communication without obliging a crew member to write it down. Salyut 4 also tested a sophisticated thermal regulation system (STR). The pressurised shell was covered with 'screen vacuum heat insulation' comprising layers of synthetic film sprayed with aluminium which

---

[20] Salyut 1's crew had spent up to 30 per cent of their time on tasks related to orienting the station.

minimised heat transfer. The station incorporated an intricate set of radiators which collected solar heat on the sunward side and radiated excess heat on the shaded side. These enabled the system to control heat transfer within a wide range of temperatures. Individual elements of the multiple-loop system, which comprised heating and cooling units, had three or more backups in order to provide a high degree of redundancy, and thus safety. The computer commanded the entire system, correlating the heaters and coolers with the station's orientation with respect to the Sun. In an emergency, heat could be shed by an evaporator which discharged water into space. While a ferry was powered down, its thermal regulation systems were put under the station's control.

Although the Korolev and Chelomei bureaux were long-standing rivals, there was cooperation. Salyut 3's water reclamation system had been fitted to maximise the use of this valuable resource; even so, water was to be taken up in small flasks by successive crews. Food was preloaded. In anticipation of equipment failure, many of the internal systems had been configured to give access for replacement. Equipment such as filters for the air conditioning system and canisters of chemicals to prevent carbon dioxide build up (which had to be replaced every few days) would likewise have to be ferried up by each new crew, together with any spare parts for repairs. If the life-support system suffered a major failure while they were onboard, the station would probably have to be evacuated and the task of effecting a repair (if this was feasible) left to their successors. Nevertheless, this new station clearly embodied the optimism of its designers. However, the gyroscopic attitude-control system had *not* been installed, so consumption of the unreplenishable propellant supply would be a major factor in planning work that required manoeuvring. A cargo–tanker was under development, but it would not be possible to use it to resupply Salyut 4 because the station did not have the plumbing to replenish its propellants. Clearly, therefore, although Salyut 4 was an improvement, it was only one step in a long development process that was advancing on several fronts.

It was a significant orbital laboratory for Earth and astrophysical observations, with the OST-1 solar telescope, which had a 25-centimetre-diameter mirror with a focal length of 2.5 metres, as its main instrument. This used the same aperture as the reconnaissance camera of the Almaz, and was contained in a similar conical mount. It had been built by the Crimean Astrophysical Observatory specifically for Salyut 4, and the cosmonauts assigned to this station had been extensively trained by the staff of the observatory to use it to make observations of dynamic processes such as the flocculi and prominences on the solar disk.[21] It could be used in conjunction with the CDS-1 diffraction spectrometer, which could produce an ultraviolet spectrum in the range 800–1,300 Angstroms with an average resolution of 2 Angstroms. Observing at sunrise and sunset, the spectrometer was to make the first comprehensive survey of the distribution of aerosols in the upper atmosphere. It could be used to measure the strength of the ozone's absorption lines (ozone

---

[21] It was similar to, but not as capable as the ATM solar telescope flown on the American Skylab a year earlier.

forms a layer in the atmosphere at a height of 20–60 kilometres) and this data later proved to be very useful.[22]

The ITS-K infrared telescopic spectrometer could be used to investigate celestial sources, but its primary objective was to study the Earth's atmosphere. The infrared sensors on previous stations had used conventional compressor-based cooling units, which had drawn significant amounts of power and had been unreliable. In this case, however, a much more advanced cryogenic system was employed which used an icy coating of solid nitrogen. This had been designed by the Kharkov Physical–Technical Institute of Low Temperatures. Not only did it draw less power, it could operate for extended periods at $-223°C$. Observing at sunrise and sunset with its diffraction slit aligned along the horizon, it determined the density and distribution of water vapour and measured the temperature of the different layers of the upper atmosphere. The photometer and spectrometers of the Emissiya experiment were clustered at the rear of the station to scan the Earth's horizon and measure the luminescence intensity of the red atomic oxygen spectral line, and to investigate processes in the atmosphere at altitudes ranging between 250 and 270 kilometres (the most Sun-sensitive part of the ionosphere, where electrons trapped in the Earth's magnetic field interact with the rarefied gas of the upper atmosphere). Combined with data from geophysical ground stations, this data was to facilitate forecasting of short-term variability in the upper atmosphere and, eventually, contribute to an understanding of global climate change.

In addition to this comprehensive atmospheric study, Salyut 4 was equipped for astronomical observations. The Filin X-ray spectrometer was sensitive in the range 1–60 Angstroms. The main telescope, mounted outside the station, was bore-sighted to a smaller sighting telescope inside. The spectrometer was essentially autonomous, so whenever it detected a signal the operator would use the sight to note the target's position against the star field. The RT-4 telescope was sensitive to the 'soft' end of the energy range sampled by the Filin. Its 20-centimetre-diameter parabolic mirror illuminated a photon counter, but its field of view was not wide enough for it to be used to scan the sky for new sources; it was used to make further studies of sources discovered by the Filin. Silya used a light nuclear-isotope spectrometer to record the isotopic and chemical composition of cosmic rays, thereby establishing a new field of investigation.

Finally, the Spektr experiment employed a variety of apparatus to measure the density, composition and temperature of the neutral gas and plasma encountered by

---

[22] Following the British Antarctic Survey's discovery in 1985 that the ozone layer above its base was severely depleted, NASA examined data accumulated by its Nimbus 7 weather satellite dating back to its launch in 1978, and then announced that there was a seasonal 'ozone hole' over the southern pole. There did not seem to be any comparable variability at other latitudes, but by including Salyut 4's data this retrospective analysis was able to be extended to 1975. This data could not confirm the depletion of polar ozone, because the station's orbital inclination was too low to view over the pole, but it set limits to the variability of ozone in sub-polar latitudes, which assisted in characterising the peculiar phenomenon over Antarctica.

the station in its orbit – a spacecraft becomes electrically charged as it flies through the ionosphere, and this interferes with communications systems; the results of this study were to improve the design of future stations.

The new station offered its crew a degree of comfort. There was a table in the main compartment, between the control panel and the telescope, and this supplied hot and cold water for food preparation. The telescope took up most of the space at the rear of the compartment, so the medical and exercise equipment was on the roof above the telescope and the toilet was in the floor at the rear. The cosmonauts were to sleep on the roof above the control panel. Despite the facilities, life would still be spartan, especially as the duration of missions increased.

Early in the new year, Salyut 4 raised its orbit to about 350 kilometres, which was almost 100 kilometres higher than its predecessors. This consumed considerable propellant, but reduced the frequency of firing its engines to overcome orbital decay. It was hoped that it would sustain three months of occupancy over a period of six months.

FIRST CREW

Soyuz 17 was launched on 11 January 1975. Alexei Gubarev and Georgi Grechko were both making their first flight, and their objectives were announced to be "observation of geological and morphological objects and of atmospheric formations" to gather data "in the interest of the national economy", together with studies of "physical processes and phenomena in outer space", further study of the adaptation of the human body to weightlessness, and testing of the new station's systems. The next day, when 100 metres from the station, Gubarev took control. The docking was achieved so easily that the crew's heart rates were lower than they had been during simulations. They found the station a little chilly, so Grechko turned up the heaters to warm it up to a comfortable 23°C – it would later be noticed that crews tended to overheat a station during their first few days. Then followed the lengthy process of activating the systems. An air hose was fed from the transfer compartment into the descent module of the powered-down ferry to ensure that it was properly ventilated (on previous long missions it had been found that the air in the ferry became stale), baseline biomedical data was sampled (Polynom-2 was used to assess the condition of their cardio-vascular systems), and then the two men settled into a daily routine, this time adopting the same duty cycle. However, their sleep cycle was scheduled to exploit the periods during which they had sustained radio con-

Alexei Gubarev (left) and Georgi Grechko in the Salyut 4 trainer.

tact with the ground, and the fact that the orbit precessed led to a corresponding migration in their sleep pattern. Over an extended period, starting work slightly earlier every day – in effect, operating a 'day' of less than 24 hours – proved to be very tiring. In general, they operated a six-day working week, and then took one day off, during which they were free to do whatever the wished, although the facilities for entertainment were rather limited.

They had Tchibis rubberised leggings from which air could be pumped to make a negative pressure that drew blood into the lower body. This proved to be useful in the initial phase of adaptation to weightlessness, because it prevented blood pooling in the torso, and immediately prior to returning to Earth, when it could increase the capacity of the circulatory system. It was a cumbersome garment which prevented movement, so there was little to do whilst using it except to undertake upper-body exercises. Each cosmonaut used the Tchibis an hour a day in the first week or so, and thereafter once per week until the final ten days of the mission, at which time the cycle returned to an hour a day. They were to follow a strict exercise cycle to reduce decalcification of their bones, muscular atrophy, and the reduction of cardiovascular capacity, which were the most debilitating effects suffered by their predecessors due to prolonged exposure to weightlessness. This cycle was repeated every four days: the first built up speed and strength, the second concentrated on exertion, the third emphasised endurance, and the fourth was 'anything goes' in which they could do whatever they liked. In addition to the standard treadmill, they were provided, for the first time, with a stationary bicycle (the VTL veloergometer). Most cosmonauts initially find exercise to be psychologically advantageous because it serves as a tonic, but they suffer in the process because, in the absence of a brisk wind to enhance evaporation, surface-tension ensures that sweat sticks to the body and pools in hollows (such as the chest) forming a puddle of moisture a centimetre or so in depth. Clothes worn during vigorous exercise immediately became saturated (which was a real problem because the wardrobe was rather limited) and every exercise period had to be followed by an all-over towel wash to clean up.

In weightlessness, the fully relaxed human body assumes a crouching position, with the knees bent, the shoulders hunched over, the elbows bent, and the forearms floating in front of the chest. The doctors had devised two types of load-inducing suit. Gubarev evaluated Atlet, in which elasticated straps linked his shoes to a waist corset and braces ran over his shoulders. The Penguin suit assessed by Grechko had elasticated straps sewn into its fabric to impose compressional loads on the muscle groups likely to suffer from lack of use in weightlessness. Both suits were designed to make the wearer's muscles work to straighten out. Because they were to be worn as everyday working clothes, they were made as comfortable as possible.

In the two weeks prior to the crew's arrival, the OST-1 had been operated by remote control. Unfortunately, a sensor in the mechanism that aligned the secondary mirror had malfunctioned, disabling its aiming system and preventing its further use. As this was the main instrument, its repair was crucial. Although they could not see the mechanism, the cosmonauts realised that by listening to the servo, and timing its travel, they could estimate its position sufficiently accurately to align the mirror, so they set it up and locked it into place. Although this rendered the telescope useable, the fact that the secondary mirror was fixed meant that the station itself would have to rotate to scan the field of view across a target, but this was feasible. Without their ingenious intervention, the telescope would have remained unusable, and a significant part of the scientific programme would have been lost. Nevertheless, the telescope was 'high maintenance' because its reflective coating tarnished after several weeks of exposure to atomic oxygen in the ionosphere. A remote-control apparatus had been incorporated to respray the surface. Exploiting the near-vacuum of space, an electric current was passed through a tungsten wire to melt a sample of aluminium in such a way that a vaporised spray of this metal would recoat the mirror. This itself was an experiment, and when restoring the mirror's reflectivity proved to be straighforward this meant that future stations would be able to be outfitted with more sophisticated mirror-based apparatus.

Two weeks into their mission, *Izvestia* reported that Gubarev and Grechko were in excellent spirits and were working well together as a crew. In fact, they were working so hard that they had to be told to take a rest, and they had developed eager appetites – they were eating four meals a day, instead of the assigned three, which was consuming the limited supplies faster than expected. They eased off for several days to catch up on much-needed sleep. In their third week, they reported that they had fully adjusted to their environment, and felt at home. On 2 February, they broke the 23-day record set by the ill-fated Soyuz 11 crew – this fact was allowed to pass without comment, however, no doubt in order not to evoke unpleasant memories. A week later, Soyuz 17 undocked and returned to Earth. With Salyut 4 operating at a much higher altitude than its predecessors, the departing ferry made its descent in two stages. Firstly, immediately after undocking over the Soviet Union, it performed a breaking burn that resulted in a 200-kilometre perigee over the South Atlantic, and once there it followed through with the standard de-orbit procedure, leading to a late afternoon recovery. Both men were found to be fit and in excellent spirits. They had set a new record of 30 days. For the first few days, they spent time in a lower-body positive-pressure apparatus similar to Tchibis, but with the pressure set to prevent blood pooling in the legs. Increasing the flow of blood in the upper torso helped the cardiovascular system to recover. Both men were able to walk without difficulty but Grechko suffered intermittent chest pains due to the fact that his heart had migrated slightly in the 'absence' of gravity. It was concluded that the exercise program had been inadequate to prepare their bodies for readaptation to gravity, so future crews would increase their exercises in the final week and also assess a variety of stamina-increasing drugs.

ABORT!

Vasili Lazarev and Oleg Makarov set off to Salyut 4 on 5 April 1975. All went well for the first five minutes, at which time the core stage failed to release properly. This had never occurred before, and it constituted an emergency.

The sequence of events during the staging operation was for the exhausted stage to separate from the interstage, so that the upper stage's engines could ignite, then for the interstage to be jettisoned. In this case, however, when the core was within seconds of shutting down, vibrations triggered a relay which detonated three of the six bolts on the *upper* rim of the interstage, partially separating it from the upper stage, and severing the cables that were, moments later, to command the firing of the bolts on the *lower* rim. When the upper stage's engines ignited, therefore, the core was still attached by the latches on one side of the rim of the interstage, and it was deflected by the efflux. The asymmetric loading in turn canted the upper stage over onto its side. The computer noticed the misalignment and shut off the engines. The vehicle was already at an altitude of about 180 kilometres, so it was above most of the atmosphere. Although it was travelling at 5 kilometres per second, this was short of orbital velocity. The escape tower for aborts in the lower atmosphere had already been jettisoned. The procedure for escaping a malfunctioning rocket on the verge of space called firstly for the second stage to be jettisoned, then for the descent module to separate from the rest of the spacecraft stack. This had been done several times by such capsules carried by errant Moon-bound rockets a decade earlier, but never with a crew onboard. The procedure worked, and the module's momentum carried it on a high arc. After following a suborbital trajectory, it plunged into the atmosphere 1,600 kilometres downrange. The ballistic re-entry resulted in very high deceleration forces – the peak of around 15 g imposed almost enough force on the chest to stifle breathing. As Lazarev and Makarov were recovering from this ordeal, they hardly noticed the jolt as the capsule landed in snow, but they were alarmed to feel it slither down a slope for some distance before the parachute snagged on a tree. When they clambered out, they were shocked to see a sheer cliff further down the hill. The Sun was about to set, so they hurriedly lit a fire to keep warm until the recovery forces could locate them. Their capsule contained survival rations for several days, but it had no heating. In fact, the radar network had tracked their descent to a spot just southwest of Gorno-Altaisk, in the Altai mountains of western Siberia, and the rescue forces soon arrived. Riding rockets is an inherently dangerous business.

A view of Soyuz 18 in the assembly building, featuring its launch escape system.

As the spacecraft had not made it into orbit, the designation Soyuz 18 was not assigned, and it became simply "the 5th April anomaly". Yet again, therefore, it was proving difficult to revisit a station. Phillip Clark's analysis suggests that a 60-day mission had been intended.

A SECOND TOUR

After the fault with the rocket had been rectified, it was decided to send up another crew, and so Pyotr Klimuk and Vitali Sevastyanov, who had served as Lazarev and Makarov's backups, set off on 24 May. When Soyuz 18's automatic system closed

A view of the docking collar of Salyut 4.

within 150 metres of Salyut 4, Klimuk took control, flew a straight-in approach, and docked at the first attempt. This marked a significant milestone, because it was the first time that a station had been revisited. As they had already had a long day, once they had verified that the station was functioning properly they retired to get a good night's sleep. Their experiment programme was a continuation of that performed by their predecessors. This time, however, they were to attempt to schedule their work so that several days would be spent on a given type of experiment, before moving on to another one. In part, this was an effort to economise on the use of propellant in reorienting the station to suit different types of work, but it would also prove to be a great improvement in working methodology, because using the same apparatus on consecutive days eliminated the time that had previously been spent first setting up and then stripping down apparatus from one day to the next. But their first task, reflecting the novelty of settling into a station that was already well into its service life, was to undertake maintenance. The design effort to make apparatus accessible now paid off. They changed the air filters, installed a gas analyser, replaced one of the six condensers in the water reclamation system with a hand-operated pump, and repaired the broken scientific apparatus (such as the Silya spectrometer) for which they had ferried up spare parts.

Pyotr Klimuk (left) and Vitali Sevastyanov in the Salyut 4 trainer.

Although biological experiments were set up with plants and insects (these had been evolved on the ground from those which had been returned by the previous crew), and solar and astronomical observations were made, the main target of their studies was the Earth. They gathered extensive spectrometric data that detected pollutants in the upper atmosphere, their work being greatly assisted by the Kaskad attitude control system that provided automatic and continuous maintenance of the orientation so that the instrument scanned the horizon. And Earth-resources imagery was taken. One specific objective was to take pictures to assist in planning the route for the 3,500-kilometre-long railway that was being built from Lake Baikal, running across Eastern Siberia to the river Amur on the Manchurian border – these pictures revealed unsuspected tectonic features which would have to be taken into account in designing the many bridges and tunnels.

In mid-June, when the Crimean Astrophysical Observatory reported an increase in solar activity, Klimuk and Sevastyanov were asked to turn to solar observations. They secured spectrograms and photographed a large prominence on the limb of the solar disk. This eruption was significant because it took place towards the minimum of the solar cycle. As the increased solar activity was likely to induce aurorae, over the next few weeks the cosmonauts made a special effort to observe such displays. Looking obliquely down, they had a magnificent view of streamers that zig-zagged for thousands of kilometres in the ionosphere. The 30-day endurance record of their immediate predecessors was exceeded a week later. By the end of June, even though Salyut 4 was six months old, it was in excellent condition. At a press conference for the buildup to the Apollo–Soyuz docking mission, Alexei Leonov reported that his colleagues had done so well that they had been ordered to remain in space for the full two-month tour, which meant that they would be in space at the same time as his Soyuz 19 flight. Konstantin Bushuyev, the technical director for the Apollo–Soyuz project, said that simultaneous missions were feasible because Yevpatoria would look after Salyut 4 while Kaliningrad switched to Soyuz 19. Sevastyanov's 40th birthday was on 8 July, and Klimuk's 33rd was two days later. They celebrated with a feast of spring onions which had grown in the Oasis. In fact, the onions represented a rare success, as, in general, plants had tended to develop well for the first few weeks only to wither and die.

Pyotr Klimuk (left) and Vitali Sevastyanov on the TV downlink.

Soyuz 19 was launched on 15 July with Alexei Leonov and Valeri Kubasov. A few hours later, the last Saturn IB rocket launched Apollo with Tom Stafford, Vance Brand and Deke Slayton. The next day, in a brief communication with Salyut 4, Leonov offered to fly across and repair any broken apparatus, a remark which reflected his frustration at having been prevented by one reason or another from visiting three of the Salyut stations. Sevastyanov observed that it was the second time that seven men had been in orbit at the same time on three spacecraft – the first time had been in 1969, when Soyuz 6, Soyuz 7 and Soyuz 8 had flown, with Kubasov on Soyuz 6. The historic docking was made on the 17th, the hatches were opened and Stafford and Leonov shook hands. After two days of docked activities, the vehicles undocked and in due course returned to Earth. It was to be an all-too-brief lull in the Cold War.

For almost a week, the plane of Salyut 4's orbit kept it in continuous insolation, because it no longer intersected the Earth's shadow. While this guaranteed power, it risked the powered-down ferry overheating, so additional hoses were run in through the hatch to help to ventilate its descent module.

Radio Moscow announced that the cosmonauts were 'rounding off' their work and would soon prepare Salyut 4 for autonomous operation. After having examined the Soyuz 17 crew, the doctors had suggested that during their final ten days Klimuk and Sevastyanov should increase their exercise programme and increase their intake of water and salt, to help to rehydrate their bodies and increase the capacity of their cardiovascular systems in the hope that this would better prepare them for return to Earth. They loaded 50 kilograms of materials, films and logbooks stuffed into storage bags. This was the most that the descent module could return as cargo, and it had to be stored carefully in order not to upset the capsule's centre of mass. On 23 July, in the midst of their preparations, Klimuk and Sevastyanov exceeded the 56 days spent in space by the second Skylab crew. Only the still-extant 84-day record of the third and final Skylab crew remained to be beaten. That, however, would have to be left to another station, as Salyut 4's environmental system had deteriorated to the extent that the portholes had fogged over with condensation and the walls were coated with a rich green mould. Klimuk and Sevastyanov had suggested curtailing the flight prior to the Apollo–Soyuz mission, but had agreed to stick to the plan, and they were not sorry finally to be leaving.

Soyuz 18 used its engine on 24 July to boost the station's orbit to 350 by 370 kilometres – it was the first time that a ferry had been used to raise a station's orbit, as previously the station had used its own engine. Finally, on 26 July, after 63 days in space, Soyuz 18 returned to Earth. Refusing assistance, the two men staggered to the medical tent. For Sevastyanov, who had been debilitated by 18 days on Soyuz 9, this was a satisfying outcome; for the doctors it was dramatic proof that the exercise regime (and in particular the readaptation process during the final week) had worked. This was excellent news for cosmonauts hoping to fly even longer missions. It was announced that the success of the flight had shown that orbital stations manned by *rotating* crews was feasible. This was a crucial milestone because, as Boris Petrov later put it, "the construction of an orbital station served by replacement crews was man's main road to outer space". A few days later, Valentin Glushko was quoted by *Izvestia* as saying that the psychological issues associated with weightlessness were the main problem facing further development of orbital stations. A tour of several months in a tin can would be an ordeal by any standards, no matter how interesting the assigned work.

In fact, Klimuk and Sevastyanov had been extremely productive: in addition to 9 days starting up, maintaining, and powering down the station, and 10 days off, they had spent 15 days on geophysical and atmospheric studies, 13 days on astrophysical work, and 10 days on biomedical activities. Highlights from their video downlink had been featured on the domestic television every evening. Analysis of the photographs returned by their predecessors had shown that the makeshift repair to the secondary mirror of the OST was satisfactory, so they had taken about 600 pictures of the Sun using it. They had also expanded the Earth-resources programme and snapped 2,000 photographs. It was hoped that comparing pictures taken during winter with those taken in summer would reveal new information. Many of the new pictures had been taken to reveal details of mountain ranges, currents and shelves in shallow seas, and alluvial deposits at river mouths, and later analysis of their photography revealed a pair of closely spaced tectonic faults near the River Ob in Siberia, between which a rich oil field was found. Surveys at the intersections of other newly discovered faults revealed coal and metallic ore deposits. The photographs also revealed two dozen previously unrecognised ring structures suggestive of impact events. The ultraviolet and infrared observations of the distribution of water vapour, ozone, and nitrogen oxides in the upper atmosphere established baseline data for subsequent studies. The ITS-K data showed that the upper atmosphere ranged from 400°C to 1,700°C, with the heating being due to absorption of most of the Sun's ultraviolet radiation. This was the first time the atmosphere had been studied on a global basis with sufficient spatial and temporal resolution to permit trends to be recorded in detail. Specialised satellites would be deployed in the years to come to extend this pioneering work.

By any measure, Salyut 4 had been a tremendous success, and although it could not support any more crews its usefulness was not over.

A TESTING TIME

In early November 1975, Salyut 4 made an adjustment to its orbit which presented favourable launch opportunities every two days. After having had its orbit boosted by Soyuz 18, it was now back down to a 350-kilometre circular orbit. An unmanned spacecraft lifted off on 17 November and, significantly, was named Soyuz 20 rather than being assigned to the generic Cosmos series as most other crewless Soyuz craft had been. It made a slow rendezvous and then docked. Whereas spacecraft delivering crews usually made the docking on their 17th orbit, the slow automated rendezvous had delayed it to the 34th orbit, which would eventually become the norm. Soyuz 20 was immediately powered down. It had verified the approach sequence for the long-awaited cargo–tanker version of the Soyuz, and an opportunity was taken to verify that its propulsion system would survive an extended period in space.

Soyuz 20 returned to Earth on 20 February 1976. Detailed examination revealed that some of the apparatus had deteriorated as a result of the thermal stresses due to flying in and out of sunlight, and this prompted a 90-day limit being imposed on the time that a vehicle could spend in space. Clearly, if a crew was to occupy an orbital station for longer than this, a way would have to be found to replace its ferry. The simplest means of doing this would be to have a station with two docking ports, as in any case a second port would be necessary to accommodate the cargo–tanker (there would be no point in sending up supplies unless there was a crew onboard the station to unload them). Commenting on Soyuz 20's docking, Konstantin Feoktistov noted that future stations would have "several docking units".

At the end of 1974, the orbital space station programme had been in dire straits. The first station, in 1971, had made it to orbit; the first crew sent to it had docked but been unable to gain entry; the next crew had been killed on their way home after spending 23 days in space; the second station had been lost prior to reaching orbit;

the next two had been crippled by malfunctions soon after entering orbit; Salyut 3 had successfully hosted a 16-day crew, but its second crew had had to abandon their approach. In just 14 months, Salyut 4 had advanced the programme tremendously by hosting two crews during 30-day and 63-day tours, and by playing a part in testing the resupply ferry.

The next step was now clear: to launch a station with "several docking units", at least one of which would incorporate the pipes needed for a tanker to replenish the station's propellants. Crews on very long-duration missions would have their ferry regularly replaced; this could be done either by having it fly up automatically, or as a result of colleagues making brief 'taxi' visits. It would even be feasible for successive crews to continuously occupy the station. The technology required for this ultimate goal was obviously available.

Table 1.5 Salyut 4 docking operations

Spacecraft	Docking	Port	Undocking	Days		
Soyuz 17	12 Jan 1975	0400	front	9 Feb 1975	0908	28.21
Soyuz 18	25 May 1975	~2300	front	26 Jul 1975	1356	61.62
Soyuz 20	19 Nov 1975	~1800	front	16 Feb 1976	~0200	88.33

Dates are Moscow Time.

Table 1.6 Salyut 4 crewing

Cosmonaut		Arrival	Departure	Days
Alexei Gubarev	CDR	Soyuz 17	Soyuz 17	29.55
Georgi Grechko	FE	Soyuz 17	Soyuz 17	29.55
Pyotr Klimuk	CDR	Soyuz 18	Soyuz 18	62.97
Vitali Sevastyanov	FE	Soyuz 18	Soyuz 18	62.97

THE MILITARY'S TURN

Salyut 5 was launched on 22 June 1976, and its telemetry immediately revealed it to be a reconnaissance platform. Over the next two weeks, it manoeuvred into a circular orbit at 275 kilometres. Soyuz 21 followed on 6 July with Boris Volynov and Vitali Zholobov, both military officers. Volynov had flown on Soyuz 5 in 1969. Although this was Zholobov's first trip into space, he had served with Volynov as backup for Soyuz 15. They docked the next day without incident.

The military station had no requirement for replenishment, ferry exchange or crew handover, so it had none of the features predicted for Salyut 4's successor. In fact, it was identical to Salyut 3 and its main instrument was an improved version of the folded-optics telescopic camera. In addition to military reconnaissance, the crew had attended courses in geology in order to observe geographical and morphological structures. The handheld RSS-2M spectrometer was to be used both to make further measurements of the distribution of aerosols in the upper atmosphere, to monitor ecological conditions on the surface, and to measure the distribution of water in the Volga basin – the Ministry of Land Improvement and Water Conservation had plans to divert some of the outflow of rivers in the northern Soviet Union to areas further south, and these observations were to help in assessing the degree to which southern rivers could accept increased capacity (and this study later influenced planning for several hydro-engineering projects in that region). In addition, the vast timber forest on the Augara river in the Ust-Ilim was to be studied to assess the number, type, age and quality of trees to assist in planning a major woodworking complex. A mission objective was to comprehensively photograph Soviet territory south of latitude 51° – the northerly limit of the orbital track – to facilitate the production of extremely detailed maps on the scales of 1:1,000,000 and 1:500,000 which would be used to identify tectonic structures that might have deposits of oil, gas and ores. They were also to take pictures to help to plan routes for the oil and gas pipelines to operate oil fields developed using data from Salyut 3 and Salyut 4. All of this was very practical work of immediate benefit to the national economy.

Vitali Zholobov (left) and Boris Volynov.

Several purely scientific materials-processing experiments were also to be carried out. The Sfera experiment was designed to study the process of smelting metals by passing ingots of bismuth, lead, tin and cadmium through a furnace in an effort to create perfect spheres upon solidification; Kristall would immerse a seed-crystal in a solution of potash and alum to observe how it grew in microgravity; Diffusia was to produce a homogeneous alloy of dibenzyl and toluene; Potok was to evaluate using capillary pumps to circulate liquids in microgravity (such pumps would not require electricity); and to evaluate a technique for assembly work in space Reaktsia was to use an exothermic chemical reaction to smelt nickel-manganese solder that would be used to join two stainless steel pipes.

Aviation Week & Space Technology argued that the scientific work on Salyut 5 was "window dressing" to mask its real mission of reconnaissance. Certainly, neither the television nor printed media carried the daily reports of the crew's activities that they had during the Salyut 4 flights. Drawing a parallel with Salyut 3, the American magazine predicted that Volynov and Zholobov would return to Earth after several weeks. Whilst meteorological observations were not a key part of their programme, in early August the cosmonauts reported that the clear boundary that had separated the summer anticyclone in western Europe from the cyclones in the east – which had been stable since they arrived – had begun to break up. Such synoptic observations often provided timely input to the weather forecasting service. The prolonged period of extreme weather caused by these uncommonly stable atmospheric fronts yielded droughts in the west and torrential rains in the east. Seeing the spawned vortices give rise to intense thunderstorms, the cosmonauts predicted a period of very changeable weather across the whole of Europe, and were soon proved right.

Izvestia reported on 18 August that the cosmonauts seemed to be suffering from "sensory deprivation" and that psychologists monitoring their health had suggested that music be played to them over the voice uplink. Two days later, *Radio Moscow* reported, in a very different tone, that solar radiation levels were "favourable" for "prolonged flight", prompting speculation that this crew might, after all, try to set a new endurance record. Then an early morning announcement by *Tass* on 24 August reported that the cosmonauts were in the process of returning to Earth. Later that evening, Soyuz 21 undocked and touched down in darkness 250 kilometres from the usual recovery zone in Kazakhstan. As they clambered out of their capsule unaided, Volynov and Zholobov were given a hearty greeting by astonished workers from the local Karl Marx collective. Uncharacteristically, the only mention on *Radio Moscow* was a sentence at the conclusion of its next broadcast – usually a returning crew was headlined. The doctors pronounced them to be "generally in satisfactory condition" after 49 days in space, which contrasted with the description of previous endurance crews as being in "good" health. In fact, they had been recalled about 10 days before the end of their planned two-month mission, just before they were due to start their intensive readaptation exercises. As a result, their physical condition was worse than that shown by the Soyuz 18 crew, who had been in space for longer. Nevertheless, they recovered in about two weeks. It was not officially acknowledged that the flight had been curtailed, but the landing in darkness so far from the usual zone suggested that the descent had been prompted by some kind of emergency condition. It seemed implausible that "sensory deprivation" could have merited such a sudden departure, and the fact that two men were in reasonable physical condition ruled out a medical emergency. Western observers concluded that the station must have been rendered uninhabitable by a systems failure. When the next spacecraft was launched six weeks later into a different orbit, this seemed to confirm that Salyut 5 had been written off.

A NEW CAMERA

Soyuz 22 was launched on 15 September 1976, with Valeri Bykovsky and Vladimir Aksyonov. It was announced to be a joint effort with the German Democratic Republic. The spacecraft, which had solar panels and so could remain in orbit for a week, was the backup built for the Apollo–Soyuz mission. Its androgynous docking system had been replaced by the six-channel MKF-6 multispectral camera, a bulky 200-kilogram package supplied by Carl Zeiss. The primary objective was to evaluate this camera, and some of the imagery was to be correlated with data collected by aircraft and survey teams already working on an Earth-resources programme. From its low orbit – more or less circular at 260 kilometres – the camera was able to image an area of 110 by 150 kilometres with a resolution of 15 metres, and there was a 60 per cent overlap between adjacent frames in order to provide stereoscopic views. An advanced form of the MKF was under development for an orbital station, and it said a lot for its importance that a Soyuz flight had been laid on specifically to try it out.

SALYUT 5 REPRISE – ALMOST

Vyacheslav Zudov and Valeri Rozhdestvensky set off in Soyuz 23 on 14 October to "continue scientific and technical work and to undertake experiments in conjunction with the orbital station Salyut 5", so the old station was evidently not uninhabitable. Unfortunately, during the 18th orbit it became apparent that there was a malfunction in the automatic rendezvous system.

Vladimir Aksyonov (left) and Valeri Bykovsky.

The station's transponder was confirmed to be working, but the Soyuz was unable to compute an approach. Vladimir Shatalov later emphasised that the crew's analysis of the system's difficulties had helped ground controllers to determine what had gone wrong, but he did not go on to say *what* this had been. Whatever the problem, it had prevented the spacecraft from completing its rendezvous. All the non-essential systems were powered down, and the cosmonauts settled down to wait for the next day's landing window. The capsule descended into a blizzard in darkness, was blown off course, and splashed down into Lake Tengiz. Although all crews were trained for water recoveries, none had been required to test the procedure, and it would have been difficult to pick a worse time to do so. The capsule was 8 kilometres offshore, in shallow freezing water that was so clogged by ice that the hastily requisitioned small boats were unable to reach it, and thick fog impeded the recovery helicopters. The capsule's batteries had been meant to supply power only for the descent, which took under an hour after the service module had been jettisoned, so the two men were forced to spend the rest of the night with the heaters off in

The Soyuz descent module is designed to float.

48 Getting started

order to sustain a lamp. It was dawn before a helicopter attached a line and dragged the capsule to the shore. It was fortunate that such an ordeal had not befallen Volynov and Zholobov in their weakened condition.

ANOTHER TRY

The failure of Soyuz 23 to reach Salyut 5 prompted the mounting of another flight; reboarding the station was evidently a high priority. On 7 February 1977, Soyuz 24 set off with Viktor Gorbatko and Yuri Glazkov. Gorbatko had flown on Soyuz 7 in 1969, but this was Glazkov's first flight. Although the rendezvous was concluded in darkness, using a spotlight to illuminate the station, the docking was made without incident. It was usual practice to open up the hatch and transfer to the station within an hour or so, but on this occasion the cosmonauts remained in their ferry and slept until the next day. It has been claimed in the West that Volynov and Zholobov had evacuated Salyut 5 when an "acrid odour" had polluted its air, and that the station's air was replenished from a tank in the ferry in this initial period of docked activities. Alexei Yeliseyev later acknowledged that the previous crew had not deactivated the life-support system, and that the air purifier had been turned on shortly before the docking. When Gorbatko and Glazkov opened the hatch the following morning, they reported that the station was "cosy and warm" and "comfortable", and that it was "easy to breathe". After testing the systems, they transferred replacement apparatus and performed maintenance. A few days later, in a press conference with journalists at Kaliningrad, they noted how "pleasant" it was to work on the station.

Alexei Leonov (left) with Viktor Gorbatko and Yuri Glazkov.

They started their science programme with a study of diffusion in microgravity, using a hermetically sealed capsule containing two substances in an originally solid state which were heated to 70°C to create a melt, permitted to diffuse for three days, and then cooled to resolidify to preserve the state of the mix for examination back on Earth. This was followed by another new experiment to investigate the diffusion of two organic compounds.

On 16 February, *Tass* made the surprising announcement that the flight was half over. The next day, both men developed slight head colds, an ailment rather more serious in space than on Earth because, in the absence of gravity, mucus collects in the nasal passages. Instead of taking the medication recommended, however, they spent fifteen minutes "sunbathing" in front of a small porthole, which proved to be a remarkably successful treatment. Intriguingly, on 21 February they tested a method of replacing the air in a station by venting air from one end of the main compartment while releasing air from tanks in the orbital module of the ferry, an operation which was televised.[23] As the air was exchanged, the crew reported "a light breeze" blowing through the compartments. Three days later, they wrapped up their programme and prepared to return to Earth. The results of their researches were transferred to the Soyuz 24 descent module, the spacecraft's systems were tested, and the station was returned to its autonomous operating mode. They returned to Earth on 25 February, after 18 days. Vladimir Shatalov noted that they had undertaken much "preventative maintenance" in order to ensure that Salyut 5 would be able to be remotely operated in its automated regime. The next day, it released its descent capsule containing "the materials of various researches". This had been loaded by the cosmonauts with items for which there was no room in their Soyuz, immediately before they departed. It is likely that the reason for revisiting the station was to retrieve the film which the first crew, making their hasty escape, had not had time to load into the capsule.[24]

In March, Salyut 5 recircularised its orbit at about 255 kilometres and the most recent backup crew of Anatoli Berezovoi and Mikhail Lisun began preparations for a short visit, but their impromptu flight was cancelled.

Despite rendezvous failures, the orbital station programme had clearly recovered from its inauspicious start, and was capable of supporting *routine* operations. With the 'civilian' and the 'military' versions evidently being alternated, it was universally expected that the next one would be a scientific facility with several docking ports which would be capable of accommodating resident and visiting crews, and could be replenished to sustain prolonged operation.

Table 1.7 Salyut 5 docking operations

Spacecraft	Docking		Port	Undocking		Days
Soyuz 21	7 Jul 1976	2133	rear	24 Aug 1976	1812	47.86
Soyuz 24	8 Feb 1977	2130	rear	25 Feb 1977	0912	16.49

Dates are Moscow Time.

Table 1.8 Salyut 5 crewing

Cosmonaut		Arrival	Departure	Days
Boris Volynov	CDR	Soyuz 21	Soyuz 21	49.27
Vitali Zholobov	FE	Soyuz 21	Soyuz 21	49.27
Viktor Gorbatko	CDR	Soyuz 24	Soyuz 24	17.72
Yuri Glazkov	FE	Soyuz 24	Soyuz 24	17.72

[23] This was probably a repeat of the procedure undertaken prior to their entry into the station. In all, 100 kilograms of air was replenished. This test served only to reinforce speculation that the first crew had been forced to leave in a hurry when the cabin was contaminated by an "acrid odour".

[24] As an interesting aside, this 'secret' capsule was publicly auctioned in New York in 1993.

2

Routine operations

SALYUT 6

Salyut 6, launched on 29 September 1977, incorporated the best elements of all the previous stations. It combined Salyut 4's power system (a trio of panels providing a peak of 4 kW), thermal regulation system, and navigational systems (Delta to depict the station's orbit on the control panel, and Kaskad to control its orientation) with Salyut 3's environmental systems and gyrodynes. The greater degree of automation would enable the crew to devote more time to experiments. The addition of the rear docking port required that the propulsion system be redesigned. Chelomei's Almaz had its docking port in the rear because the crew's capsule was on the front, and a twin-chamber engine had been designed to fit around the transfer tunnel. When it had been given the design and told to get it into service as soon as possible, the Korolev Bureau had replaced this untested engine with a Soyuz service module. Chelomei had continued to develop his own configuration and, after the loss of Salyut 2, the twin-chamber engine had flown on Salyuts 3 and 5. The Korolev Bureau now adopted this peripherally mounted engine in order to leave the axis free for the rear docking port, but instead of leaving the engine and its propellant tanks exposed, as on Almaz, the entire assembly, together with other apparatus, was encased in an unpressurised bay of the same diameter as the main section of the stepped-cylinder configuration. On earlier Korolev stations, the main engine had burned nitric acid and hydrazine fed by a hydrogen peroxide powered turbine, and the attitude-control thrusters had vented cold hydrogen peroxide. In this revised design, the orbital manoeuvring engines and the attitude control thrusters burned unsymmetrical dimethyl hydrazine in nitrogen tetroxide, which were drawn from a common set of tanks which were pressurised by bladders inflated by high-pressure nitrogen gas instead of a turbine. As previously, these reactants ignited on contact, which eliminated the need for an igniter. Unifying the propellant system greatly simplified the task of replenishing the station. It was not announced, but only the rear port had the plumbing to accommodate a tanker, so operations would have to be carefully coordinated to ensure that this port was free when a resupply ferry arrived. Salyut 6 introduced an improved transponder system for automatic dockings. Like its predecessor, this Igla system would turn the station to face the appropriate port towards the inbound ship to simplify the manoeuvres in the final approach. It provided the ferry with range, closing rate, line-of-sight angular velocity, and perpendicular deviation from the straight-in vector. It paused when the range was 200 metres, rolled the ship to the proper alignment, and then flew straight in to dock.

The main compartment was laid out like previous Korolev stations, except that there was now an axial tunnel behind the massive conical housing of the primary instrument, leading to the rear docking port. As on Salyut 4, the forward transfer compartment could serve as an airlock, and it had an inward opening side-hatch for extravehicular activity. It contained two semi-rigid spacesuits that were designed to facilitate greater mobility than previous suits. The entire backpack was hinged, and a cosmonaut entered the suit by the large oval hinged hatch in its rear, which formed a hermetic seal when swung shut. In an emergency, it could be donned in five minutes. The integrated backpack contained the life-support systems, which were accessible for maintenance while the hatch was open. It could support several hours' work in space, but required an umbilical for communications and power, so a spacewalking cosmonaut could not possibly float off. As the suits had to accommodate successive crews, they were adjustable. One of the portholes in this compartment was designed to pass ultraviolet sunlight, so that this would act as a germicide to sterilise the air. It would also permit the cosmonauts to recover from any vitamin-D deficiency.[1]

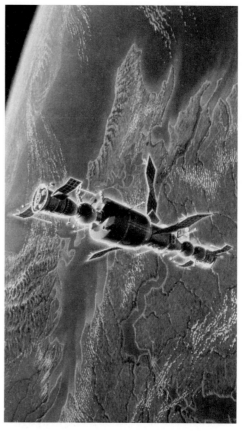

A painting of Salyut 6 by A. Sokolov. Courtesy of *Tass*.

The primary instrument was the BST-1M telescope for infrared, ultraviolet and

---

[1] Even better, synthesis of excess vitamin-D would assist calcium absorption in the intestines, and reduce the build up of calcium leached from bone tissue in response to prolonged weightlessness.

submillimetre astronomical observations. With a 1.5-metre-diameter mirror, this was larger than most ground-based instruments. Although it operated in the vacuum of space, its sensors had to be cooled to $-269°C$. This was achieved using an improved version of the closed-cycle cryogenic unit tested on Salyut 4 with the ITS-K. Helium was produced using a compressor, a pair of refrigerating units and intermediate heat exchangers, and chilled by being fed through an expanding-throttle valve. Although this apparatus consumed 1.5 kW, once the sensor had been chilled to its operating temperature it could be maintained for an extended period at little extra cost. However, the telescope could be used only when in the Earth's shadow, and for the rest of the time its cover had to be closed to protect it.

The second major instrument was the 6-channel MKF-6M multispectral camera for Earth-resources observations. This was an improved form of the camera that had been tested on Soyuz 22. At Salyut 6's higher orbit, which was more or less circular at 355 kilometres altitude, however, each photograph captured an area of 165 by 220 kilometres with a resolution of 20 metres. Simultaneous imagery was secured in six spectral bands (four visual, and two in the infrared) in cassettes with 1,200 frames. Because film deteriorated rapidly in the increased radiation environment in orbit, the exposed film would have to be returned before it could fog over, so long-term use of the camera would rely on spacecraft visiting on a regular basis. Salyut 6 also had a KATE-140 stereoscopic topographic mapping camera (so designated because it had a focal length of 140 millimetres) that operated in the visual and infrared spectrum. From the station's altitude, its 85-degree field of view could photograph an area of 450 by 450 kilometres with a resolution of 50 metres, and either produce individual frames or generate an extended strip image. Each cassette had sufficient film for 600 pictures. Once it had been set up, the camera could be operated either by the crew or by remote control. The imaging capability of this station, therefore, was formidable. The Ministry of Agriculture had planted specific cereals, vegetables and grasses at test sites at Salsky near Rostov in the Ukraine and at Voronezh near Lake Baikal to assess the capabilities of the cameras.

As Salyut 6 could be resupplied, a variety of biological and materials-processing apparatus was to be delivered to augment that fitted at launch, so that, over time, its science capability could be greatly expanded. It was to become a space laboratory to support long-term projects. The Biosfera experiment, for example, was to start with visual observations of the oceans. What was observed was to be measured against a chromatic chart. Compiling observations of the optical properties of the atmosphere in different conditions would reveal the optimum conditions for observing a variety of phenomena. Later, hand-held cameras were to be used to test this analysis using various films. Later still, spectroscopic studies were added. All of this was designed to improve the effectiveness of an orbital station. As Biosfera data accumulated, it would map the boundaries between fertile and arid terrain, determine the capacity of rivers and run-off from glaciers and snowfields, and assess oceanic and atmospheric pollution. Its overall purpose was to provide a global overview for assessing various Earth-based ecological projects.

The station had been made as comfortable as possible, because cosmonauts were to spend periods of several months living in its cramped confines. The

54 Routine operations

suggestions by the inhabitants of earlier stations had been heeded, and particular attention had been paid to soundproofing the many motors and pumps to reduce the general level of background noise. Previous crews had derived comfort from hearing the various noises, because they were indicative of the state of the different systems, but they had nevertheless argued for a much lower level of ambient noise. Similarly, although it was perfectly feasible to attach sleeping bags to the ceiling, the Almaz crews had felt better with specific sleep facilities, and so recognisable 'cots' had been installed to offer a sense of functionality, although one was on the roof and the other was on a wall. Also exploiting experience on Almaz, a shower had been installed. Located at the rear of the compartment, this was a deployable polythene tube that relied upon forced airflow and a suction device to direct and collect a spray of hot water, which was recycled, purified, mineralised and later used to rehydrate prepackaged food. An improved menu enabled the cosmonauts to have a different meal each day of the week, and an improved diet provided a daily intake of 3,200 calories. Many items of food had been broken into bite-sized chunks to make them easier to handle in space, and flaky items such as cakes had been coated with an edible film to lock in crumbs. Some 65 types of food had been prepared prior to launch. To prevent dehydration (caused by increased urination as the body tried to remove what it perceived to be an excess of fluid in its upper torso) each cosmonaut was required to ingest 2 litres of water per day – this included that used to rehydrate food. All consumables could be replenished by a cargo ferry as required. In the centre of the main compartment, between the bulk of the telescope and the control panel recessed into the floor, was the large table for food preparation, and to either side of the table were lockers. For relaxation, there was a stock of novels, a library of cassettes, and a chess set with magnetic pieces on a metal board. The gymnasium had a veloergometer, a treadmill and a mass-meter (a spring-loaded apparatus on the ceiling used by the cosmonauts to weigh themselves). In the roof there were two small airlocks which could be used either for scientific apparatus or to eject rubbish. Most of the scientific apparatus, the shower, toilet and gymnasium were in the aft section of the main compartment, the front housed the control panel and a variety of storage lockers. Most of the 20 portholes were dedicated to specific equipment. There was one colour and two monochrome television cameras mounted internally and a monochrome one looking out from each docking unit. The colour camera (the first on a Salyut) was fixed above the forward hatch of the main compartment, looking aft to enable the flight controllers (and the psychologists) to monitor the crew's activities in the brief periods when the station was within communications range.

Ignoring the differences between the two variants of the Almaz produced by the Chelomei and Korolev Bureaux, and the improvements made with each successive vehicle, the introduction of the two-ended version offered such a significant increase in capability that it can justifiably be said to be a new *generation* of orbital station, marking the switch from engineering development to routine operations. Its systems had been designed to operate for 18 to 24 months, during which time it was hoped that it would be able to host a 90-day, a 120-day and a 175-day crew, each of which would be visited by brief Intercosmos missions as required to ensure that the docked Soyuz ferry never expired its 'service life'. Although a continuous orbital presence was the ultimate goal, given the state of the technology this was too much to expect at this stage, so the immediate objective was to occupy Salyut 6 for at least 50 per cent of its design life; anything beyond this would be a bonus.

FRUSTRATION

Soyuz 25 set off on 9 October with Vladimir Kovalyonok and Valeri Ryumin, both of whom were making their first flight. The spacecraft followed the usual approach and, the next day, was close enough to Salyut 6 to start the automated rendezvous system. When this paused at 200 metres, Kovalyonok took over and soft docked. Unfortunately, although the probe retracted, the latches in the annular collar did not

Valeri Ryumin (left) and Vladimir Kovalyonok.

engage, preventing the establishment of a hermetic seal in the transfer tunnel. While the ferry remained loosely docked, the flight controllers considered possible failures and recovery procedures. Finally, Kovalyonok was ordered to withdraw and try the docking sequence again, which he did with no better result. He undocked again and made a third attempt, this time with greater force, but this, too, failed to engage the latches. Although the cosmonauts requested permission to fly around to try to dock at the rear port, this was denied because the engineers thought the fault was in the ferry's collar rather than the station's (unfortunately, because the docking system on the orbital module could not be returned to Earth for study, there was no way to be sure why its latches had failed to engage). The dejected crew were told to withdraw and return to Earth at the next day's

The Kaliningrad control room during the Soyuz 25 mission.

landing opportunity. This failure to dock was a poor start to operations with the new station.

SECOND TIME LUCKY

Yuri Romanenko and Georgi Grechko were launched in Soyuz 26 on 10 December, and the following day they docked at Salyut 6's rear port without incident. This was Romanenko's first flight. Having served with Romanenko as backup for Soyuz 25, Alexander Ivanchenkov had hoped to fly next, but he had been replaced by Grechko, who had spent a month on Salyut 4 and was very familiar with the mechanics of the docking system.

Georgi Grechko (top) and Yuri Romanenko prepare to depressurise Salyut 6's forward compartment.

On 20 December, Romanenko and Grechko sealed themselves into the forward transfer compartment, donned the spacesuits and depressurised the compartment to make the first Soviet extravehicular activity since Yeliseyev and Khrunov transferred from Soyuz 5 to Soyuz 4 in 1969, which was itself the first since Alexei Leonov's pioneering spacewalk in 1965. Although a hatch had been provided specifically for spacewalks, in this case Grechko swung in the drogue assembly of the front docking unit. He examined the conical receptor and reported that it seemed to be in perfect condition. It was, he said, "without a single scratch" just as if it was newly delivered from the factory. Then he poked his head and shoulders out of the tunnel to inspect the external facilities. He examined the collar of the docking unit to see if any damage had been caused by Soyuz 25's repeated attempts to link up, methodically examined the various joints, sensors, guide pins, fasteners and seals, used special tools handed out by Romanenko to exercise the mechanisms, inspected the radar transponder, and used a portable television camera to show everything to the engineers at Kaliningrad. Romanenko remained in the compartment and held Grechko's feet, to anchor him so that he would have both hands free. Having verified that there was no damage to the docking unit (in other words, that the problem had indeed been in the ferry's docking assembly) Grechko moved out and affixed the Medusa experimental cassette (which held biopolymer compounds) onto an attachment on the outer surface of the forward compartment. When Grechko re-entered, Romanenko, against instructions, stuck his head and shoulders out to admire the view. Seeing that his commander's tether was loose, Grechko instinctively grabbed for it, but there was no need because their suits were permanently attached to the station by umbilicals. They neglected to tell the flight controllers of this breach of the rules. After their return to Earth, however, Grechko mischievously gave the impression in an interview of narrowly managing to save his commander from drifting away to his death!

There was some concern among the flight controllers while the airlock was being repressurised, as telemetry indicated that the valve used to vent the final vestiges of air to enable the inward-opening hatch to be opened had not shut, in turn suggesting that the compressed air that was flooding into the airlock was leaking to space. The fact that this ate into the limited reserve of pressurised air was not the main issue – unless they could pressurise the airlock, Romanenko and Grechko would not be able to re-enter the main compartment, and with their ferry at the other end of the station they would have nowhere to go. Depressurising the entire station would be difficult. This state of anxiety lasted about half an hour, until it was established that the valve had closed and the airlock was repressurising. It was found that a cable in the airlock had created the spurious signal. Later, Grechko said that his impromptu spacewalk was a "tremendous pleasure". Vladimir Shatalov pointed out that external activities would be an integral part of orbital operations, for which all cosmonauts would train in a neutral buoyancy tank.

Salyut 6 cosmonauts followed a different daily cycle from that used on Salyut 4. Adapting the cycle in order to make full use of the periods when the station was in radio contact with Kaliningrad had disrupted the circadian rhythm, which over a long period had been debilitating. Salyut 6 crews adopted Moscow Time and followed a regular schedule. At 0800 they would wake up, shave and exercise. They would have breakfast at 0845, and at 0945 would have a medical check and inspect the station's systems. They would take their early break at 1045, start work at 1100, and at 1230 have lunch and exercise. Work would resume at 1500 and run through to 1800, when they would have their evening meal, after which they would do whatever was most urgent until 2300, when they would retire. This liberation from the constraint of the communications network was possible because sophisticated data recorders were to store experimental data while out of radio contact, and then 'dump' it either to the ground stations or to the communications ships, which then relayed it to Kaliningrad via the Molniya satellites.[2]

Two weeks into the mission, *Tass* announced that while Grechko was behaving in

---

[2] A system of geostationary relay satellites was planned to maintain continuous contact with future stations.

58 Routine operations

Yuri Romanenko (left) and Georgi Grechko.

an "even tempered" manner, Romanenko was much more "volatile", and whereas Grechko tended to relax in the forward transfer compartment making sketches and taking pictures of the Earth, Romanenko spent much of his free time monitoring the control panel. They were clearly different personalities but, as an improvised team, they were well matched. They welcomed in the New Year with fruit juices, and since their orbit took them across the International Date Line sixteen times per day, it was fortunate that alcohol was prohibited.[3] Because they were working more effectively than expected, and were ahead on the flight plan, on 3 January 1978, having already completed their assigned technology experiments, they asked Kaliningrad to suggest more work. They observed the Earth, studying a strong ocean current skirting South America whose boundaries were easily identified by the sharp contrast in the colour of the water, and ventured that their atlas of ocean currents was incorrect. They also observed glacial melting in the southern hemisphere, and reported that whereas some glaciers were sky blue others displayed a striped pattern. Most of the glacial studies were on Soviet territory however, to help to forecast glacial flow into rivers in order to predict water levels, and to measure the water flowing into the deserts of Soviet Central Asia. One highlight was spotting for the first time the snowy peak of Mount Fujiyama near Tokyo – it had previously been obscured by cloud. Continuing earlier studies, they reported on noctilucent clouds in southern sub-polar latitudes.

VISITORS!

History was made on 11 January when a spacecraft approached an already occupied station. Soyuz 27 docked at the front port without incident. Some engineers had feared that because a ferry hits the docking port at about 0.3 metres per second with a force of about 20 tonnes, the propagation of the brief but very sharp 40 g shock through the station's structure might crack the hermetic seal of the collar of the ferry docked at the other end. Romanenko and Grechko had sealed themselves into their

---

[3] Romanenko and Grechko were the first cosmonauts to celebrate New Year in space; only the final Skylab crew had had this dubious pleasure before.

Vladimir Dzhanibekov (left) and Oleg Makarov.

ship, therefore, just in case. As soon as it was confirmed that the seal had held, they re-entered the station and prepared to greet their visitors. The usual post-docking procedures had to be followed though, so it was over two hours before the valve was opened to equalise the pressures so that the front drogue assembly could once more be swung back. After a month with only each other for company, the residents were naturally delighted to welcome their visitors. What Vladimir Dzhanibekov and Oleg Makarov made of the smell that wafted through when they opened the hatch is not recorded, but one consequence of weightlessness is that stomach gases do not rise to the gullet; instead they pass through the intestines and give rise to intense flatulence of an aromatic nature. The olfactory senses of the residents would have become desensitised, and the air quality would have been very striking to the newcomers, the first people to visit an already inhabited station. A television camera recorded them being enthusiastically dragged in through the hatch, hugged and offered the traditional bread and salt, which was washed down with a toast of cherry juice. The newcomers performed a standard set of cardiovascular tests with the Polynom-2 multifunction apparatus as part of the effort to study the process of adaptation to the absence of gravity during the first week in space. Each visiting crew was to undertake a similar battery of tests in order to provide a statistically valid database.

As this was the first time that an orbital complex had been assembled using three vehicles, an engineering objective was to determine the dynamic

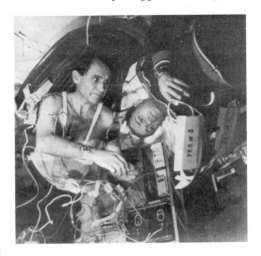

Oleg Makarov (left) and Vladimir Dzhanibekov.

characteristics of the structure, which was now 30 metres long with a mass of just over 32 tonnes. Adding a ferry changed the centre of gravity, and it was important to verify that the Kaskad attitude-control system could accommodate this change of mass distribution. It was also necessary to determine the stresses placed on the structure – in particular the docking collars – under different loads. The Resonance experiment, in its simplest case, required a cosmonaut to jump on the treadmill at intervals determined by a timing signal transmitted by Kaliningrad so that instruments at various points within the complex could measure the propagation and damping of the resulting vibrations. This would identify any resonance which might induce metal fatigue. Although the designers had made every effort to eliminate dangerous vibrations in ground trials, it was essential to verify that the structure was indeed sound. Dzhanibekov also did preventive maintenance on the electrical systems, and replaced a voice transmitter that had failed. Soyuz 27's orbital module was stocked with fresh food, newspapers, books, mail and a variety of other supplies, including the joint Soviet–French Cytos experiment. A variety of micro-organisms had been delivered in a thermostat which kept its contents at 8°C in order to inhibit their development in transit. The French experiment used paramecium protozoa and the Soviet one used an aquatic proteidae. These were placed into the apparatus, which was maintained at 25°C to facilitate their development, and chilled after 12 hours for return to Earth for analysis. It was a simple test designed to determine the effects of microgravity and space radiation on the kinetics of cell division – this being the very basis of the life process. It was part of an investigation of the long-term ability of organisms to live in space.

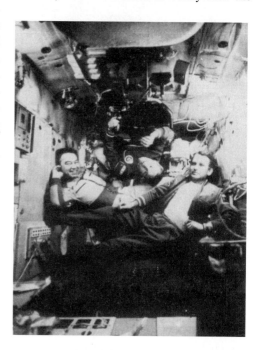

Georgi Grechko (left), Vladimir Dzhanibekov and Oleg Makarov.

Although a ferry can be launched at any time when the station's orbital plane passes through the cosmodrome, ideal recovery windows occur only at two-monthly intervals. This interval is due to the fact that the Earth is not perfectly spherical but oblate, with the gravitational attraction of its equatorial bulge acting on a satellite's angular momentum in such a way as to make its orbital plane precess. The rate of this rotation is dependent on the inclination of the orbit. At the inclination used by the second-generation Salyuts this was 58 days. This arcane orbital dynamic dictated the timing of the visiting flights to these stations. Each recovery window lasted for

Georgi Grechko (left), Oleg Makarov, Vladimir Dzhanibekov and Yuri Romanenko.

about ten days. Visiting ferries set off early in a window, and returned towards its end, which placed a limit on the time that such visits could last. Visits using ideal recovery conditions could be mounted only at two-monthly intervals, and could last only about a week. Dzhanibekov and Makarov were to return to Earth in Soyuz 26, so they transferred their Kazbek contoured couch liners and Sokol pressure suits from Soyuz 27, stowed the various packages that they were to take away, and then reset the adjustable weights on the couches to correct the capsule's centre of mass. On 15 January, Dzhanibekov powered up Soyuz 26 to verify that it had not been damaged during its month in space, ran through all the procedures required to return to Earth, tested its attitude-control thrusters, and briefly fired the main engine. The next day, he and Makarov undocked and returned to Earth, leaving Romanenko and Grechko with a brand new ferry with which to continue their planned 90-day mission. Alexei Leonov emphasised that Dzhanibekov and Makarov had done an enormous amount of work. Valeri Kubasov added that the ability to visit an occupied station not only meant that a scientist could be sent up to undertake a specific experiment, but also that an engineer could install new (or repair faulty) apparatus, in parallel with the ongoing mission.

The failure of Soyuz 25 to dock at Salyut 6's front port had seriously disrupted the planned operations. While the station was occupied, it required replenishment at roughly two-monthly intervals. It had expended considerable propellant climbing to its operating altitude. Also, much of the apparatus for science experiments was to be installed in orbit, so the first cargo delivery had been scheduled early. A tanker could replenish the station's propellant only through the rear port, so the station could not be resupplied until this was clear – the exchange of Soyuz ferries had achieved this. In effect, therefore, Soyuz 27 had been an extra flight inserted into the schedule to restore the docking sequence.

Romanenko and Grechko took the next day off, even although this was not a scheduled rest day, in order to recover from the intensive joint programme. To help them to relax, Kaliningrad relayed radio coverage of a Soviet–Canadian ice hockey match. They spent the next two days photographing vast tracts of the Central Asian Republics with the MKF-6M camera in a search for mineral deposits. Meanwhile, at the cosmodrome, the first supply ferry had been set up on the pad.

RESUPPLY

Launched on 20 January, the new vehicle was named Progress. As expected, it was a modified Soyuz. Because it was not required to return to Earth, the descent module with its heavy heat shield and landing system had been replaced by a cylindrical compartment with tanks for 'wet' cargo and the navigation and control packages that were normally in the descent module. The 'dry' cargo was in containers in the orbital module. This new configuration was about half a metre longer than the Soyuz ferry. The only difference in the carrier rocket was the omission of the escape system – in an accident the spacecraft was expendable. Unlike the Soyuz, Progress had a unified propulsion system based on that introduced by Salyut 6. It did not have solar panels but because it was not required to support a crew it could sustain up to a week of independent operations. It pursued a two-day rendezvous which resulted in a much slower closing rate that simplified the task of the automatic approach system. This procedure had been demonstrated by Soyuz 20 during its approach to Salyut 4 three years earlier. To enable the flight controllers to supervise the final approach and, if necessary, take control, there was a television camera on the front of the vehicle that provided a view similar to that through the Vzor periscope of a Soyuz, and the ship had a spotlight to assist in docking in darkness. One of Kaliningrad's control rooms monitored Salyut 6, and the other the manoeuvring spacecraft. It had originally been intended that the cosmonauts would retreat to their ferry, just in case the impact of the automatic craft damaged the station, but after the success of Soyuz 27's

Progress 1 approaches Salyut 6.

docking they had been given permission to remain in the station. However, they manned the controls, ready to manoeuvre the station to compensate for deviations in the ferry's approach, but the newcomer made a perfect straight-in approach and docked on its first attempt. Although Romanenko and Grechko had been scheduled to take the rest of the day off, they were given permission to open the hatch immediately to retrieve their gift packages. Alexei Yeliseyev, a cosmonaut now serving as a flight director, pointed out that although there were still "many problems" to be solved in operating an orbital station, this docking proved that they had finally overcome the restrictions imposed by the limited amount of apparatus and consumables that could be installed in a station prior to launch. This achievement was "a really big step forward" in the exploration of space.

Of the Progess's 7-tonne mass at orbital insertion, 2,300 kilograms was cargo. Of this, 1,000 kilograms was propellant for the station. The orbital module (which had a volume of just 6 cubic metres) was criss-crossed with a metal framework supporting large containers which could be released simply by turning a number of locking bolts through 180 degrees. Each such container carried a variety of individual items. Bulky apparatus was bolted directly onto the frame. The compartment was so full that the docking probe could barely be swung back into the compartment. An inventory and a locator had been stowed near the hatch to assist in unloading. The cargo included compressed air, food, water and other consumables. The two men squeezed into the compartment and closed the hatch behind them so that metal, wood and rubber chips from the packaging could not drift into the station. They wore face masks to prevent inhaling fragments and goggles to protect their eyes. With two people working so hard in such a small volume, the compartment was soon humid and stuffy. A small vacuum cleaner was used to clean up the debris prior to reopening the hatch. After being transferred into the station, every item was either stored in a locker or installed on the appropriate experiment site. The main experiment this time was the Splav-1 furnace. Spare parts (in this case an orientation sensor) were included as required. Every cargo would include items for routine maintenance, such as the 1-metre-long cylindrical canisters of potassium superoxide for the air regulation system. Much of the food was fresh, and apples, onions, caviar and garlic were particularly welcome. Water was supplied in small spherical flasks, each of which contained 5 kilograms of water.[4]

A set of pipes had been installed to feed propellant through the orbital module to the docking collar. Previous stations had not been capable of replenishment, so this transfer system had never been tested. The first step was to verify the integrity of the pipes by pressurising them with nitrogen, a process which involved proving the hermetic seal of each manifold and valve in the path. This could be controlled either by Kaliningrad or by the crew, and on this occasion the cosmonauts supervised the task. The next step was to reduce the pressure in the station's tanks. Propellant was fed to the engines by metallic bladders within the tanks. These bladders were driven by nitrogen gas at a pressure of 20 atmospheres. To enable the pump in the ferry to force propellant into the tanks against the bladders, the nitrogen pressure had first to be reduced. A 1 kW compressor had been incorporated into the station to drive the nitrogen back into its bottle to relax the bladders, but the nitrogen was stored at 220 atmospheres and the compressor was power hungry. This preliminary operation had to be timed not to place too great a load on the station's power supply, which meant that it could only be done while in sunlight, and it was usually staggered over several days. Whilst this was a slow process, it was simple and reliable, and the use of the bladder guaranteed that there would be no gas bubbles in the propellant supply (any gas in the flow would disturb combustion and cause the engine to misfire). Once the pressure was down to 3 atmospheres, the ferry's pump (pressurised by a bladder at

---

[4] Manual transfer of water was soon deemed to be inefficient, and on later stations the Progress was able to pump water directly into the station's tank.

only 8 atmospheres) forced propellant into the station. After the oxidiser had been transferred, the pipes were vented to clean them before the fuel could be transferred. If the station had used a lot of propellant, it might be necessary to top up each of its independent pair of oxidiser and fuel tanks. Clearly, given these constraints, it could easily take a week to pump onboard the tonne or so of propellant the ferry carried. Afterwards, high-pressure nitrogen cleaned the pipes, to ensure that when the ferry undocked it would not leak corrosives onto the rear of the station.

Even after it had been completely unloaded, the ferry was retained to serve as a 'tug' to raise the station's orbit, further reducing its propellant consumption. Before such manoeuvres, all loose apparatus had to be strapped down. Before the ferry was jettisoned, dirty clothing, empty water flasks, food containers, clogged air filters and saturated carbon dioxide scrubbers were stowed in the racks in its orbital module. Care had to be taken not to offset the vehicle's centre of mass, or its attitude-control system would not be able to orient it accurately once it resumed independent flight. It undocked on 7 February, dropped some 14 kilometres behind the station, and then activated its backup rendezvous system and initiated a second approach, but as soon as it was evident that this was working it was commanded to withdraw, and was de-orbited over the Pacific Ocean the next day and – as the communiqué put it – "ceased to exist" when it burned up on re-entering the atmosphere.

By early February, the cosmonauts admitted that they were suffering from acute home sickness and were looking forward to finishing their mission. On 10 February they celebrated the 150th anniversary of the birth of Jules Verne by noting during their daily television broadcast that whereas one of the famous writer's characters had circled the world in 80 days, they did so in almost as many minutes. They also observed that they had circled the Earth more than 1,000 times. The next day, they broke the 63-day endurance record established in 1975 by Klimuk and Sevastyanov on Salyut 4. Dr Anatoli Yegorov, the head of the medical team in Kaliningrad, said that their latest medical check-up had confirmed that both men had attained "a stable adaptation" to weightlessness. They received a radio message from Thor Heyerdahl, who was sailing across the Indian Ocean in *Tigris*, a reed boat that demonstrated the technological gulf between past and present modes of transport. After chatting with the previous holders of the record, Romanenko and Grechko resumed photography of Siberia with the MKF-6M. Among other things, this imagery helped in assessing the extent to which silkworms infested the forests – diseased trees could be spotted more easily from space using multispectral techniques than they could be by visual examination on the ground – and enabled pesticides to be more efficiently applied. In contrast to observations by the crews of earlier stations, 90 per cent of the pictures taken by Romanenko and Grechko were at the specific request of specialists.

In mid-February the Splav furnace was installed in one of the scientific airlocks. This apparatus took the form of a broad disk attached to a thin cylindrical stem, and it was put in the airlock so that it could radiate excess heat to vacuum. It operated at 1,100°C, and its electronic control system could maintain this temperature to within 5°C. It incorporated a set of molybdenum reflectors to focus the heat onto a shaft containing three ampoules. The fused ampoules were automatically transferred to a 'cooling chamber' with a linear thermal gradient that ensured optimum conditions for the formation of a monocrystal. When this had cooled to 650°C, it was transferred into a third chamber which was maintained at this temperature to sustain the three-dimensional crystallisation process over several days, with the growth of the crystal recorded by a time-lapse camera. Although the Kaskad attitude-control system was deactivated so that the experiment would not be disturbed, the cosmonauts reported that they could see tiny imperfections in the crystal due to the vibrations caused by their moving around in the station. One experiment which the cosmonauts welcomed was Svezhest. Although this made the station slightly more comfortable by ionising the air, they also liked it because it released a refreshing scent of pine. By now they were growing tired, but their productivity was still 10 per cent in excess of that on the flight plan – a similar trend had been observed in the case of Salyut 4, when Klimuk and Sevastyanov developed a 'second wind' towards the end of their tour.

A CZECHOSLOVAKIAN VISITOR

Despite the frustration of Soyuz 25, the ferry exchange and the replenishment flight had demonstrated that the technology to sustain an orbital station for an extended period existed. Although the automated docking system made it possible to send up an unmanned replacement ferry with extra supplies and have the expired ferry return with the results of experiments, it had been decided to have a pilot deliver the new vehicle and take away the old one, and to offer the second couch on such flights to a guest nominated by the individual member states of Intercosmos, the international space research organisation run by the Soviets in partnership with fraternal socialist states. With Romanenko and Grechko's record-breaking flight going so well, it had been decided to fly the first of these Intercosmos missions just before the residents were due to pack up and return to Earth. On 2 March, therefore, Soyuz 28 lifted off with Alexei Gubarev and Vladimir Remek. To mark the fact that a Czechoslovak was to be the first person to fly in space who was neither a Soviet nor an American, the launch ceremonies were attended by Boris Petrov, head of Intercosmos, and Jaroslav Kozesnik of the Czechoslovak Academy of Sciences. Georgi Beregovoi, now heading the Gagarin Cosmonaut Training Centre, announced that researchers from Poland and the German Democratic Republic would visit Salyut 6 later in the year, and that candidates from Bulgaria, Hungary, Cuba, Mongolia and Romania were to train for later flights. The programme had clearly taken a giant step forward during the time that the last Almaz had been in orbit – the difference in capability between Salyut 4 and Salyut 6 was marked. The docking of Soyuz 28 at the rear port was a welcome reunion for Grechko and Gubarev, who had spent a month together on Salyut 4. The next day, the residents exceeded the 84-day endurance record for a single mission which had been established by the final Skylab crew, so they now held the principal record. The flight controllers read up various congratulatory messages, including one from the former record holders.

The visitors carried out the standard cardiovascular tests in order to add another

Routine operations

Vladimir Remek (rear) and Alexei Gubarev.

chapter to the file on adaptation to weightlessness. Comparing data from crews had shown that the Tchibis lower-body negative pressure suit inhibited blood pooling in the torso, and accelerated adaptation to weightlessness. They also performed Opros, which involved filling out a comprehensive questionnaire concerning their eating and sleeping habits, level of physical fitness and posture, their sense of smell, vision and hearing and other factors in an effort to correlate their physical and mental state.[5] In general, visitors would not be allowed to interfere with the residents' programme, but Romanenko and Grechko had already completed the majority of their own work and they helped their guests with some of *their* experiments. Remek's Czech-built Oxymeter correlated the concentration of oxygen in body tissue with adaptation to weightlessness. Splav experiments yielded an extremely pure semiconductor crystal and a glass-melt more homogeneous than anything which could be produced on Earth but, while it yielded a large crystal, there were physical deformities (such as a helical surface) caused by micro-accelerations due to ongoing operations on the station. One advantage of having so many visits was that once film which had been returned to Earth by the first visitors had been developed, prints could be sent up with the next visitors to indicate sites for further study, and this feedback increased the residents' sense of achievement. The operational

The Kaliningrad control room on the day of Soyuz 28's launch.

---

[5] The visitors filled in the form on a daily basis, and the residents did so once a week.

constraint that the visiting mission fit within a single recovery window required that such missions be brief, which meant that the visitor could not be assigned complex tasks. Nevertheless, the public relations value of having an operational facility accommodating a succession of foreign researchers was tremendous – especially at a time when the great rival, America, was unable to mount *any* sort of space mission. Vladimir Shatalov reflected that the two crews had worked well together. Nevertheless, as Remek would later wryly reflect, the back of his hand had soon turned red because whenever he reached out his hosts would slap it and yell, "Don't touch that!"

Vladimir Remek (left), Alexei Gubarev, Georgi Grechko and Yuri Romanenko.

Gubarek and Remek landed on 10 March. They did not exchange ferries because the main mission was drawing to a close – if Soyuz 27 had not been added to the schedule, this Intercosmos flight would likely have occurred earlier, and *would* have swapped ferries.

END OF A MARATHON

On 16 March Romanenko and Grechko returned Salyut 6 to its automated regime. The visitors had taken away most of their results, so they stowed what remained in Soyuz 27's descent module, powered up the ferry, tested its engine, and returned to Earth. Following a medical examination at the recovery site, the doctors announced that (considering what they had done) both men were in excellent health. This was a little misleading in Romanenko's case, however. For the last few weeks he had been suffering severe toothache. All that the medics had been able to recommend was that he wash out his mouth with warm water and keep warm. By the time he returned to Earth, the nerve was exposed and giving him great pain. A few days later, Dr Anatoli Yegorov ventured that cosmonauts should be able to "live and work in space for a year or longer". Dr Oleg Gazenko congratulated Romanenko and Grechko on having completed all their tasks "with unflagging interest and even with inspiration". They had extended the record to 96 days, and Grechko's accumulated total was now

Mi-8 helicopters land on the snowy steppe to recover Georgi Grechko and Yuri Romanenko.

125 days. Grechko found readaptation to be easier than after his first flight. During their first few days, they had difficulty coordinating their actions, such as when picking up a cup of tea, and when they awoke they attempted to swim out of their beds. Dr Robert Dyakonov, their personal physician, explained that this was because they had grown so accustomed to the absence of gravity that most actions now required deliberate effort – in effect, they were "still up there in space, not only physically but also mentally", and even tuning a radio dial needed concentration.

Boris Petrov said that no further missions would be sent up to Salyut 6 until the results of this latest flight had been assimilated. So far, the new station had been an outstanding success. From this point on though, in an engineering sense, they would be pioneering.

Georgi Grechko (left) and Yuri Romanenko after their mission.

A SECOND LONG-DURATION MISSION

In early June 1978, after three months in a virtually powered down state, Salyut 6 recircularised its orbit at 350 kilometres and reactivated its environmental systems to prepare for its next crew. Soyuz 29 was launched on 15 June. It was commanded by Vladimir Kovalyonok, who had suffered the frustration of Soyuz 25's failure to link up. His flight engineer, Alexander Ivanchenkov, had been dropped from Soyuz 26 to make way for Grechko, and this was his first flight. They had been thrown together by fate to be the second resident crew. This time, Kovalyonok had the satisfaction of hearing the latches engage with the station's front port. They wore Penguin suits from the start, so as to exercise their muscles during the initial phase of adaptation to weightlessness, and Dr Dyakonov noted that they adapted much more rapidly than their predecessors. Unlike Romanenko and Grechko, who had opted to miss sleep in order to get ahead of schedule in activating the station, and had suffered as a result, this crew had been told not to overwork during the initial phase of adaptation. Their research programme was a continuation of that before: Earth-resources, atmospheric phenomena, astrophysical studies and materials-processing. As part of their training they had flown at high altitude in an aircraft equipped with an MKF-6 camera, and specialists had pointed out the types of feature they were to seek when they made similar observations from orbit.

At the suggestion of Romanenko and Grechko, it had been decided to alter the work schedule. Whereas their predecessors had used a fixed schedule that matched their natural sleep patterns but fragmented their day, Kovalyonok and Ivanchenkov were to adopt a more conventional '9 to 5' working day and switch to a 5-day week, doing whatever they wished in free time. This crew-orientated approach contrasted with the way that the Skylab programme had evolved, where a study of how long it had taken the first crew to perform each task had enabled mission planners to set a schedule that assigned a coordinated series of tasks to each astronaut for essentially each minute of each day of the later missions. This mission-orientated approach had prompted the third crew to rebel, to gain a degree of control over their schedule and to create time off – time off was time wasted to the mission planners. However, on Romanenko and Grechko's flight (which had broken the Skylab record) there had been no such revolt because the flight controllers had left the cosmonauts alone. It was hoped that the second expedition would be able to further extend the record, and Romanenko was placed in charge of the support team to ensure that the controllers were in tune with the crew's mood. The majority of NASA's astronauts were pilots first and foremost, and served in the control centre only to communicate with a flight crew using familiar vernacular. There was less demarcation at Kaliningrad. Many of the recently recruited cosmonauts were engineers from the design group which had developed Salyut, and veteran cosmonauts were serving as flight directors.

Once Salyut 6 had been fully reactivated, Kovalyonok and Ivanchenkov carried out maintenance chores. This included replacing the pressure indicator in the airlock that had caused Romanenko and Grechko consternation after their spacewalk, and replacing a faulty ventilator in the Splav furnace. They then oriented the station to exploit gravity-gradient stability, turned off the Kaskad attitude-control system, and started an

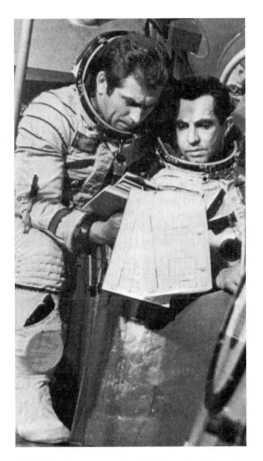

Vladimir Kovalyonok (left) and Alexander Ivanchenkov.

Alexander Ivanchenkov (left) and Vladimir Kovalyonok.

extensive programme of materials-processing experiments in the repaired furnace. The station turned out to be remarkably stable in this position – its dorsal solar panel acted like a rudder and the drag of the rarefied ionosphere locked it in trail, so that the station travelled belly forward.

Later in the month, the station flew in continuous sunlight. This condition was due to a combination of the station's orbit precessing around the Earth, the Earth's travel around the Sun, the offset of the Earth's axis to the plane of its orbit, and the inclination of the station's orbit. For a week or so, at six-monthly intervals when the angles reinforced, a spacecraft in a highly inclined orbit spent most of its time in sunlight. This put a greater demand on the thermal regulation system, but maximised power generation and facilitated not only prolonged observations of the atmosphere but also the potential to study surface relief at very low angles of illumination – and the cosmonauts reported that under perfect conditions they could see undulations in the floor of shallow seas which they were able to correlate with surface currents.

A POLISH VISITOR

No sooner had the residents settled down to life in space than their first Intercosmos visitors were making preparations to join them. Soyuz 30 was launched on 27 June with Pyotr Klimuk and Polish cosmonaut–researcher Miroslaw Hermaszewski. This automatically docked at the rear port the next day. Most of Hermaszewski's luggage was biomedical apparatus to perform a range of studies into space ergonomics and psychology. Smak investigated changes in the sense of taste in space (some favourite foods had turned out to be unpleasant in space, while others had proved to be much tastier). It measured the electrical stimulation of the taste buds, so that the subjective sense of taste could be correlated with specific data. Hermaszewski also had several materials-processing experiments, one of which (Sirena) was to make a monocrystal of cadmium-mercury-telluride (CMT) semiconductor, a material that was extremely sensitive to infrared radiation, for study by the Warsaw Institute of Physics.

A standard assignment on Intercosmos visits was for the guest to take MKF-6M pictures of his own country for Earth-resources studies. On this occasion, however, a solid bank of cloud intervened. The weather over the Ukraine and Kazakhstan was ideal, and these were imaged instead. There was a basic conflict between materials processing and imaging in that the furnace benefitted if the station was placed in the gravity gradient whereas the porthole through which the MKF-6M viewed had to be maintained facing the Earth. Dynamic scheduling to exploit weather windows risked impairing a crystal. Consequently, once a materials experiment requiring several days had started, this restricted opportunistic imaging. Weather forecasting was therefore an important element of planning work on the station (although, of course, a specific picture could be taken only when the ground track and lighting conditions were both appropriate). Nor was it just a matter of snapping pictures with a handheld camera. The MKF-6M had to be primed and the station oriented to aim the camera directly at the ground when that the photographs had to be taken. It could easily take several days before these factors combined to allow a given image sequence to be taken, and it was a lucky visitor who managed to finish his programme. In the gravity-gradient, however, all of the portholes in the forward transfer compartment faced the horizon and gave a panoramic vista for observations using hand-held instruments. Persistent poor weather over a guest's home country was not a disaster, however, because the residents could usually fill in gaps in required coverage later.

Miroslaw Hermaszewski (left) and Pyotr Klimuk.

On 5 July Klimuk and Hermaszewski returned to Earth in their own spacecraft. Hermaszewski's haul included two crystals of CMT, one with a mass of almost 50 grams.

A NEW FURNACE

Progress 2 was launched on 7 July, and on 9 July docked at the newly vacated rear port. It had been loaded with only 600 kilograms of propellant so as to deliver more cargo. In addition to the usual parts for the environmental systems and consumables to last the two cosmonauts 50 days, it delivered several new experiments, including a new furnace (Kristall) and electrocardio monitors which could be worn continuously and transmit data as telemetry. While the cosmonauts were busy with the unloading operation, the flight controllers remotely supervised the preliminary procedures for replenishing the propellant tanks. Previously, the fluid transfer had been managed by the crew, but in this case (and in the future) it was to be done remotely.

Unlike Splav, which had to be installed in a scientific airlock to radiate excess heat into vacuum, the Kristall furnace could be mounted inside the station. It was installed in the rear transfer tunnel. Splav maintained a temperature gradient during crystallisation, but Kristall employed uniform heating, and exposed its sample to a steady-state thermal zone at a temperature in the range 400–1,200°C. It could be used in four different ways: one produced monocrystals directly from the gaseous phase by sublimation, and evaporated the sample and then transported the gas to the cooling zone where it settled onto and enhanced a seed; a second method made homogeneous multi-layered films using chemical gas-transportation; a third made a monocrystal by using a 'moving solvent' to make a high-temperature solution; and in the fourth method the temperature was slowly reduced in the crystallisation zone – one side of the sample was cooled while the other was kept hot, forming a seed on the cold side which, as it was slowly stretched across the zone, produced incremental crystallisation. Kristall controlled its temperature more accurately than did Splav, so its results were expected to be more homogeneous. During its first test, a crystal of gallium arsenide – a semiconductor used in the construction of highly efficient solar collectors – was grown by the high-temperature solution technique.

EXTERNAL ACTIVITY

Whereas Grechko had merely swung in the drogue to inspect the assembly, on their spacewalk on 29 July Kovalyonok and Ivanchenkov opened the larger side hatch of the forward compartment. Ivanchenkov used a colour television camera to show his commander using special tools to retrieve experiments mounted outside the station. In addition to the Medusa cassette that Grechko had deployed, he retrieved cassettes that had been mounted before launch, one of which was a micrometeoroid detector and others that contained materials such as titanium, steel, glass, ceramic, paint, rubber and sealant compounds used in spacecraft construction which had been exposed to the space environment for a prolonged period in order to determine how they degraded. This done, he installed several new cassettes and a radiation detector. When the station flew into the Earth's shadow, Ivanchenkov aimed a lamp to enable Kovalyonok to continue working. His final act was to evaluate the effectiveness of a set of handholds positioned to help a spacewalker move about on the outside of the station. To their surprise they completed all their assigned tasks in less time than in training in the hydrotank. Although the flight director politely suggested that they re-enter the station, Kovalyonok said that he wanted to admire the view because he had been cooped up inside for almost six weeks – the view through his curved visor was incomparably better than that through a flat porthole. As he watched, a meteor burned up in the atmosphere far below. Konstantin Feoktistov later pointed out that the 6-square-centimetre micrometeoroid detector they retrieved had 200 tiny craters in it – more than predicted. Analysis revealed that many of these were the results of impacts with chips of paint and aluminium flakes from solid-fuel rocket motors. The Medusa experiment revealed that exposure to sunlight had caused components of the nucleic acids to develop into substances similar to nucleotides; this was a step in the process

Television of Vladimir Kovalyonok outside.

Vladimir Kovalyonok snaps Alexander Ivanchenkov in the hatch.

of creating nucleic acids, which were vital for the development of organic life.

Progress 2 left on 2 August. Once it had withdrawn, Kaliningrad methodically tested its systems for signs of deterioration prior to de-orbiting it. Progress 3 arrived on 10 August. Since it followed so soon after its predecessor, no propellant had been loaded into its cargo tanks, so its orbital module was full to capacity. Viktor Blagov, the deputy flight director, reported that whereas the first two resupply craft had served to replenish the station's consumables, in this case it was to build up a stockpile for future activities – it delivered 280 kilograms of food, 200 kilograms of water and 450 kilograms of compressed air. Ivanchenkov had asked Grechko to send a guitar, and this had been done. Fur boots had also been sent up because they had complained that their feet were cold (a consequence of the reduced blood circulation in the lower body). While the cosmonauts unloaded the cargo, the station was left in the gravity gradient so that the furnaces could process the newly delivered samples. After being unloaded in record time, Progress 3 was discarded on 21 August in order to free the rear port for the next visitor.

AN EAST GERMAN VISITOR

A week later, Soyuz 31 docked at the busy rear port. It delivered Valeri Bykovsky and Sigmund Jähn, the first cosmonaut from the German Democratic Republic. The delivery of fresh onions, milk, lemons, apples, honey, pork, peppers and gingerbread assured them of a warm welcome. They had a heavy biomedical, biotechnology and materials-processing programme. One biological experiment was designed to reveal how the space environment affected the operation of basic biopro-

Valeri Bykovsky (left) and Sigmund Jähn.

cesses in cells in a tissue culture. Another, supplied by the Institute of Aviation Medicine, investigated how micro-organisms could be 'sewn' together with organic polymers, to determine whether different geometries could be formed in space (on Earth, floccule structures formed) of pharmaceutical benefit. Of two Splav experiments, one boiled beryllium-thorium to produce homogeneous glass of higher quality than was possible on Earth, and the other used a special quartz matrix to produce a monocrystal of a bismuth-antimony semiconductor. The Kristall furnace grew a lead-telluride crystal, using the sublimation technique for the first time, and bismuth-antimony by stretching a seed – this being processed with the material sandwiched between two plates, within an ampoule, and it produced a tree-structured crystal that was five times larger could be made on Earth (this was done so that the result could be compared with one made in the Splav). This kind of research was readily accommodated by the limited facilities of the station but, as yet, there was no capacity for visitors to devise wholly novel experiments and send up heavy apparatus. However, Progress 3 had delivered a pair of modified cameras for an experiment to test different types of film for external and internal use, and Jähn methodically worked through this Reporter sequence. Another aspect of the Intercosmos programme began to develop when Jähn reran the Polish Smak experiment – while each visitor could introduce new biomedical tests, existing ones would often be repeated to improve the database on their results.

Salyut 6 with Soyuz 31 at the rear, taken by Sigmund Jähn as he left in Soyuz 29.

As he was to take away the old ferry, Bykovsky transferred the couch liners and pressure suits from Soyuz 31 to Soyuz 29, then stowed 100 accumulated samples from the furnaces and a cassette of film from the MKF-6M. A deceptively trivial test was to take a sample of the station's air. Even though the environmental system purified the air, there was concern that, after such a long period of inhabitation, toxic compounds might accumulate. If this was so, then the next cargo ferry would have to carry sufficient compressed air to purge the cabin. In the event, the test revealed no chemical, biological or bacterial pollutants.

SWAPPING ENDS

On 3 September Soyuz 29 returned to Earth, leaving Soyuz 31 at the rear port. The next cargo ferry would not be able to dock until this had been moved. A ferry had never been relocated from one port to the other before. On 6 September Kovalyonok and Ivanchenkov verified that Soyuz 31 was fully functional, and the following day

Sigmund Jähn signs his name in chalk on the Soyuz 29 descent capsule.

they powered down most of the station's systems, just in case they were unable to redock and had to return to Earth, undocked and pulled straight back 150 metres. As the two vehicles flew in formation, Salyut 6 was commanded to rotate through 180 degrees to present its front port to the ferry, a procedure which used less propellant than would flying the ferry around to line up to approach the opposite end – all the ferry had to do was to pull back, wait and then move back in. As Kaliningrad wished to monitor the station, the manoeuvre was done during a 35-minute communications session from the South Atlantic, northeast across Africa, and over the Soviet Union. About 30 minutes after undocking, Soyuz 31 redocked at the front without incident, marking a significant milestone in the programme. Although this transfer could have been undertaken at any time, it was done while the recovery window remained open. Transferring a new ferry to the front port within days of the old one leaving was to become standard practice. But swapping ends imposed a significant overhead on the operation of the station. The prospect of a docking failure meant that the station had to be returned to its automated flight regime, and it took several days to place it into hibernation and fully restart it. No power-hungry apparatus could be left running, so materials-processing had to be carefully scheduled; with smelts often running as long as a week, an imminent ferry transfer might affect operations over a ten-day period. All of this was necessary because the plumbing to replenish the station's fluids was available only at the rear port.

A view of Salyut 6 snapped as Soyuz 31 was transferred from the rear to the front docking ports.

RECORD BREAKERS

On 20 September, Kovalyonok and Ivanchenkov exceeded the 96-day record set by their predecessors. Professor Nikolai Gurovsky, head of the Ministry of Health's space medicine department, reported he was fully satisfied with their state of health, Dr Robert Dyakonov said that they were sustaining a high workload, and Professor Oleg Gazenko, a member of the Kaliningrad medical team, announced that they were to return as soon as convenient after achieving the 120-day milestone. It had been observed that their choice of activity selected for time off had changed. Initially they had opted for extra medical tests, then turned to visual observations, and later began work with the BST-1M telescope. The objective of the experiment with a schedule involving a 5-day week with weekends off, had been to find out whether they could use the 'free' time to recover both their physical and mental state after a full week of intensive work, but it was difficult for the cosmonauts to disengage from their daily activities and do something completely different, and in any case their circumstances were not conducive to the pursuance of hobbies. Until a large space station became available with a large staff, there was little hope of realistically mimicking terrestrial daily life.

FINAL RESUPPLY

Progress 4 set off on 4 October, the 21st anniversary of the launch of Sputnik 1. It docked two days later, and immediately raised the station's orbit to 370 kilometres. It delivered a full load of propellants and consumables, spare parts, samples for the furnaces, film, clothing, mittens to keep warm, a new tape player, and a selection of cassettes to relieve the monotony of the exercise regime. At the specific request of the cosmonauts, it brought a set of screens which could be erected to create two tiny 'rooms' for improved privacy. When Ivanchenkov opened a box of chocolates from his wife, the sweets sprayed out and drifted away, and rediscovering them in nooks and crannies in the days to come became something of a treat. Unloading Progress 4 was to be their last major task. After replenishing the station's propellants, the ship left on 24 October. The heavy labour of shifting cargo had helped to develop their stamina. One of the items delivered was eleutheroccus, a drug that acts as a tonic and stimulates the body to work harder, and they took this daily to prepare for gravity. For most of the final month, they used the Tchibis suit twice a week to help increase the capacity of their cardiovascular systems, and in the final week they used it every second day. They had worn the Penguin load-suits throughout in order to keep their muscles in trim, and had rigorously followed the exercise schedule, even occasionally resisting the temptation to postpone sessions to finish another task, and as the flight drew to its conclusion, they increased this regime to three hours per day. In addition, they drank large quantities of salinated water to restore the fluids lost in adapting to space. It was starting to look as if there was no limit to the time a human body could spend in weightlessness, so long as appropriate precautions were taken for returning to the Earth.

Alexander Ivanchenkov taking pictures from Salyut 6's forward compartment.

At the end of October, Kovalyonok and Ivanchenkov began the lengthy process of accounting for and stowing the results of their researches in Soyuz 31. Although the visiting crews had taken much of it away, a great deal of material remained. As the duration of flights increased and crews became more productive, returning their results would become a major issue.[6] It had been a highly productive mission. Some 18,000 photographs had been taken using the MKF-6M camera. To survey the same terrain conventionally would have taken several years by aircraft, or a lifetime in the field. Continuing previous programmes, pictures had been taken of specific targets in Belorussia, the Ukraine, the Crimea, the Caucasus, the Caspian Basin, Kazakhstan, the Central Asian Republics, the Urals, Siberia and the Soviet Far East. Among their specific results were the discovery of an underground water reserve on a peninsula on the eastern shore of the Caspian (in an area that surface surveys had pronounced to be utterly devoid of water), a comprehensive glacial and snow-cover survey of the Pamirs, a survey of pasture in Turkmenia, an assessment of irrigation in Uzbekistan and a geobotanical map of the Lake Balkash area. In addition, the Salsky test site had been imaged regularly to continue to refine the ability of the multispectral cameras to distinguish a wide variety of cereals, vegetables and grasses in different conditions. Infrared studies of the atmosphere had been made using the BST-1M, and hundreds of atmospheric and surface phenomena had been sketched.

Kovalyonok and Ivanchenkov had attempted to grow higher plants in Phyton, a cultivator of nutrient medium with a lamp to simulate sunlight. It was set up so that the root grew 'down' in the dark beneath a black sheet, while the shoots grew 'up' towards the lamp. The objective was to make a plant flower and yield a seed, to test the growth cycle from seed, to mature plant and back to seed in weightlessness. The fact that plant growth seemed to be intimately tuned to gravity was fascinating for botanists. This supplemented the experiments that investigated the development of

---

[6] One option was to send up an automated Soyuz for the residents to load and dispatch, because an unoccupied ferry could return several hundred kilograms of cargo to Earth, but, strangely, this was rarely done.

individual cells. Arabidopsis – also known as wallcress, a common weed – had been selected for its very short life cycle. However, although it did grow rapidly, it died prior to flowering. Onion quickly grew seedstalks (which was surprising because on Earth it does not try to seed so soon) but did not actually produce its seed. A hardy strain of wheat characterised by a short stalk and a very heavy yield of grain was tried, but it grew much more slowly than expected and when it was returned it had only just started to produce grain. The pleasure of witnessing a plant run through its complete growth cycle in space would therefore fall to a future crew. Apart from the plant experiments, the mission had been a great success.

Soyuz 31 returned to Earth on 2 November 1978. The recovery was carried live by domestic television. Despite having set a new record of 140 days, they emerged from the capsule unaided. Intoxicated by the brisk wind on the sundrenched steppe, Kovalyonok knelt to pick up a lump of dirt to celebrate his return. The doctors were delighted to find that they were in better health than previous crews who had made long flights. Significantly, because they had been in space for longer than 120 days, which was the life cycle of the erythrocytes that carry oxygen to body tissue, their blood stream contained only red cells generated in space, and analysis revealed that although these were rather smaller than normal, they functioned properly. On their second day back, they were able to go for a walk in the park around the Cosmonaut Hotel at the cosmodrome, and after 10 days they had recovered sufficiently to play tennis. It was now evident that readaptation had been successfully decoupled from the time spent in weightlessness, suggesting that a flight of a year or more would be possible. This would not be possible on Salyut 6, however, because its expected life of 24 months was already more than half over. The original plan had called for only one more expedition, to extend the endurance record to six months. But a fault had developed, and the station's future was in doubt. A few days later, *Tass* ominously reported that because Salyut 6 had been occupied for 60 per cent of its time, it had successfully *completed* its primary mission.

WRITE-OFF?

In mid-November, Konstantin Feoktistov said that in the coming months Salyut 6 would be thoroughly assessed "to determine its possible further use". What he did not reveal was that while Progress 4 had been replenishing propellant, a problem had developed in one of the tanks in the unpressurised compartment at the rear of the station. An intensive effort was underway to understand the failure, and to devise a remedy. Telemetry analysis implied that the pressurised bladder within the tank had developed a leak. Bubbles of nitrogen gas mixed in with the propellant would make the engine misfire, so the engineers had recommended not using it until the problem could be fixed since there was a risk of the highly-corrosive unsymmetrical dimethyl hydrazine seeping through the pressurisation system into the other bladders. But the unified propulsion system incorporated two *independent* pairs of tanks, and if the damaged tank could be drained and isolated then the engine would be restored to use. However, the repair operation could not start until a crew was onboard, and until an automated ferry docked to drain the propellant from the damaged tank. An operation such as this would not have been attempted on an earlier station – it would simply have been written off. The decision to rescue Salyut 6 therefore indicated how far the programme had come in such a short period. Although the station was regularly reported to be operating normally, it had not escaped Western observers that it had not manoeuvred since Progress 4 left and had decayed to 307 by 330 kilometres, but this was interpreted as indicating that it was low on propellant and it was supposed that a tanker would be sent up to restore it to a circular orbit at 350 kilometres prior to its reoccupation. In mid-January 1979, Vladimir Shatalov reported that specialists were still considering the possibility of sending another crew. A month later the fleet of communications ships left port for their operating points, which suggested that a launch was imminent, and while this was true, it would not be the expected tanker.

REPAIR CREW

When Soyuz 32 set off with Vladimir Lyakhov and Valeri Ryumin on 25 February, Salyut 6's orbit had decayed to 296 by 309 kilometres. The ferry docked at the front port the next day, and on opening the hatch the newcomers were greeted by a faint scent of pine, which was the residue of the Svezhest experiment. Before powering down the Soyuz, they used its engine to boost the orbit to 308 by 338 kilometres. This was feasible only because it had not taken as much propellant as usual to reach the station. Raising the orbit further would not be possible until a tanker arrived, and its engine used. This was Lyakhov's first flight. Ryumin had suffered the frustration of Soyuz 25's failure to dock. As a member of the team that designed the station, he was well qualified to attempt to isolate the damaged tank.

Valeri Ryumin (left) and Vladimir Lyakhov.

Although Salyut 6 had been in orbit for 18 months, its systems had proved to be extremely reliable. Nevertheless, some of its apparatus was nearing the end of its recommended service life, and this would have be replaced before the station would be able to support another long-duration crew. Numerous spare parts and tools had been ferried up by Soyuz 32, but most of whatever would be required would have to be shipped up on the next cargo ferry, which could not be done until the cosmonauts submitted a list of requirements. Accordingly, once they had reactivated the station, they inspected each piece of apparatus for potential failures, then replaced filters in the environmental unit, ventilators in the thermal regulation system, elements of the

80 Routine operations

Vladimir Lyakhov (left) and Valeri Ryumin in the Salyut 6 trainer.

veloergometer and the treadmill, miscellaneous cables and all the lights. The ability to replace expired items, repair faulty apparatus and replenish consumables showed the maturity of the programme. Their inspection confirmed that (apart from the fault in its propulsion system) the station was in good shape. In early March, when all the maintenance that could be attempted had been completed, the station was orientated nose-down in the gravity gradient. Unlike their predecessors, Lyakhov and Ryumin had been given primarily visual observation tasks. With the portholes in the forward transfer compartment taking in the entire horizon, they could make comprehensive observations without expending propellant. They installed a sextant which they had brought with them, for use by later crews to determine the station's position relative to the stars rather than the Earth.

When Progress 5 arrived on 14 March, the station's orbit had decayed to 293 by 325 kilometres. It brought food, water, film, ampoules for the furnaces, clothing and various other items including improved smoke alarms, a linen dryer, a tape recorder, the Koltso walk-around communications system, a replacement Stroka teletype (the original had broken), a nickel–cadmium storage battery (to augment the existing ones, which had degraded – as, indeed, had the solar panels, about which nothing could be done), a monochrome video monitor to set up a television uplink, and parts to repair the water reclamation system. In addition, reflecting the confidence that the efforts to repair the engine would be successful, it delivered some new scientific equipment, including the Yelena-F gamma-ray detector, an improved Kristall furnace (the other had broken after performing 40 smelting operations), the Biogravistat and Phyton-2 plant cultivators, the Spektr-15K spectrometer, and the Duga electrophotometer for an investigation of luminous phenomena in the upper atmosphere. They opened 27 containers and unpacked over 300 individual items, each of which had to be stored in its proper place in the station. In the process, as the assorted stores and its packing floated around in the main work compartment, the station took on the appearance of a warehouse. Nevertheless, they finished this task in just four working days.

The repair of the propulsion system started on 16 March. The first step was to drain the damaged fuel tank. The Soyuz 32–Salyut 6–Progress 5 complex was placed in an end-over-end spin, so that the centrifugal force would separate the liquid–gas mix in the damaged tank. The fuel was then pumped into the second fuel tank of the integrated propulsion system until this was full, and the remainder was pumped into an empty tank in Progress 5. Finally, the damaged tank's valves were opened to vent the residue behind the bladder. Progress 5's thrusters were then used to stabilise the complex. The pipes were exposed to vacuum for a week to ensure that all the highly corrosive unsymmetrical dimethyl hydrazine was removed, and nitrogen was blasted out as a further guarantee. The damaged tank was filled with nitrogen and then sealed off. The entire repair operation had been completed without incident. It was quite an achievement. After recircularising the station's orbit at its operating altitude of 350 kilometres, Progress 5 left on 3 April.

The station's radio and television systems were overhauled and upgraded so that it could receive a video uplink from Kaliningrad. This was not as easy as might at first be expected, due to interference from radiation at orbital altitude and from the station's own systems. Nevertheless, this marked a significant step forward, because scientists would be able to demonstrate their requirements, the crew would be able to see the results of their Earth-photography, and be able to observe their colleagues demonstrating procedures for impromptu spacewalks, and, of course, crews on long missions would be able to see their families as well as talk to them.

With the station once again fully operational, Lyakhov and Ryumin settled down to a comprehensive programme involving Earth-resources, upper atmospheric and astrophysical observations, materials-processing (some of the samples for which had been supplied by France) and biological experiments (including starting a new line of research by placing the egg of a quail into an incubator for an embryological study). The Yelena-F was set up to measure gamma-rays and charged particles in near-Earth space. Because this drew just 10 watts, it was to carry out a long-term survey of the radiation in the station's environment. Not unexpectedly, it recorded its highest readings near the South Atlantic Anomaly, a zone where the inner radiation belt dips down towards the ionosphere.

SO NEAR AND YET SO FAR

On 10 April Nikolai Rukavishnikov and Georgi Ivanov, a Bulgarian cosmonaut, were launched in Soyuz 33. The following day, when the final transfer orbit brought the spacecraft within 6 kilometres of the station, the Igla system was activated to make an automated approach. With 1,000 metres to go, as the ferry aligned itself with the station, the engine shut off part way through a six-second burn and the vehicle shook violently, and Rukavishnikov instinctively reached out to steady the shaking control panel. Ryumin, observing from the station, reported what he thought was an unusual lateral plume from the engine, but as this was *behind* the ferry it was difficult to be sure. Although Rukavishnikov later admitted to being "scared as hell" by the engine failure, at the time he asked permission to resume the approach using only the small thrusters, but was told not to do so. While he may well have been able to complete the docking, this would have been to no avail, as there would have been no point in leaving a damaged ship at the station

Georgi Ivanov.

for the residents.[7] The newcomers settled down to await the next landing opportunity. At launch, Ivanov's pulse had not risen above 74 beats per minute. Continuing this relaxed cool, he slept. Rukavishnikov, however, remained awake. There was concern that the shaking that accompanied the engine's misfire might have damaged the adjacent backup engine, which would be required for the de-orbit burn. Whereas the main engine was designed to be restartable, its backup was intended for one long firing at full thrust to terminate orbital flight. It had never been used previously. Although it was set to shut down automatically, it failed to do so and Rukavishnikov shut it down manually. The orbital and service modules were jettisoned without incident, and the capsule landed in darkness within 15 kilometres of the target. By the time the recovery team arrived, Rukavishnikov had scrambled out. He announced that he felt as if he had been in space for a month!

In determining what had gone wrong, the engineers had to rely on the telemetry, and their task was complicated by the fact that the propulsion unit was not heavily instrumented. Nevertheless, within days they announced that they had identified the fault, which was a part of the engine itself, rather than the control system as initially suspected. Fortunately, a pressure sensor in the combustion chamber had shut down the engine when it sensed that it had not achieved the correct operating pressure, thereby eliminating the danger of the chamber exploding from uncontrolled mixing of propellants. This malfunction was all the more surprising because such engines had been fired in excess of 1,000 times since their introduction in the mid-1960s. It took a month to fully investigate the problem, but as soon as the exact cause of the failure had been identified steps were taken to modify the spacecraft which had already left the production line.

As regards Soyuz 32, the concern was not that its engine might suffer this rare malfunction but that it had been in space for ten weeks. It was to have been returned by the Intercosmos crew, and it would now expire its 90-day service life long before the next visiting crew could be sent up. However, its status would become an issue only if an emergency necessitated its use. Its unserviceability did, however, rule out sending up the next Intercosmos crew, because *they* would not be able to return in the expired

---

[7] This must have been particularly frustrating for Rukavishnikov, as it was the second time he was prevented from boarding a station, having been unable to open Soyuz 10's hatch to enter Salyut 1.

vehicle. The Hungarian mission in June was postponed so that its launch window would be used to send up an unmanned Soyuz with a modified engine. The Soyuz 33 drama had completely disrupted the tightly interleaved schedule of visiting missions.

WORKING ON

Lyakhov and Ryumin took these events stoically. There was nothing they could do, and they were safe so long as they remained where they were, so they worked on. They observed ocean currents, made semiconductor crystals in the Kristall furnace, and installed the Biogravistat cultivator. Even the planned Bulgarian programme had not really been lost. The equipment had been delivered by Progress 5, so they would be able to perform much of Ivanov's work, including the spectroscopic survey of the Earth by the Spektr-15K for the Balkan experiment. Using this spectrometer and an infrared camera, they discovered that they could detect a thin film of oil that formed over a large shoal of fish swimming in a warm ocean current, and they reported their observations to fishermen. On 1 May they took a holiday and watched the military parade through Red Square on their video uplink.

Progress 6 arrived on 15 May. The automated ferry was not affected by the failure of Soyuz 33's engine because it incorporated the unified propulsion system engine. Once again, it carried primarily dry cargo, including, in addition to the usual parts for the environmental system, replacement hardware (higher-power television lights, and a control panel for the navigational system), samples for the furnaces and for the Biogravistat and Phyton-3 cultivators. The replenishment of the propellants was carefully monitored by Kaliningrad to ensure that there was no further problem.

On 5 June, as Lyakhov and Ryumin passed their 100th day in space by loading assorted rubbish into Progress 6's orbital module, the final preparations were made to launch Soyuz 34 the following day. Once the modified engine was confirmed to be functioning properly, Progress 6 departed – this had been retained to ensure that the station would have a proven propulsion system (despite the station's engine having been repaired, there was some reluctance to rely on it). Soyuz 34 pursued a 48-hour rendezvous and docked at the rear port without difficulty. This assured the station's crew of a serviceable ferry for their return to Earth. Opportunity had been taken to stock it with a variety of additional supplies, so Lyakhov and Ryumin resumed their cargo transfer routine. A week later, Soyuz 32's descent module was loaded up with 180 kilograms of results and sent back to Earth using the window that the Hungarian mission would have used. Although considered unsafe for returning cosmonauts, the fact that it was being trusted to return the results of their experiments indicated that the ship was fully expected to return safely; and it did. In fact, because the residents had not received either of their assigned Intercosmos visits, each of which was to have relieved them of about 50 kilograms, this was the *only* means of returning their haul. Opportunity had also been taken to send back a variety of 'expired' apparatus, all of which would otherwise have been jettisoned with Progress 6, for a reassessment of the schedule for routine maintenance. And of course, the descent module was itself a useful result because analysis of how well its systems had survived 108 days would be able to be compared with those of Soyuz 20, the degradation of which had led to the imposition of a 90-day service life. Soyuz 34 was moved to the front port on 14 June to clear the way for the next cargo ferry. This was not scheduled for six weeks, but the transfer had to be done while the recovery window was open. The station's propellant reserve was so great that Progress 7, which arrived on 30 June, delivered a heavy load of new scientific equipment, including Isparitel for materials processing, Resistance for precisely measuring the atmospheric drag on the station (which was a force that might interfere with crystal growth) and the KRT-10 radio telescope. In the mail bag, Lyakhov and Ryumin found a guidebook for the Moscow area, sent by one of their colleagues to help alleviate symptoms of homesickness.

The Vaporiser experiment used the Isparitel apparatus supplied by the Institute of Electrical Welding in Kiev. This was installed in the scientific airlock. After some teething problems, the low-voltage electron beam melted tiny granules of aluminium and silver and then sprayed this vaporised melt onto small carbon and titanium disks to condense as a thin film. The duration could be varied between one second and a maximum of 10 minutes to deposit a layer ranging from only a few micrometres to a few millimetres in thickness. The disks were to be returned for analysis to determine whether such coatings could be applied to protect stations from deterioration in the space environment.[8] The Deformatsiya engineering experiment used optical sensors to measure the distortion of the complex's structure resulting from solar heating. The three-vehicle train was set in a fixed attitude with respect to the Sun that provided a 300°C temperature differential between the illuminated and the shaded sides, to bake one side and freeze the other. This attitude could not be held for long without risk of damage, but the test indicated that even several hours of exposure deformed the axis by just a few tenths of a degree.

On 15 July, Lyakhov and Ryumin passed the 140-day record of their immediate predecessors. Their objective was to achieve the psychologically important 'half a year', and this would require them to remain onboard for at least another month. By this point, they required only about seven hours of sleep, rather than the 10 hours with which they had begun, but even at this late stage their working efficiency was extremely high. In addition, whereas their predecessors had undertaken an intensified exercise regime in the final weeks to help prepare to return to gravity, Lyakhov and Ryumin had followed this throughout, so they were in much better shape. Only the stamina-inducing drugs, vitamins, saline rehydration and the Tchibis cardiovascular conditioning would be required to prepare for the return to Earth.

A RADIO TELESCOPE

The KRT-10 incorporated a directional antenna and five radiometers (four horns in the 12-centimetre band and a spiral antenna operating in the 72-centimetre band). As

---

[8] This test followed directly from the successful respraying of the mirrors in the OST-1 on Salyut 4.

delivered, the 350-kilogram unit comprised a control system and a parabolic antenna with an integral three-legged mount to hold the radiometers at its focus. A video had also been sent to show the cosmonauts how to operate it. The controller was to be retained inside, but the rest of the structure had to be set up outside, and this was to be done without requiring the crew to make a spacewalk. In mid-July, they 'docked' the tightly-furled assemblage to the rear docking drogue and left it projecting axially into Progress 7's orbital module (its docking probe had been dismounted). When the ferry's docking latches retracted, the escaping air pushed it smoothly away from the collar, exposing the antenna. The ferry was commanded to halt a few hundred metres away to enable Kaliningrad to monitor the deployment of the 10-metre-diameter dish, which unfolded as intended, totally obscuring the station behind. Due to the need to recharge the station's batteries, the radio telescope could be used only on alternate days. Over the next few weeks, it was operated as an interferometer in conjunction with the 70-metre dish of the deep-space communications site in the Crimea. These observations demonstrated mixed results, but it was only an early test of an orbital telescope (the very act of trying to use it had generated useful design feedback). For celestial sources, it had been used in two modes: in one case the station had been stabilised so that the telescope gave continuous coverage of a given object, and in the other mode the station was turned to scan the telescope along the plane of the Milky Way to search for new sources. When pointed at the Earth, it was able to discern the structure of meteorological features, map sea state, measure water salinity and soil humidity, and even although its surface resolution was a poor 7,000 metres, it was able to monitor an eruption of Mount Etna, in Sicily.

When the KRT-10 antenna was jettisoned to free the rear port, it wobbled as it drifted away and became entangled in the apparatus that projected from the station's rear. The station was 'rocked' to attempt to shake the antenna free, but it remained stuck fast. It was decided that the cosmonauts would have to cut it free of whatever it had fouled. Since they were likely to be the final occupants, Lyakhov and Ryumin had been assigned a brief spacewalk at the end of their mission to retrieve the various packages that had been mounted near the hatch by a previous crew. Although the doctors expressed concern that the men might be too tired to undertake such a complex unscheduled operation, this contradicted the earlier assessment that they were working extremely efficiently. When asked, Lyakhov and Ryumin said they would extend their excursion to give it a try. While they loaded their experiment results into Soyuz 34 to ensure that these would not be left aboard the station if it proved impossible to re-enter following the spacewalk, their colleagues worked in the hydrotank to develop procedures to cut the dish free. The big day was 15 August. As Ryumin pulled on the airlock hatch, he found that it was jammed, and it took considerable force to release it. Just after they finally emerged, the station moved into the Earth's shadow and Kaliningrad told them to wait by the airlock. Back in sunlight, half an hour later, Ryumin left Lyakhov at the airlock and made his way slowly to the rear of the main compartment using a set of handholds and 20-metre umbilical provided for such a contingency. Once in place, he paused to rest. After studying the antenna to decide how he would cut the frame free, he waited to re-establish communication with Kaliningrad prior to proceeding. Although all that was required was a few quick snips with his metal cutters, this was a potentially dangerous operation because he had to crawl over the rim and down in between the dish and the rear of the station. As a general rule, mission planners tend to treat items not designed for use by spacewalkers as inherently dangerous. Ryumin was reminded of this when he severed the first strand of the fouled mesh and this set the sharp-edged structure rotating towards himself, but he persisted and managed to push it away. On his way back, he examined some of the portholes, whose glass had gradually become tarnished by exposure to space. He rubbed one with his glove in an effort to collect a sample of this dust for analysis, but it appeared to be embedded in the glass rather than resting on its surface. Lyakhov, meanwhile, had retrieved all the experiment packages from around the airlock.

Postage stamps issued to celebrate Valeri Ryumin's success in releasing the dish of the KRT-10 radio-telescope.

As with the isolation of the damaged propellant tank, this impromptu spacewalk had demonstrated just how rapidly the programme was maturing. It was not clear whether Salyut 6 would be able to sustain another crew, but if the rear port had been left fouled this would have precluded docking cargo ferries, which would have meant that the station would have been able to be occupied only until its consumables were exhausted – if indeed it was deemed useable at all, with the vast but lightweight dish hanging off its back end. Real-time planning in Kaliningrad, and determined action in space, had paid off.

HALF A YEAR

During their long mission Lyakhov and Ryumin had undertaken a wide variety of experiments. In addition to the ongoing multispectral Earth-resources and large-scale mapping, 54 samples had been processed in the various furnaces, and ocean currents and plankton blooms had been seen (Lyakhov was awarded a certificate proclaiming him to be an "outstanding worker" by the trawler skippers that he had steered onto rich shoals of fish). The long-duration missions were becoming so productive that the researchers on Earth were beginning to be swamped by the flood of data from the science programme, and it was evident that special support teams would have to be formed for future stations. As previously, the great disappointment was the plant experiments. It was speculated that the mechanism that disposed of waste products did not operate in microgravity, with the result that the plant

Valeri Ryumin (left) and Vladimir Lyakhov in recliners after having crawled out of the Soyuz 34 capsule, while technicians inflate the medical tent.

poisoned itself. Only the root vegetables had been successful. Despite the limited return-cargo capacity of a Soyuz carrying a crew, the broken Kristall furnace was dismantled and stowed to enable its designers to examine it, as was one of the regeneration canisters from the environmental system that had started to corrode. On 19 August, after 175 days in space, Lyakhov and Ryumin finally returned Salyut 6 to its automatic flight regime, and departed. Because Soyuz 33 had pre-empted the Intercosmos series, they had been obliged to fly the entire mission without the relief of company, so the crowd at the recovery site must have seemed quite overwhelming.

A month later, Konstantin Feoktistov announced that Salyut 6's systems were being studied to assess its future utility. In July, Progress 7 had boosted it up to the unprecedented altitude of 400 kilometres. By the end of November it had decayed to its nominal operating altitude at around 360 kilometres. It had been decided that even though it had already operated far beyond its design life, it should be able to support another expedition in the new year.

Valeri Ryumin (centre) and Vladimir Lyakhov are interviewed immediately after returning from 6 months in space.

A NEW FERRY

Meanwhile, other developments were bearing fruit. A new variant of the Soyuz had been tested under the anonymity of the Cosmos series. Cosmos 869 and Cosmos 1001, launched in November 1976 and in April 1978 respectively, had performed a simulated rendezvous and then returned within a few days, but, earlier in the year, Cosmos 1074 had been powered down for 60 days to assess how its new systems degraded. On 16 December, this new configuration, launched crewless, was openly declared as Soyuz-T 1. After a two-day flight it approached Salyut 6 by descending from above and ahead of the station, rather than from below and behind it as was usual, but it did not dock. Instead, it withdrew. It re-rendezvoused the next day, and docked at the front port. After raising the station's orbit to 370 by 382 kilometres, it was powered down. In March 1980, having spent 100 days in space, it returned to Earth. In each of these test flights, the 1,260-kilogram orbital module was jettisoned prior to the de-orbit burn in order to reduce the overall propellant requirement by 10 per cent, which meant that the cargo capacity could be correspondingly increased.

Konstantin Feoktistov pointed out that although this new Soyuz was externally similar to the station ferry then in use, it incorporated major internal modifications. The most significant upgrade was the installation of a digital computer to control the spacecraft's attitude and execute all of the manoeuvres required to reach and return from a station, do so automatically, and in the most propellant-efficient manner. The use of advanced electronics also meant that more sophisticated telemetry could be produced to help to diagnose faults. The Argon 16 computer introduced by Salyut 4 had been installed to monitor the sensors and give real-time status data to the crew. The inertial platform of the navigational system enabled the new vehicle to operate completely independently of Kaliningrad. It also incorporated the unified propulsion system that was already used by the Progress. Solar panels had been reintroduced so that if a spacecraft was unable to dock it would not be forced to beat a hasty retreat (which had happened four times). It was not announced, but this new version could accommodate three pressure-suited cosmonauts, which restored the capacity lost by the revision following the loss of the Soyuz 11 crew in 1971.

YET ANOTHER EXPEDITION

The remarkable longevity of Salyut 6 enabled a fourth expedition to be mounted. In this case, the tanker, Progress 8, was launched first, and it docked at the rear of the vacant station on 29 March 1980. The mission had been assigned to Leonid Popov and Valentin Lebedev, but in March Lebedev had badly injured a knee trampolining. Their backups were deemed to be insufficiently experienced to serve as a resident crew – Vyacheslav Zudov had flown only the aborted Soyuz 23 and Boris Andreyev had yet to fly. Valeri Ryumin, who had been coaching the new crew on the lessons learned

Leonid Popov (left) and Valeri Ryumin.

during his six-month tour, realised that he was best placed to restore Salyut 6 to life for another long-term mission, and he offered to take Lebedev's place, and so it was that when Soyuz 35 docked at the front port on 10 April, he had the déjà vu experience of reading the note which he had left to welcome anyone who might visit his old haunt. As a joke, he then inflated a plastic cucumber, presented it to the ever-watching flight controllers, and reported that it had grown in his absence. They were to continue the weekend-off work schedule, and use Moscow Time, but whereas the intense exercise regime had previously been enforced daily, this time every fourth day was to be a lazy day, to break the monotony. Ryumin's recent rapid recovery had convinced the doctors that they understood the requirements for a tour of duty of six months. They were delighted to hear that Ryumin readapted to weightlessness almost immediately.

In addition to the usual consumables and replacements for the thermal regulation and atmospheric systems, Progress 8 had delivered another nickel–cadmium battery, a new Splav (Splav-2), a Kristall (Kristall-3) and the Malakhit plant cultivator. After a fortnight of maintenance work, they made a start on their experiments. They had been assigned a comprehensive smelting programme designed to test a wider variety of substances and greater range of crystallisation times – extending from a typical 10 hours to as long as 120 hours in the case of the CMT samples, during which time the station would be in the gravity gradient. They were also to make observations of the oceans and their biological productivity. The Yelena-F gamma-ray detector, which continuously sampled data on the near-Earth environment, was restarted (and when it failed soon afterwards they took it apart and repaired it). The Malakhit cultivator was to study the growth of orchids, which had been chosen because they thrived in a dry atmosphere, and so might survive in the artificial environment within the station. The plants delivered after they had started to flower immediately wilted. However, those planted in space grew roots so long that they protruded from the apparatus! Nevertheless, although they flowered, they did not seed. Samples were to be sent to Earth with visiting crews at 60 days, 110 days, and at the end of the mission, which was scheduled to last six months. Progress 8 left on 25 April, and Progress 9 took its place four days later. It brought the Lotos experiment, which was to make shapes of structural elements from polyurethane foam, and a new motor for the centrifuge in the Biogravistat. It vacated the rear port on 20 May.

A HUNGARIAN VISITOR

It must have been with yet another sense of déjà vu that Ryumin awaited the launch of the first Intercosmos crew, which would bring the Hungarian cosmonaut who was to have visited him the previous year. This time there were no problems, and when Soyuz 36 docked at the rear on 27 May the residents welcomed Valeri Kubasov and Bertalan Farkas.

The main experiment was Interferon, to evaluate the possibility of manufacturing interferon (a chemical, produced by human cells, that inhibits viruses) in commercial quantities. One aspect of this test involved injecting a vial of white blood corpuscles

Valeri Kubasov and Bertalan Farkas (with mustache) before and after their mission.

into an interferon-yielding substance to find out whether the rate of production was different in microgravity. A related blood analysis experiment studied whether the generation of this chemical by the body was influenced by weightlessness. Another experiment tested whether existing interferon pharmaceutical preparations (delivered both in the liquid state and in a lyophillised gel) were rendered more or less effective in weightlessness. The results were so interesting that several follow-up experiments were devised. Of the Earth-observations, Refraction used the VPA-1M to measure the optical polarisation of the horizon, Zarya used the Spektr-15K spectrometer to derive temperature and density profiles of the atmosphere by measuring absorption lines at orbital sunrise and sunset, and Utrof used the MKF-6M camera to compile a geomorphological map of the Carpathian Basin and the Tisza River Basin, to study the ecology of Lake Balaton and to assess the effect of the Kishkere reservoir on soil salination in the inland waterways linked to the Danube.

HOTEL SALYUT!

On 3 June the visitors retreated to Soyuz 35, undocked and returned to Earth. They had been so active that they had caught no more than three hours of sleep a night, and they were exhausted. The following day the residents relocated Soyuz 36 to the front port, and the day after that Soyuz-T 2 lifted off. Yuri Malyschev and Vladimir Aksyonov wore the new Sokol pressure suit, which, in addition to being lightweight at just 8 kilograms, was more comfortable than the old one. This was Malyschev's first flight, but Aksyonov had been flight engineer on the solo Soyuz 22 mission that tested the MKF-6 camera. After a 24-hour rendezvous, Soyuz-T 2 initiated its final approach. The Argon computer was to select the optimum sequence of manoeuvres, but when it was 180

Vladimir Aksyonov (left) and Yuri Malyschev.

metres out Malyschev saw it start a manoeuvre with which he was not familiar, so he took over and manually docked. A later analysis established that the route selected by the computer should have resulted in a successful docking, and that the intervention had been premature. Malyschev and Aksyonov spent three days on routine biomedical tests, took some pictures using the MKF-6M and made Spektr-15K observations, and then departed having achieved their primary objective of evaluating the performance of the new spacecraft. This was the first time that two visiting missions had been mounted in a single recovery window. Several days later, Salyut 6 suffered a serious inconvenience when the Kaskad attitude-control system malfunctioned. Popov and Ryumin repaired it, but doing so took considerable time, and the subsequent manoeuvring trials consumed a great deal of propellant.

On 1 July Progress 10 delivered a colour monitor for the video uplink, a Polaroid camera (one of the few objects onboard of Western origin), a new image converter for the Duga electrophotometer (to replace the one that projected an inverted image) and replacement parts for the BST-1M. It replenished the station's propellants, adjusted the orbit, and left on 18 July. Popov and Ryumin watched the opening ceremony of the Moscow Olympics on their new uplink and then sent a pre-scripted message to be beamed onto a large monitor in the Lenin Stadium. A few days later, the portholes began to fog up, suggesting that the environmental control system was deteriorating. They took pictures to document the degradation of the outer surface of the glass (the mark left on one porthole by Ryumin's glove during his spacewalk a year earlier was still clearly visible).

On 24 July, Soyuz 37 settled into the busy rear port of 'Hotel Salyut' to deliver Viktor Gorbatko and Pham Tuan. As a Vietnamese citizen, Tuan's main assignment was the Kyulong experiment. This used the MKF-6M and KATE-140 cameras and the Spektr-15K spectrometer to make the first comprehensive study of hydrological features on the Central Vietnamese Plateau, and tidal flooding and silting in the deltas of the Mekong and Red rivers. The Halong experiment (named after a bay on the Vietnamese coast) used the Kristall furnace to produce bismuth-tellurium-selenium and bismuth-antimony-telluride alloys, and a gallium-phosphide semiconductor. The Azolla experiment tested a fast-growing nitrogen-rich water fern (azolla pinnata was a commonly used fertiliser in rice growing) to determine whether it would be suitable for use in a closed-cycle hydroponics farm on a future station, but it fared no better than any other higher plant to date. One great advantage of the frequency

Viktor Gorbatko (left) and Pham Tuan.

92 Routine operations

of flights was that a hypothesis could be tested quickly. The inability of plants to survive had perplexed the botanists. However, they were delighted several weeks later when an arabidopsis flowered. After it had begun to grow, it had been set beside a porthole to illuminate it with sunlight, in the hope that this would be more satisfactory than the cultivator's lamp. Furthermore, it demonstrated that the plant had not been confused by the orbital illumination cycle. The next objective was to make one complete the cycle and produce a seed.

Gorbatko and Tuan departed in Soyuz 36 on 31 July. The next day, Popov and Ryumin once more transferred their new ferry to the front port. They then resumed observations. After having tried so hard to hold the station still in order to minimise disturbance to smelting operations, a new experiment involved rotating it to subject a sample in the Splav to directional solidification. The comprehensive tests had shown that the semiconductor monocrystals grown in microgravity were of "immeasurably better" quality than those made on Earth. Another novel experiment involved using a lens to project an image of the Sun onto a screen, so that, as the Sun dropped below the horizon, one camera filmed the Sun while another filmed its image on the screen to accurately measure the distortion of the solar disk by atmospheric refraction. For most of August and September, the station was left in the gravity gradient so as to combine smelting with observations of the Sun on the horizon using the BST-1M. As a result, little propellant was used and there was no need to send up a new cargo ferry. In mid-September, its orbit having decayed considerably, Salyut 6 boosted itself back up to 335 by 352 kilometres – this was the first time that its engine had been used since the damaged fuel tank had been isolated.

Soyuz 38 arrived on 19 September with Yuri Romanenko and Arnaldo Tamayo Méndez, a Cuban. As usual, the visitor's main task was the study of his home. The Tropics experiment involved using the MKF-6M to map Cuba's natural resources – in fact, this was Tropics-3, and it continued observations already made earlier of other locations. As always, the orbital imaging was coordinated with data-gathering by airborne and ground teams. Unfortunately, although the weather in the Caribbean was excellent throughout the flight, the timing of the flight meant that there were few suitable daylight passes. However, the residents secured the necessary pictures after Tamayo Méndez had left, and these established that at both its eastern and western extremities, the island is criss-crossed by intersecting networks of faults.

One novel Cuban contribution to the study of adaptation to weightless-

Valeri Ryumin (left), Arnaldo Tamayo Méndez, Vladimir Romanenko and Leonid Popov.

ness was the Support experiment. This involved Tamayo Méndez wearing a special adjustable shoe for six hours each day. It was designed to place a load on the arch of his foot to simulate the force the foot feels when standing on Earth, to test whether it affected the adaptation of the vestibular system. This was based on the idea that on Earth the state of the muscles in the foot contributes directly to the sense of balance, and that there was a vestibular reaction to this sensory input that was being denied in space. Such a device was also to be used to investigate the recovery of locomotive stability on returning to Earth, because cosmonauts tended to adopt a distinctive gait and posture whilst recovering from long flights.

It must have been an interesting experience for Romanenko to revisit Salyut 6 after so many years of use, but he was probably glad to be making only a brief stop. He and Tamayo Méndez departed on 26 September, and the fact that they did so in their own spacecraft indicated that the residents expected to follow them quite soon.

COMPLETING A SECOND SIX-MONTH TOUR

As Salyut 6 began its fourth year in orbit, Progress 11 slipped into the rear port. It delivered sufficient propellant to sustain the station for a period of automated flight – just in case it proved possible to mount another expedition. On 1 October, Popov and Ryumin exceeded the 175-day endurance record that Ryumin had established the year before. During their residency, they had devoted some 35 per cent of their work to observations of the Earth. In addition to a large number of synoptic pictures with hand-held cameras, almost 3,500 MKF-6M and 1,000 KATE-140 pictures had been taken. The MKF-6M had regularly recorded the Salsky test site. They had reported on various crop yields and forestry in preselected areas, and on plankton and large fish shoals, and their observations had been exploited to great effect. Of three faults which they had identified in Hungary, one had already been found to have oil and gas reserves. They had collected considerable data using the Spektr-15K, RSS-2M and VPA-1M for atmospheric studies, and infrared data using the BST-1M had revealed that there were small-scale variations in the submillimetre flux of the atmosphere in the zones in which cyclonic activity originated. The Yelena-F gamma-ray detector had charted streams of high-energy electrons flowing in the South Atlantic Anomaly, and revealed that the background flux was highly dependent on latitude (its strength varied by a factor of 10 between the equator, where it was weakest, and the latitude extremes of Salyut's orbital inclination). They had performed variations of the Vaporiser experiment, spraying small disks of glass, carbon, titanium and various other metals with a variety of thin coatings of gold, silver, copper and aluminium, in all producing 196 samples. Future crews would experiment with polymer coatings. In addition, they processed almost 100 samples in the furnaces and created semiconductors and a variety of alloys. Overall, they spent about 25 per cent of their working schedule on routine maintenance and repairs, although most of the latter had been carried out at the start of the mission, to prepare the station for their own period of residency. In addition, they devoted a few days to unloading each cargo ferry and then packing it with rubbish for its departure. Their working efficiency had been extremely high, but

Valeri Ryumin (left) and Leonid Popov, with their Sokol pressure suits.

had fallen markedly when they had entertained visitors. For Ryumin, who had spent 12 of the last 20 months onboard Salyut 6, the contrast between his tours had been striking – on his first tour he had received no visitors, but then had played host four times on the second.

Popov and Ryumin landed on 11 October. Almost as soon as the recovery team arrived, they were astounded to see the two men emerge from the capsule and walk without assistance to the recliners. Ryumin recovered much more rapidly than after his previous flight, and he had actually gained a little weight (2 kilograms). Although the rigorous exercise regime had maintained the capacity of his cardiovascular system and limited the degradation of his calf muscles, the idle back and chest muscles had atrophied. However, by moving so much cargo he had increased his arm and hand muscles. This 185-day mission banished all doubt that it would be feasible to fly for a year or more.

REPRISE?

Salyut 6 had been left with Progress 11 still docked, to ensure an independent means of orbital manoeuvring. Before departing, Popov and Ryumin had undertaken routine maintenance on equipment nearing the end its service life, and had made such repairs as they could. Its age notwithstanding, the station was in reasonably good condition and, with further routine maintenance, it might possibly support another expedition. One factor limiting further use, was that the hydraulics within its thermal regulation system were starting to degrade. The station had been designed to operate for up to 24 months, and thus far had been in use for *three years*. The hydraulics had not been designed to be serviced, so the station would be rendered uninhabitable if the thermal regulation system failed. Another factor in deciding whether to send up another crew was that the station was constantly short of power due to the inevitable degradation of its solar panels, and the research would have to be planned with this in mind. The decision therefore depended on the cosmonauts' debriefing, and on analysis of the station's telemetry.

Having assessed the situation, and compiled a list of all the systems that required attention, in early November it was decided to send a maintenance crew for a brief visit. Interviewed by *L'Humanité* in Paris, Georgi Beregovoi reported that a *three-man* crew would be sent, which was the first indication that the new Soyuz had this capability. However, as the ferry could not reach the standard operating altitude of 350 kilometres when carrying three cosmonauts, the station's orbit had been allowed to decay below 300 kilometres. In mid-November, Progress 11 boosted the station to 300 by 315 kilometres, which was the highest that the heavily loaded ferry could reach.

Soyuz-T 3 set off on 27 November 1980 with Leonid Kizim, Oleg Makarov and Gennadi Strekalov. Konstantin Feoktistov had been involved in the detailed planning for this mission and had hoped to make the flight, but had been medically grounded, and Strekalov took his place. Kizim and Makarov had served as the backup crew for Soyuz-T 2, so they were well acquainted. When the spacecraft arrived the next day, the Argon computer was allowed to choose its own approach route, and it docked at the front port without incident.

Gennadi Strekalov (left), Oleg Makarov and Leonid Kizim wearing the new Sokol suits.

The complex was immediately put into the gravity gradient to save propellant, and the crew set to work. It was usual to avoid heavy work during the early phase of adaptation to weightlessness, but they ignored this. And because they were so busy they also let the strict exercise regime lapse, but because they did not intend to stay for long this would not pose a significant risk. Despite the problem of gaining access to the thermal systems, the objective was to perform maintenance by replacing the pumps that circulated fluid through coolant loops to radiators on the surface. If they could achieve this, they were to move on to other tasks. Having gained access to the enclosure, they had to saw through a metal supporting bracket to reach the pumps. This was tricky, as they had to hoover up the metal shards from the saw. They had brought a complete assembly preloaded with glycol (in effect, antifreeze), and were conscious that if they spilt any of this fluid they would have to evacuate the station. The pipes through the shell had to be depressurised before the new assembly could be installed, and in the meantime the temperature in the station had to be able to be maintained at the required level. This was essential not just for the crew's comfort, but also to preserve the service life of the many rubber gaskets that sealed the holes through which pipes and cables ran to externally mounted apparatus. With the new pump assembly safely installed, the system was as good as new. Having crossed off the most urgent chore, they moved down the list in short order. They dismantled the telemetry system, and although this was another system that had not been designed to be serviced they were able to replace the faulty components. Several timers were replaced, as were a commutator in the power distribution system and a transducer in a compressor in the unified propulsion system's replenishment unit that had failed while Progress 11 was pumping in propellant when the station was unoccupied. The Resonance and Amplitude experiments were run to verify the structural integrity of the docking collars. Everything checked out.

The most significant new experiment installed involved using the KGA-1 camera built by the Physical-Technical Institute in Leningrad. It used a helium-neon laser to make holographic images. Such imaging is extremely susceptible to interference from vibration. On Earth a massive supporting structure must be used, but it was hoped that this would not be an issue in space. Romanenko was to have tested it, but it had not been ready in time for his flight and had been reassigned to this impromptu crew. It was tested by recording a dissolving salt crystal in the absence of convective flow, but was later to be used to record the deterioration of the station's portholes, which were otherwise difficult to photograph.

After boosting the apogee of the station's orbit to 370 kilometres, Progress 11 departed on 9 December. The next day, Salyut 6 was returned to autonomous flight and Soyuz-T 3 returned to Earth. The recommendation of the service crew was that Salyut 6 could support another expedition. The decision made, Progress 12 was sent up on 24 January 1981. It was hoped that its cargo, together with the stockpile of consumables already on the station, would be sufficient for a two-man resident crew for three months, to enable the final two Intercosmos missions to employ successive launch windows.

GREMLINS

In late February, the motor of one of Salyut 6's solar panels failed, further reducing the already low power supply unless the station was orientated so that the jammed panel faced the Sun, but this would bake one side of the station and freeze the other, and thereby overtax the thermal regulation system. The power deficiency meant that the heaters had to be turned down, with the result that the internal temperature fell to 10°C, which caused water vapour to condense and introduced the risk of a short circuit. In early March, the station was continually reorientated to trade power for thermal control, in an effort to increase the internal temperature.

ONE FINAL TOUR

Soyuz-T 4 slipped into the front port on 13 March with Vladimir Kovalyonok and Viktor Savinykh. Because of the unexpected longevity of the station, this team had been formed only when the repair crew reported that a further occupancy should be possible. In 1978, when Savinykh had joined the cosmonaut corps, Kovalyonok had been on Salyut 6. The third seat had been removed so that extra equipment could be carried. The first task was to attend to the stuck solar panel, and to

Vladimir Kovalyonok (left) and Viktor Savinykh.

Vladimir Kovalyonok (left) and Viktor Savinykh prepare the shower in the Salyut 6 trainer, behind which is the conical bulk of the primary instrument.

general delight, once the controller was replaced (an operation that took only a few hours), the panel rotated to face the Sun. With the heaters restored, the cabin was soon a comfortable 22°C. On spotting a crater in a porthole caused by the impact of a micrometeoroid, they set up the KGA-1 and made a hologram of it.

Savinykh was able to benefit from Kovalyonok's experience, and they rapidly adapted to weightlessness. Progress 12 was hastily unloaded so that it could leave, and much of its cargo was strapped to the cabin walls using a series of bungee cords. It delivered a new plant cultivator that was basically a Biogravistat with a magnetic field across its centrifuge compartment. In a television broadcast Popov and Ryumin had related their difficulty in growing higher plants and had appealed for ideas, and when applying a magnetic field was suggested this Magnetobiostat had been built to test the idea. It was planted with shoots and roots of crepis. Arabidopsis was placed in the Phyton, and orchids in the Malakhit. After boosting the station's orbit back to 345 by 358 kilometres, Progress 12 left on 19 March.

When Soyuz 39 arrived on 23 March, it brought Vladimir Dzhanibekov and the Mongolian cosmonaut Jugderdemidiyin Gurragcha. The visitors undertook the usual biomedical programme, and tested the Mongolian-supplied cervical shock absorber. Worn continuously during the first three days (except when sleeping), this applied pressure to the topmost part of the spinal column, restricted head movement, and simulated local loads experienced on Earth. It was hoped that this would inhibit the motion sickness commonly suffered in the first stage of adapting to weightlessness. The primary task was Earth-resources

Jugderdemidiyin Gurragcha (rear) and Vladimir Dzhanibekov.

photography. Mongolia is a vast, inaccessible and largely unexplored country, and orbital observation was the most cost-effective method of surveying. The KATE-140 and MKF-6M were used to locate geological faults, identify subterranean water deposits, assess crops and forests, and compile a map of soil conditions.

98 **Routine operations**

A view from Soyuz 39 of Salyut 6 with Soyuz-T 4 at the front port.

Spectroscopic observations were made with the Spektr-15K for the ongoing Biosfera project. The Emission experiment involved using dielectric detectors to record the flux of cosmic rays. One of the detectors was exposed in a scientific airlock, and another was kept inside the station to measure how well the hull blocked this radiation. The visitors returned to Earth on 30 March. They did not swap ferries as none of the Intercosmos crews were familiar with the new Soyuz-T spacecraft.

As soon as things returned to normal, Kovalyonok and Savinykh set about routine preventive maintenance on the water reclamation system and the fans in the thermal regulation system, replaced a switching unit in the cryogenic system of the BST-1M, and serviced the MKF-6M camera and the Yelena-F gamma-ray detector. Most of April was devoted to Earth-resources photography of the Soviet Union and making observations to complete the unfinished Intercosmos projects. At one point, with the station in the gravity gradient to observe the Earth panoramically through the various portholes, they saw noctilucent clouds extending over almost the whole of one hemisphere, and they used the BST-1M to take data. One significant result of using the BST-1M was the discovery of intense submillimetre emission high above thunderstorms.[9] During this period, while waiting for the final Intercosmos mission, smelting was done while the crew slept and on two occasions six-day experiments were run in the Splav to create CMT.

---

[9] In 1994, a NASA research aircraft equipped with a low-light camera observed thin jets (nicknamed 'sprites') projecting above violent storms to an altitude of 50 kilometres which give rise to red glows. This is a result of establishing a temporary discharge path from a thunderstorm to the ionosphere – a form of lightning to space rather than to the ground.

Leonid Popov looks on as Dumitru Prunariu signs the Soyuz 40 capsule.

Soyuz 40 arrived on 15 May with Leonid Popov and the Romanian cosmonaut Dumitru Prunariu. Although Prunariu's flight had been pending for some time, the scheduling chaos after the fault with Soyuz 33's engine had led to him training with a succession of potential commanders, and he had been paired with Popov only at the start of the year. His experiments included one to determine the capacity of human lymphocytes to synthesise interferon. Minidoza measured the energy spectrum of the heavy nuclei in cosmic rays. Two novel smelting experiments were performed, one forming a crystal of germanium in the Kristall furnace by exploiting capillary action within a molybdenum matrix, and the other a disk-shaped silicon monocrystal to test the making of solar cells in space. The Pion apparatus exploited the fact that there are no convection currents in fluids in space to crystallise materials both in and out of solution. It was set up for the Struktura experiment that investigated heat exchange and mass transfer during crystallisation in an aqueous solution, and the film of the process was returned when the visitors left on 22 May. Standing by the side of the charred capsule, which was the final one of the old series, Popov praised its manufacturer on behalf of all the cosmonauts who had flown it during the previous decade.

Alone once again, Kovalyonok and Savinykh undertook a comprehensive medical check, and then began to prepare to depart, which they did on 26 May 1981 after spending 75 days in space. A few days later, it was announced that no more cosmonauts would be sent to Salyut 6.

A CLEAR SUCCESS

Magnificently successful, Salyut 6 had been occupied for a total of 684 days. It had hosted five long-duration crews, four of which were able to set successive endurance records, and 11 shorter visits. It had been sustained by propellant, consumables and experiment apparatus delivered by a dozen Progress ferries. Remarkably, it had been able to accommodate the entire first round of Intercosmos missions, and of the nine such flights only one had been frustrated. Although its rôle as a habitable station was

Routine operations

over, Salyut 6 could still serve a useful purpose, and Kaliningrad had a brand new project planned.

Table 2.1 Salyut 6 docking operations

Spacecraft	Docking		Port	Undocking		Days
Soyuz 25	10 Oct 1977	0709	front	11 Oct 1977	~0800	1.03
Soyuz 26	11 Dec 1977	0602	rear	16 Jan 1978	1422	36.35
Soyuz 27	11 Jan 1978	1706	front	16 Mar 1978	1100	63.75
Progress 1	22 Jan 1978	1312	rear	7 Feb 1978	0855	15.82
Soyuz 28	3 Mar 1978	2010	rear	10 Mar 1978	1325	6.72
Soyuz 29	17 Jun 1978	0058	front	3 Sep 1978	1123	78.43
Soyuz 30	28 Jun 1978	2008	rear	5 Jul 1978	1315	6.71
Progress 2	9 Jul 1978	1559	rear	2 Aug 1978	0757	23.66
Progress 3	10 Aug 1978	0300	rear	21 Aug 1978	–	11
Soyuz 31	27 Aug 1978	1937	rear	7 Sep 1978	1353	10.76
Soyuz 31	7 Sep 1978	1421	front	2 Nov 1978	1046	55.85
Progress 4	6 Oct 1978	0400	rear	24 Oct 1978	1607	18.50
Soyuz 32	26 Feb 1979	0830	front	13 Jun 1979	1251	107.18
Progress 5	14 Mar 1979	1020	rear	3 Apr 1979	1910	20.37
Progress 6	15 May 1979	0919	rear	8 Jun 1979	1100	24.07
Soyuz 34	8 Jun 1979	2302	rear	14 Jun 1979	1918	5.84
Soyuz 34	14 Jun 1979	~1950	front	19 Aug 1979	1208	65.68
Progress 7	30 Jun 1979	1418	rear	18 Jul 1979	0650	17.69
Soyuz-T 1	19 Dec 1979	1705	front	24 Mar 1980	0004	94.29
Progress 8	29 Mar 1980	2301	rear	25 Apr 1980	1104	26.50
Soyuz 35	10 Apr 1980	1816	front	3 Jun 1980	1447	53.85
Progress 9	29 Apr 1980	1109	rear	20 May 1980	2151	21.45
Soyuz 36	27 May 1980	2256	rear	4 Jun 1980	1938	7.86
Soyuz 36	4 Jun 1980	1808	front	31 Jul 1980	1455	56.86
Soyuz-T 2	6 Jun 1980	1858	rear	9 Jun 1980	1224	2.73
Progress 10	1 Jul 1980	0853	rear	18 Jul 1980	0121	16.69
Soyuz 37	24 Jul 1980	2302	rear	1 Aug 1980	1943	7.86
Soyuz 37	1 Aug 1980	~2010	front	11 Oct 1980	0930	70.56
Soyuz 38	19 Sep 1980	2349	rear	26 Sep 1980	1534	6.66
Progress 11	30 Sep 1980	2003	rear	9 Dec 1980	1323	69.72
Soyuz-T 3	28 Nov 1980	1854	front	10 Dec 1980	0910	11.59
Progress 12	26 Jan 1981	1856	rear	19 Mar 1981	2114	52.09
Soyuz-T 4	13 Mar 1981	2333	front	26 May 1981	–	74
Soyuz 39	23 Mar 1981	1928	rear	30 Mar 1981	1122	6.66
Soyuz 40	15 May 1981	2150	rear	22 May 1981	1337	6.66
Cosmos 1267	19 Jun 1981	1052	front	permanently docked		–

Dates are Moscow Time.

Table 2.2 Salyut 6 crewing

Cosmonaut		Arrival	Departure	Days
Yuri Romanenko	CDR	Soyuz 26	Soyuz 27	96.42
Georgi Grechko	FE	Soyuz 26	Soyuz 27	96.42
Vladimir Dzhanibekov	CDR	Soyuz 27	Soyuz 26	5.96
Oleg Makarov	FE	Soyuz 27	Soyuz 26	5.96
Alexei Gubarev	CDR	Soyuz 28	Soyuz 28	7.93
Vladimir Remek	CR	Soyuz 28	Soyuz 28	7.93
Vladimir Kovalyonok	CDR	Soyuz 29	Soyuz 31	139.60
Alexander Ivanchenkov	FE	Soyuz 29	Soyuz 31	139.60
Pyotr Klimuk	CDR	Soyuz 30	Soyuz 30	7.92
Miroslaw Hermaszewski	CR	Soyuz 30	Soyuz 30	7.92
Valeri Bykovsky	CDR	Soyuz 31	Soyuz 29	7.87
Sigmund Jähn	CR	Soyuz 31	Soyuz 29	7.87
Vladimir Lyakhov	CDR	Soyuz 32	Soyuz 34	175.06
Valeri Ryumin	FE	Soyuz 32	Soyuz 34	175.06
Leonid Popov	CDR	Soyuz 35	Soyuz 37	184.84
Valeri Ryumin	FE	Soyuz 35	Soyuz 37	184.84
Valeri Kubasov	CDR	Soyuz 36	Soyuz 35	7.86
Bertalan Farkas	CR	Soyuz 36	Soyuz 35	7.86
Yuri Malyschev	CDR	Soyuz-T 2	Soyuz-T 2	3.93
Vladimir Aksyonov	FE	Soyuz-T 2	Soyuz-T 2	3.93
Viktor Gorbatko	CDR	Soyuz 37	Soyuz 36	7.86
Pham Tuan	CR	Soyuz 37	Soyuz 36	7.86
Yuri Romanenko	CDR	Soyuz 38	Soyuz 38	7.86
Arnaldo Tamayo Méndez	CR	Soyuz 38	Soyuz 38	7.86
Leonid Kizim	CDR	Soyuz-T 3	Soyuz-T 3	12.79
Oleg Makarov	FE	Soyuz-T 3	Soyuz-T 3	12.79
Gennadi Strekalov	FE	Soyuz-T 3	Soyuz-T 3	12.79
Vladimir Kovalyonok	CDR	Soyuz-T 4	Soyuz-T 4	74.77
Viktor Savinykh	FE	Soyuz-T 4	Soyuz-T 4	74.77
Vladimir Dzhanibekov	CDR	Soyuz 39	Soyuz 39	7.86
Jugderdemidiyin Gurragcha	CR	Soyuz 39	Soyuz 39	7.86
Leonid Popov	CDR	Soyuz 40	Soyuz 40	7.86
Dumitru Prunariu	CR	Soyuz 40	Soyuz 40	7.86

Routine operations

Table 2.3 Salyut 6 spacewalks

Date	Hours	Activity
20 Dec 1977	1.5	Romanenko and Grechko opened the front docking port, inspected the drogue and the external rendezvous apparatus to verify that there was no damage, and then deployed the Medusa cassette.
29 Jul 1978	2.1	Kovalyonok and Ivanchenkov exited the airlock hatch, retrieved the Medusa and other cassettes, installed new ones, retrieved a passive meteoroid detector, and deployed a radiation monitor.
15 Aug 1979	1.4	Lyakhov and Ryumin exited the airlock, and Lyakhov remained at the airlock and collected experiment cassettes while Ryumin made his way to the rear to cut loose the fouled KRT-10 radio telescope dish.

ANOTHER ALMAZ?

With the 'military' and the 'civilian' stations being alternated, it seemed reasonable to assume that while the lessons of Salyut 6 were assimilated, the next Almaz would be flown as Salyut 7. Even though the Chelomei Bureau had successfully flown its Almaz as Salyut 3 and Salyut 5, it had been forced to rely on Korolev's Soyuz for crew transport. Without the large TKS ferry that it had designed to complement the reconnaissance platform, there was no way to deliver bulk cargo. Throughout, therefore, Chelomei continued with the development of his own spacecraft, so that it could finally fly the missions originally proposed for the military. The first step was to test the re-entry characteristics of the capsule. A Proton rocket put Cosmos 881 and Cosmos 882 into low orbit on 15 December 1976, both of which returned after a single orbit, although they did not use the usual recovery site in Kazakhstan. As the main compartment of the TKS was not included, a pair of capsules were stacked for two tests using a single launch. These verified the attitude-control system, de-orbit engine, heat shield and parachute system. Cosmos 997 and Cosmos 998 repeated the test in March 1978. On 5 January 1979, however, a Proton with two more capsules malfunctioned, and while the escape system pulled the top one clear, demonstrating that this system worked, the one underneath had to be abandoned. A repeat in May 1979 with Cosmos 1100 and Cosmos 1101 certified the capsule, but in keeping with the secrecy of the military programme, the fact that these were tests of a spacecraft for human use was not announced.

Following the initial test, another Proton launched the full TKS configuration as Cosmos 929 on 17 July 1977. During the next month, this performed a number of minor manoeuvres while holding a low perigee to enable radars to determine its drag characteristics. On 17 August, one component of the sophisticated telemetry stream was noticed by Western radio monitors to have ceased – the capsule had returned to Earth. The following day, the TKS raised itself to 310 by 330 kilometres, and on 19

December it climbed to 440 by 448 kilometres, which was higher than that used by Salyut stations. On 2 February 1978, after six months in space, it de-orbited itself over the Pacific. In addition to showing that it had a restartable engine, by making a total change of velocity of 300 metres per second it demonstrated that it had a large propellant tank. Clearly, it was not testing a mere rocket stage for geostationary or planetary probes, and the Royal Aircraft Establishment's Satellite Tracking Group speculated that it was a test of a 'tug' which would deliver large laboratory modules to a future space station; this proved to be a prescient prediction.

With the TKS and its capsule successfully tested, Chelomei had hoped to launch an Almaz with a docking port at each end. The crew of three would have gone up on a TKS, and a succession of such ships would have delivered replacement crews for extended operations. But Salyut 6's longevity prompted repeated postponements, and final cancellation. However, a TKS was launched as Cosmos 1267 on 25 April 1981 and returned its capsule a month later. The TKS then made a slow rendezvous with Salyut 6 and on 19 June it docked at the front port. Over the next 12 months, during which the complex remained unoccupied, Cosmos 1267 periodically fired its engine to maintain a nominal 350-kilometre circular orbit while the degradation of its systems was monitored. In late 1981, *Aviation Week & Space Technology* reported a Department of Defense claim that Cosmos 1267 incorporated a ring of "infrared-homing guided interceptors" and had turned Salyut 6 into an "orbital battle station". In fact, the long slim cylindrical objects on the surface of the vehicle were externally mounted propellant tanks. In response, the Soviets said that it was the prototype of a module for the assembly of a large orbital station from modular components.

3

A step towards continuous occupancy

BUILDING ON SUCCESS

On Salyut 6, each two-man crew had used up 20 to 30 kilograms of consumables per day. For a station to be continuously inhabited, the logistical system would require to deliver 1 tonne of consumables per month, which meant one Progress flight every two months. However, the longer life of the Soyuz-T would reduce the need to send up crews on brief visits to exchange ferries, which meant that the pace of operations should be considerably less hectic than that which had been sustained while Salyut 6 was occupied. One objective of the next station would be to have one crew hand the station over to its successor. This could have been done on Salyut 6 if a retiring crew had departed in its own ferry and left the new residents with their own. A handover in orbit would eliminate the time wasted in powering down and then reactivating the station, and would facilitate a lengthy period of occupation without requiring a single crew to endure the stress of an extended flight. And, if the life-support system could sustain three people, it would also facilitate partial crew exchanges, and even allow a research cosmonaut to serve with successive crews to carry out an extended study of adaptation to weightlessness.

SALYUT 7

Salyut 7 was launched on 19 April 1982. It had been expected that the next station would have a cluster of docking ports to enable it to be augmented by modules such as Cosmos 1267, but other than having a reinforced front collar to accommodate the greater loads expected to occur when manoeuvring with a large mass docked, it was structurally identical to its predecessor.[1] It had a revised form of the Delta automatic navigation system, which was deemed to be sufficiently reliable to permit the station

---

[1] Resonance tests would demonstrate that this reinforcement was unnecessary.

to be manoeuvred without crew supervision, and an improved Kaskad system which could align the station to within 1 degree. The Niva video enabled observations to be recorded at any time for downloading when in communication. The long-range Mera transponder had been added to assist a ferry manoeuvre from its final transfer orbit to where the Igla could lock on, several kilometres out. To prevent micrometeoroid impacts from degrading the outer surface of the instrument portholes, these had been fitted with covers. To ameliorate the inevitable degradation of the transducers in the solar panels, attachments had been fitted along the edges of the deployment frames so that small panels could be clipped on by spacewalkers to re-establish the required output. Umbilical connectors had been installed in the airlock to enable spacewalkers to complete their preliminary activities prior to drawing on the life-support systems of their suits, thereby not only facilitating a full five hours of external activities but also guaranteeing supplies in the event of difficulties in repressurising the airlock.

The plan was to operate Salyut 7 for four years. With crews flying ever-longer missions, there would be fewer flight opportunities. Vladimir Shatalov reported that only 16 cosmonauts had been assigned. It had been decided that two-person resident crews would be visited by teams of three, and the station's environmental systems had been upgraded to be able to support five people for brief periods. A Progress tanker could pump water directly into the tank of the Rodnik in the unpressurised engine bay at the rear of the station, which was more practical than having the crew unload lots of small flasks. The experience of living onboard Salyut 6 had prompted substantial improvement in the crew facilities. Previously, food had been prepacked irrespective of whether that combination would appeal to a cosmonaut. Food items on Salyut 7 were packed separately, and the crew could 'pick and mix' to suit their individual tastes. A small refrigerator had been added to store the fresh food that was to be delivered regularly by the Progress ferries. The stove could heat tinned food for both cosmonauts at once. About 65 per cent of each meal was reconstituted freeze-dried food. Four full meals per day would provide 3,150 calories. A new toothpaste and electric toothbrushes were provided. The electric shaver had a stubble collector (but this proved inadequate). Although the main compartment appeared to be walled off, there were spaces behind many of the panels for storage, to try to avoid the problem encountered late in Salyut 6's life where much of the apparatus ferried up over the years had to be strapped to the walls (this was to prove to be wishful thinking). The walls had been covered by brightly coloured washable panels. The lighting had been strengthened, and made more uniform. As previously, there was both a treadmill and a veloergometer (this time mounted on the ceiling) for exercise (although the fact that the treadmill faced the wall, upside down, proved disconcerting). The multi-function Aelita biomedical test kit superseded the Polynom-2M. An improved Bania shower had been installed (but, as previously, proved to be frustrating to use). Penguin suits were still to be worn, but because the elastic strapping had been removed these were simply comfortable work clothes.

Most of the scientific apparatus was to be delivered by a succession of ferries, but the bulkiest items were preloaded. This initial suite included the RT-4M mirror-based 'soft' X-ray astronomical telescope (an improved form of the telescope tested on Salyut 4) which was in the bulky conical mounting in the main compartment, the MKF-6M multispectral camera, a KATE-140 mapping camera, the SKR-2M X-ray spectrometer, the Spektr-15M and MKS-M spectrometers to study the atmosphere and oceans, the Yelena-F gamma-ray spectrometer for near-Earth radiation studies, an improved Oasis plant cultivator and the KGA-2 holographic camera.

COMMISSIONING CREW

Anatoli Berezovoi and Valentin Lebedev had begun training to commission the new station in early 1981. Soyuz-T 5 was launched on 13 May. The Mera transponder was activated while still 250 kilometres away, but did not detect the station until the range was down to 30 kilometres. The Igla locked on soon thereafter, and the Argon docked the ferry at the front port. There was a moment of concern when the hatch would not open, and Lebedev had to stand on the roof of the orbital module with his feet around the hatch's rim to prise it open. After everything had been checked out, they placed their sleeping bags on the ceiling of the station's main compartment and retired for the first night of what they hoped would be a record-breaking mission.

As usual, for the first week, as they adapted to weightlessness, they undertook only light duties. They gave the Delta and Kaskad systems a thorough test, then set them on automatic. To execute a manoeuvre, they needed only to enter the command into the computer. They redistributed the apparatus that had been set on frames for launch, and then removed the temporary brackets and struts. Wheat and peas were placed in the Oasis cultivator, flax in the Biogravistat and arabidopsis in the Phyton. Then on 17 May, after verifying its battery and transmitters, the Iskra amateur radio satellite was ejected from the scientific airlock to mark the start of the Congress of the Young Communists' League. This was the first time that a subsatellite had been released by an orbital station. To minimise propellant consumption, the programme had been organised so that the station could be maintained in a given orientation for several days at a time. The most economic attitude exploited the gravity gradient to provide stability. As much of the observational equipment was portable, it could be set up in the forward transfer compartment, which had a ring of portholes providing panoramic coverage. The station was set vertically, either up or down depending on whether the sky or the Earth was to be observed. Materials-processing experiments could be performed during these periods. But the MKF-6M and KATE-140 cameras could be used only when the belly of the station was maintained facing the Earth, the manoeuvres for which precluded materials processing. Other experiments required instruments to be aimed at the horizon and, depending on the instruments, this could be done either in the gravity gradient or by placing the station in a complex rotation. All these flight regimes had to be carefully scheduled. Spending several days on one type of work was also efficient because it eliminated the time wasted setting up and stowing equipment. Lebedev spent so much time working in the forward transfer compartment that he soon developed an impressive tan.

Progress 13 arrived on 25 May. As in the case of Salyut 6, only the rear port had the plumbing to replenish the station's fluids. The cosmonauts were to have isolated the rear transfer compartment for the docking, but they rigged the sensor so that the flight controllers would think that the internal hatch was closed, and then they went to monitor the approaching ferry through the tiny porthole beside the docking collar. Among its scientific cargo was the Magma-F furnace and 250 kilograms of apparatus for their first visitors. The Soyuz-T could not reach the high orbit carrying a crew of three, so Progress 13 lowered the station to 300 kilometres before leaving on 4 June. As the ferry withdrew, the Astra apparatus observed how its engines contributed to the gaseous environment in the immediate vicinity of the station. While assembling the Magma-F furnace, it was noted that there were errors in the instruction manual. An improved form of the Kristall used on Salyut 6, it had a microaccelerometer and a magnetometer to record effects which might influence the melting and crystallisation processes. It could operate at up to 900°C, and report its status to Earth as part of the station's telemetry stream. After initial tests, it was decided that it gave better results if it was unbolted from its mount and left to float freely, as it did not pick up the vibrations transmitted through the hull of the station.

A FRENCHMAN VISITS

Soyuz-T 6 was launched on 24 June. It carried Vladimir Dzhanibekov, Alexander Ivanchenkov and Jean-Loup Chrétien, the first French cosmonaut. Dzhanibekov had been assigned in February after Yuri Malyschev had been taken ill. This, the first of the new series of international flights, was the first involving a citizen of a state not in the Intercosmos organisation. The 1979 agreement raised the profile of a close cooperation which has seen French apparatus flown on Soviet satellites since the mid-1960s. Initial enthusiasm had turned to calls for the flight to be cancelled in response to the Soviet invasion of Afghanistan in December 1979, but preparations had quietly continued. When 900 metres out, the Argon computer turned the ship to fire the engine to cut its closing rate. As it turned back to face the station, prior to using the less powerful thrusters for the final phase of the approach, the inertial reference locked. The Argon abandoned the manoeuvre and left the ferry in an end-over-end spin. Dzhanibekov stabilised the spacecraft, completed the approach, and docked.

Alexander Ivanchenkov (left), Vladimir Dzhanibekov and Jean-Loup Chrétien.

Dzhanibekov and Ivanchenkov tested the Braslet, which involved wearing thin elastic rings around the thighs for up to an hour, several times a day, during the first few days in order to restrict the migration of blood from the legs up into the torso. It

Salyut 7 with Soyuz-T 5 attached, taken by Jean-Loup Chrétien on Soyuz-T 6.

complemented the Tchibis leggings. Chrétien had a broad programme which included Earth and astrophysical observations, materials processing, biological and biomedical experiments, the latter focusing on cardiovascular and sensory functions. Echograph utilised an ultrasound scanner to provide a sectional image of the heart and a doppler sensor to measure blood flow, cavity dimensions, mycocardiac thickness, and arterial capacity. When Grechko reported chest pains after returning to Earth, it was realised that the heart migrated up into the chest cavity in space. Chrétien initially had some difficulty locating his heart, because it was not where trials had led him to expect it to be. The Posture experiment studied how the body adopted given positions in the absence of gravity. Posture is maintained by a variety of sensory organs and muscle groups different from those used in locomotion. The experiment required Chrétien to don motion sensors and muscle monitors, and place one foot in a restraint to enable spatial coordinates to be measured. With both eyes closed, he executed a series of movements to exercise sensory and locomotive functions. The apparatus took longer to set up than expected, which prompted him to make recommendations for future apparatus, but when it was finally ready the other visitors performed the experiment to provide additional data. The Cytos-2 experiment employed an incubator to study the structure and functioning of bacteria cells in weightlessness, and the effects of a variety of antibiotics on them. Biobloc-3 studied how cosmic rays affected biological materials by sandwiching seeds between a pair of heavy-ion detectors. A number of materials-processing experiments were performed in the newly assembled Magma-F. One measured small accelerations to develop a mathematical model of the furnace, a second observed the dissolution of a polycrystalline solid alloy in its own liquid in a state of thermodynamic equilibrium, to develop a model of the solidification process and investigate capillary forces, and a third created an alloy of aluminium and indium – which are immiscible in Earth's gravity. The PCN low-light camera was used to photograph faint sky phenomena, including the zodiacal light, noctilucent clouds and lightning. Chrétien was assisted in this work by Ivanchenkov who, as a result of his previous mission, had become adept at seeing such phenomena. The Piramig camera was placed in one of the portholes in

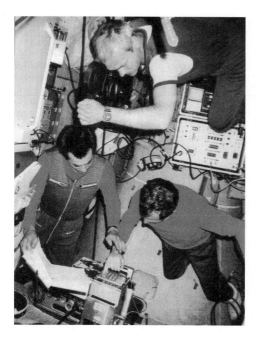

Vladimir Dzhanibekov films Alexander Ivanchenkov (left) and Jean-Loup Chrétien.

the forward transfer compartment. It could be operated only in the Earth's shadow, but fortunately at that time the orbit provided 25 minutes of darkness per cycle. With the hatches shut, the lights off, and with the attitude-control thrusters off to preclude spewing efflux across the field of view, the photomultipliers could make faint ultraviolet, visible and near-infrared observations, with exposures of a few minutes. A total of 350 pictures were secured of the upper atmosphere and interplanetary dust. Those of noctilucent clouds were so fascinating that the residents were later asked to use the PCN twice per week to photograph the horizon in the hope of capturing other transient stratospheric phenomena.

Although Chrétien started off by sleeping for the assigned period, his schedule was so demanding that within a few days he was working late and getting only a few hours rest, with the result that by the end of the visit he was exhausted. The visitors departed on 2 July. Although the residents had been assigned a long tour, Soyuz-T 5 was not exchanged. As the next visit was not due for several months, Salyut 7 raised its apogee to 344 kilometres. Berezovoi and Lebedev were so worn out playing host that they were told to take a few days off to recover. They spent a long time looking for Dzhanibekov's watch and a roll of PCN film which had inadvertently been left behind. Although sad to be alone once more, they soon resumed the familiar routine. However, they became a little grumpy, requesting Kaliningrad not to feel obliged to chat simply because they were in radio contact, as this interrupted their work, which was usually most intensive while passing over the Soviet Union, and later rebuking a call for them to start an exercise period, even although this would mean missing a run with the MKF-6M, whose target was about to appear over the horizon.

Soyuz-T 6 descends by parachute.

Progress 14 docked on 12 July. The miniature onions that it delivered for one of the cultivators were so enticing that they were eaten as delicacies. It also brought the Korund furnace, and the EFO-1 electrophotometer which monitored the attenuation of stars as they were occulted by the Earth's limb to measure the vertical structure, density and composition of the thin layers of the upper atmosphere for the Climate experiment.

INSTALLING EXPOSURE CASSETTES

On 22 July Berezovoi and Lebedev cleared the accumulated loose items out of the forward transfer compartment, and retrieved the spacesuits. As the suits were new, they had to be thoroughly inspected, the oxygen bottles filled, the lithium hydroxide canisters loaded, and the water tank filled that fed the network of fine tubes which were woven into the fabric of the undergarment that would keep the user cool. These preparations consumed most of the week. The experiments to be mounted outside, which had been delivered by Progress 14, were then prepared. The Medusa cassette contained biopolymers, but the others had samples of materials and structures being assessed for use on future spacecraft. The 20-metre umbilicals which would provide electrical and radio links were unpacked and tested. The bracket that would hold the apparatus while the airlock hatch was open (to ensure that it did not drift off) was installed. The battery of the TV camera that would be used to record the state of the solar panels was recharged. The cosmonauts were keen to extend their scheduled external activities to allow them to crawl along the length of the station to inspect its thermal insulation, but Kaliningrad insisted that they remain near the hatch, where the experiments were to be installed. Finally, they donned the suits and pressurised them slightly above the ambient level to verify their seals. There was some concern when Lebedev discovered that his suit leaked, but an inspection revealed that a strap had caught in the rear hatch when the hinged backpack had been closed, preventing it forming a proper seal. The day before the big event, just in case it proved impossible to repressurise the airlock and they had to abandon the station, their research results were stored in Soyuz-T 5's descent module, the internal hatch was sealed, and the flasks of air that would be required to repressurise the orbital module in the event of such an emergency were checked. The preparations complete, Salyut 7 was reset for autonomous operation regime. The excursion, on 30 July, took less than three hours. Lebedev positioned himself on the foot restraint immediately outside the hatch and retrieved several packages placed prior to launch. These

Anatoli Berezovoi snapped this picture of Valentin Lebedev passing out through the airlock hatch.

112 A step towards continuous occupancy

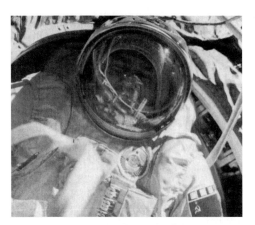

Once outside, Valentin Lebedev took this picture of Anatoli Berezovoi in the hatch.

included Etalon, a plate that had exposed optical coatings (some of which had bubbled and peeled off), a Medusa package and a Spiral package. He then attached the new packages. He was eager to move up to another anchor at the base of a solar panel to test the winch that a later crew would use to install clip-on solar panels, but the station was approaching the Earth's shadow and he stayed where he was. Berezovoi, who remained in the airlock with only his head and shoulders outside, passed Lebedev the camera to record the state of the solar panels. The hatch was closed, and the airlock repressurised without incident. Lebedev was astonished to find a 2-centimetre tear in the protective cover of his helmet, as he had no idea how it happened. The air lost when depressurising the airlock was replenished by air pumped into the station's tank from the attached cargo ferry.

BACK TO WORK

After a day's rest, the cosmonauts set up the new furnace delivered by Progress 14. The Korund incorporated a revolving sample holder which could be loaded with up to a dozen ampoules. Although each was processed in sequence, the furnace could be set up to operate either automatically or by remote control so that the retiring crew could set it up for use following their departure, when it would not be disturbed by their presence. In contrast to the Splav (in which the sample was fixed and the thermal field was varied to control crystallisation) and Kristall (in which the thermal field was held constant and the sample was moved), this new furnace was capable of both moving the sample at speeds varying from a few millimetres a day to a few centimetres a minute, and of changing the temperature between 20°C and 1,270°C at rates of 0.1°C to 10°C per minute. One of the rôles long-ago identified for the Salyut stations was testing apparatus intended for use on automated vehicles. It was hoped that this second-generation technology would facilitate large-scale manufacturing of semiconductors of exceptional quality on an automated free-flying spacecraft which would be periodically serviced. Unfortunately, the Korund overheated and switched itself off, so it could be used for only a few hours at a time. It took three months to realise that the thermal coefficients of its heating element had been computed from trials conducted on Earth, in which thermal convection had drawn off energy. In microgravity, where convection did not occur, the element had heated up so rapidly that the controller had concluded that it was overheating. Once reconfigured with appropriate coefficients, it worked perfectly.

In early August, the arabidopsis in the Phyton cultivator produced pods – it was the first time that a plant grown from seed in space had done so. A week later, the pods burst to release seeds. The peas in the Oasis were developing well, but even more mouthwatering were the tomatoes in the Svetobloc hothouse. All of this was excellent news for the botanists.

A WOMAN VISITOR

Progress 14 departed on 11 August, and on 20 August, with the station again down to a 289 by 300 kilometres orbit, Soyuz-T 7 took its place. It delivered Leonid Popov, Alexander Serebrov and Svetlana Savitskaya. The second woman to fly in space, Savitskaya was presented with a flowering arabidopsis. She was also given a floral apron and invited to play hostess for the evening meal. The males reserved the orbital module of a ferry for Savitskaya, to enable her to have her own toilet, but she chose to sleep in the main compartment with them.

The main item of the visiting crew's programme was the testing of the Tavriya electrophoresis apparatus. This was essentially a column of biological compounds to be separated into homogeneous fractions by the application of an electric field. For this test, human blood protein (albumin and haemoglobin) was separated. Savitskaya filmed the process, then extracted a sample of each separated product. Later, it was used to purify urokinase (an enzyme present in human urine) and, for the first time, interferon was processed by the electrophoresis technique. The results were very encouraging, and demonstrated a full order of magnitude improvement in purity over similar apparatus on Earth. The electric field acted regardless of molecular weight, so the apparatus could be used to separate a wide range of biological substances. It was hoped that improved equipment on later missions would produce very high quality vaccines and other pharmaceutical products on a commercial basis.

Popov, Serebrov and Savitskaya

Leonid Popov (left), Svetlana Savitskaya and Alexander Serebrov.

Anatoli Berezovoi (left), Valentin Lebedev and Svetlana Savitskaya.

Valentin Lebedev and Svetlana Savitskaya.

returned to Earth in Soyuz-T 5 on 27 August. This left Soyuz-T 7 at the rear port. It had to be moved while the recovery window remained open, so it was done two days later. A fly-around had not been done with the Soyuz-T type before, and there was some concern that the narrow beam of the Igla system might not be able to lock on at only 200 metres, but the redocking was completed without incident. As a precaution, though, the station had been placed in its automatic regime. It turned out to be much easier than in the earlier model of the spacecraft, which had not had the benefit of the Argon computer. Although they had been ordered not to re-enter the station until communications were re-established, as soon as the pressure was equalised Berezovoi and Lebedev went in to get something to eat, and then retreated to their ferry to await permission to open the hatch. With the scheduled visits over, the orbit was raised.

THE LONG HAUL

There was some flexibility in the length of the mission, and the cosmonauts had been suggesting for some time that they would like to break the 185-day record. By 10 September they were once again growing irritable with the flight controllers regarding the merits of adhering to the daily schedule. The best days, ironically, were the rest days because the controllers left them alone and they performed tricky observational tasks in uninterrupted peace. At last, on 14 September, Viktor Blagov invited them to extend the mission by two months, which they eagerly accepted. They also asked to make another spacewalk, to test the apparatus that was to be used by a later crew to install extra solar panels. No commitment was made to this new request, however. The arrival of Progress 15 on 20 September (after a two-week postponement in its launch) was welcomed because both men were thoroughly bored with reconstituted food. As usual, the routine maintenance of the thermal regulation and environmental systems followed the unloading of replacement parts. Several days later, there was a moment of concern when the cosmonauts noticed a smell of burning insulation. They closed all the hatches to isolate the compartments, and switched off the fans. With fire extinguishers at the ready, they opened a suspect panel and found a fan coated with a thick layer of dust – the dust had evidently clogged the mechanism and caused it to overheat, so they cleaned it. Because the entire incident had taken place when out of radio contact, they decided not to tell Kaliningrad.

By this point, a sense of wonderment had set in. One piece of apparatus had arrived that they had never seen before. When they asked what it was, they were told to eject it from the scientific airlock and track it using the sextant, so they did, and reported that it had drifted away in exactly the same way as did the rubbish that they regularly tossed out! Resident crews on very long flights would later grow used to unpacking apparatus that they had never trained with, and even to having people that they had never met float through the hatch of a newly arrived ferry; but it was disconcerting. Another source of frustration was apparatus that malfunctioned. Thus far, they had fixed the leaky gasket in the Oasis, repaired the Korund, and replaced burnt-out parts in the Aelita, the EFO-1, the vacuum cleaner and a transmitter in the telemetry system. But all of this paled into insignificance compared with the fault in the Delta navigation system. An astrophysical observation using the RT-4 telescope in late August had been interrupted by the Delta dropping off-line. After analysis, it was decided to re-enter its programme, so in mid-September Lebedev had manually keyed a sequence of 325 six-digit instructions. In early October he had to do it again; unfortunately, he failed to key 'enter' at the end and the sequence was lost. He redid it, but the checksum did not match! By the time the sequence was finally accepted, it was far into the night. Back in June, a test had demonstrated that the sextant and star tracker could be used to orient the station with an accuracy comparable with the Delta, but doing this was so time-consuming that little other work could be done. As confidence in the Delta diminished, the station was oriented vertically and stabilised in the gravity gradient and their work rescheduled. In October, the water reclamation system failed. When the panel was removed they discovered a massive blob of water floating inside. When Lebedev disassembled the pump, he found that its bearing had seized. By this point, they were intimately familiar with the noises on the station. A month earlier, he had warned that this pump had been running rough, but Kaliningrad had told him not to worry. Another irksome problem was that the STR had recently begun to allow the temperature to fall to its minimum level before switching on the heater. Salyut 7 had been in use for only six months, but already it was proving to be less reliable than its illustrious predecessor.

At the end of October, Berezovoi and Lebedev passed the point at which they were nominally to have returned. Ironically, after having canvassed for an extension, they now started to count the days to their departure. As they had opted to stay on, it had been decided to stock up the station for the next crew. Progress 15 had left on 14 October. Progress 16 arrived on 2 November. Problems with the Delta continued, and it was set to repeatedly recalculate the station's position in space in the hope of catching it 'fall over', but this failed to expose the fault. On 11 November, Lebedev awoke to find Berezovoi in pain. After an hour without improvement, he issued him with biseptol from the medikit. Once in contact with Kaliningrad, they described the symptoms and an injection of atropine was prescribed. They were told to prepare to return to Earth the following day, but Berezovoi's condition improved and this order was rescinded. If they had been forced to return, it would have been very bad timing, as it would have left them just two days short of the 185-day record. Time began to pass more quickly. When they found that they were running out of paper on which to record observations they acknowledged that it really was time to leave. When the

Delta failed completely in early December, it was decided to bring them home. They left on 10 December, having set a 211-day record – although without making their second spacewalk. The capsule descended in darkness in a snowstorm, was dragged along the ground by its parachute, and rolled down a slope, which was disconcerting after so long weightless. The recovery window would not open for another fortnight, so they had barely begun their drug and exercise programme, and were in pretty bad shape. The weather was so bad that the first helicopter broke its gear upon landing. It was deemed too dangerous to land another helicopter, so the intrepid spacefarers spent the rest of the night shivering in the back of the disabled chopper.

Among the many scientific results was a monocrystal of cadmium selenide with a mass of 800 grams, a hoard of almost 200 arabidopsis seeds produced in space, and 2,500 MKF-6M pictures. In recognition of his Earth-resources work, Lebedev was declared a "distinguished prospector" by the Minister of Geology.

Progress 16 departed on 13 December, leaving Salyut 7 in the gravity gradient. The Korund had been left fully stocked so that smelting could continue under remote control. With the Delta system overwriting its memory, the engineers concluded that its electronics would have to be replaced. The crippled station could be manoeuvred only while in radio range. Fortunately, it was orbiting somewhat above its usual 350-kilometre altitude, so there was no danger of its orbit decaying before the next crew could be dispatched – or so it was thought.

TRIALS AND TRIBULATIONS

On 2 March 1983 a Proton put Cosmos 1443 into low orbit, and within hours this began to manoeuvre towards Salyut 7. It was reported to be an 'operational' version of Cosmos 1267, which had docked with Salyut 6 after that station's final crew had left. As such, it was a multi-function vehicle. By serving as a tug, it would be able to manoeuvre and stabilise the ailing station. As a freighter, it carried 3,600 kilograms of cargo. As a power module, its pair of 20-square-metre solar panels added 3 kW to the station's supply. Finally, its capsule had a 500-kilogram cargo-return capability. Although now unlikely to carry a crew, Chelomei's TKS was nevertheless making a major contribution to the programme.[2] After an eight-day rendezvous, Cosmos 1443 docked at the front port. It would provide attitude-control and orbital manoeuvring until a crew could be sent to repair the Delta. Although the TKS had large propellant tanks, it could not be replenished, so it was crucial that the station's control system be repaired as soon as possible. The next favourable launch window was mid-April. On 5 April, Cosmos 1443 lowered the orbit of the complex to 300 kilometres. This was a clear sign that the next crew would comprise three cosmonauts. An attempt to launch Soyuz-T 8 on 11 April was abandoned due to a technical problem, as was the second attempt three days later, but on 20 April all

---

[2] Perversely, although it had been relegated to a cargo-return rôle, Cosmos 1443's capsule still had couches installed.

Trials and tribulations

A painting of the Salyut 7–Cosmos 1443 complex by A. Sokolov. Courtesy of *Tass*.

went to plan and Vladimir Titov, Gennadi Strekalov and Alexander Serebrov finally set off.[3] However, their problems were not over.

The boom with the antenna of the rendezvous radar, which was tight against the side of the orbital module for launch, was torn off when the aerodynamic shroud was jettisoned, but there was no telemetric indication because this was not instrumented. When the crew ran through the post-insertion checks and the radar failed to respond, they reasoned that the boom must have failed to deploy. Titov rocked the spacecraft in an attempt to release it, but to no effect. Although the rule book required the flight to be aborted, it was decided to continue. After a nominal 24-hour rendezvous, they drew within 10 kilometres of Salyut 7. Without the Igla transponder, the Argon was useless. The final approach had to be attempted by combining what Titov could see through the Vzor periscope with closure-rate data derived from tracking by ground radars. Once they were close enough for Titov to see the shape of the complex, its apparent size on the graticule on the periscope gave him a feeling for range. All went well until they were 300 metres out, when the Earth's shadow suddenly swallowed the station. Titov switched on a spot-lamp to show the rear docking port and moved in to 160 metres, which was the range at which he had trained to take command after a computer failure, so he was feeling confident. However, sensing that he was closing too

Alexander Serebrov (left), Vladimir Titov and Gennadi Strekalov.

---

[3] Irina Pronina (Savitskaya's backup for Soyuz-T 7) had been assigned, but she had been replaced by Serebrov late in training.

rapidly, he aborted the approach and pulled back. By the time he emerged from the shadow he was out of radio range, and by the time contact was re-established (on the next pass) the range had opened to 4 kilometres and they were ordered home, landing the next day.[4] Cosmos 1443, having lowered Salyut 7's orbit, now boosted it back to its usual altitude – in the process consuming more of its precious propellant.

THE SECOND EXPEDITION

In late June, Cosmos 1443 adjusted Salyut 7's orbit to 326 by 337 kilometres in preparation for another attempt to reoccupy it. Because this was near the station's nominal operating altitude, it was an indication that a two-cosmonaut crew was to be sent, and true enough, when Soyuz-T 9 was launched on 27 June it carried Vladimir Lyakhov and Alexander Alexandrov. Valeri Ryumin, now serving as a flight director, told reporters that they would not attempt to break the endurance record, and added that they would be extremely busy because they were to tackle the programme that had been assigned to a crew of three. This was because Lyakhov and Alexandrov had been assigned with Viktor Savinykh as the backups for Soyuz-T 8, and it had been decided to leave Savinykh behind in order to carry the additional propellant required to reach the station in its high orbit, and hence preserve Cosmos 1443's remaining manoeuvring capability.

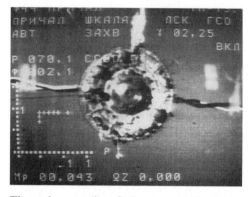

The solar panels of Cosmos 1443 can be discerned behind those of Salyut 7 in this television view from Soyuz-T 9 as it made its approach to the rear of the station.

The first task was to retrieve the new computer from a cargo rack in Cosmos 1443. This was the first time that cosmonauts had ventured into a Chelomei module. Apart from the Delta computer, it contained 600 other items including a pair of clip-on solar panels. A telescopic pole was extended through the hatch into Salyut 7, and a sled run back and forth to transfer apparatus through the hatch to assist unloading. Lyakhov's immediate reaction upon seeing the racks of cargo was that it looked like a warehouse. Alexandrov was a specialist in control systems, and in their first week he replaced the electronics in the Delta and reloaded and tested it, thereby restoring Salyut 7's operational status.

In an effort to compensate for their absent crewmate, Lyakhov and Alexandrov ran a 12-hour day, and a six-day week. At the end of the month, although they were

---

[4] Titov's heroic effort was to pay a dividend later.

working efficiently, the doctors warned them not to burn themselves out. The labour of unloading Cosmos 1443 resulted in "uncommonly good appetites", and they soon began to put on weight. The unloading continued through July, and was interleaved with installing new experiments. The Yelena-F monitored gamma-rays, and Ryabina studied the charged particle radiation in near-Earth space. Two different instruments were available for the MKS-M spectrometer – one to study the atmosphere and the other the oceans. The first phase of the scientific programme was concluded in early August. They made extensive oceanic studies, in particular the Mediterranean Sea, and also participated in an experiment in the Black Sea during which a 500-metre diameter pool of dye was released. This was part of the preparations for a 20-day study of the Black Sea region, later in the month, undertaken by the Intercosmos organisation. Their observations were correlated with data from an aircraft, a ship and the 10-channel Fragment multispectral camera on the Meteor-Priroda satellite. In addition to observations for the Intercosmos organisation, they also imaged Brazil for UNESCO. Crops in the Central Asian Republics were extensively surveyed, and images taken of the Caspian and the Volga regions, the Sea of Azov, the Crimea, the Urals and the Caucasus. They were so productive that in just two months they took more pictures than their predecessors had during seven months. In fact, during one 14-day period they snapped 3,000 pictures. They used the PCN low-light camera to record lightning bolts and aurorae, to supplement Chrétien's programme. The Astra mass spectrometer noted a tenuous stream of matter trailing the station, composed of air vented by the small scientific airlocks and the efflux from the attitude-control thrusters. If it were not for the capacity of the Cosmos 1443 capsule, the results of this intense observational programme would not have been able to be returned by Soyuz. Nevertheless, despite their hard work, the 318 kilograms of accumulated results fell short of the capsule's 500-kilogram capacity, and so they sent back the components of the faulty Delta together with a variety of spent filters, fans and scrubbers from the environmental system so that their service life could be reassessed.

Alexander Alexandrov in the Cosmos 1443 'warehouse'.

Alexander Alexandrov (left) and Vladimir Lyakhov on the TV downlink.

Cosmos 1443 had to go because the station's propellant needed replenishment, and this could not be done until the rear port was vacated, which in turn required the front port to be vacated so that Soyuz-T 9 could be relocated there. If Soyuz-T 8 had managed to dock, its crew would have had much longer to complete the programme. Cosmos 1443 left on 14 August. Its departure suggested that the freighter version of the TKS was not intended to become a permanent part of an orbital complex. While it was reasonable to jettison a cargo carrier once it had been unloaded, it would be wasteful to discard a scientific module after just a few months, so it was evident that the specialised laboratory variant would require to incorporate a second docking port in order not to inhibit ongoing operations. Cosmos 1443's capsule returned to Earth on 23 August.[5] While the module had been in place, a series of tests had been made to determine how well its thrusters could control the complex. Over the years, it had repeatedly been stated that the ultimate objective of the programme was to construct a large complex from modular components. The simplest configuration was to link up the modules in line. Studies after engineering trials by Cosmos 1267 while docked with Salyut 6, suggested that a revised control system would be able to manoeuvre a larger complex using only 20 per cent of the propellant normally required. Lyakhov, however, had found it difficult to control the station manually, and had used much more propellant than expected – hence the requirement to send up another Progress tanker as a matter of priority. The development of an automatic system which could be readily adjusted to take into account the changing dynamics as a modular complex was reconfigured was a major long-term priority.

'A SLIGHT LEAK'

On 9 September, as Progress 17 was pumping oxidiser onboard, a pipe in Salyut 7's unpressurised engine bay fractured. As soon as telemetry indicated that propellant was leaking, the crew was ordered into Soyuz-T 9 just in case it proved necessary to abandon the station. Although the nitrogen tetroxide was extremely corrosive, there was no danger of an explosion, so several hours later the cosmonauts were allowed to re-enter the station. This fault effectively disabled the main engines and one of the two sets of attitude-control thrusters, which would once again constrain the research programme. There was no announcement. Only after the crew eventually returned to Earth did Roald Sagdeev, director of the Space Research Institute in Moscow, admit that "technological problems" had developed, and in his post-flight press conference Lyakhov acknowledged that there had been "a slight leak" from one of the propellant tanks. Progress 17 raised the station's orbit to 337 by 358 kilometres, and then left on 17 September.

---

[5] Many years later, it was sold at auction in the United States.

PAD ABORT!

An attempt to launch Titov and Strekalov on 26 September almost ended in their deaths. With just seconds remaining on the countdown, a valve failed to close, and a fire broke out at the bottom of the rocket. It was fully 20 seconds before the escape tower was manually triggered, by which time the rocket had been engulfed by flame. This was the first time that a crew had 'aborted' on the pad. When the rocket finally exploded, the capsule was descending on its parachute 4 kilometres away. Titov and Strekalov had intended to relieve Lyakhov and Alexandrov, and thereby achieve the first handover of an orbital station from one crew to another. While both crews were present, Titov and Strekalov were to have made a spacewalk to affix supplementary solar panels. Lyakhov and Alexandrov would have returned in early October. Titov and Strekalov would have served until either early December or early January, when they would have handed the station over to their successors. If this had been able to be achieved, the programme would have taken a great step towards the ultimate goal of continuous occupancy. Lyakhov and Alexandrov agreed to extend their mission in order to install the solar panels. The fact that Soyuz-T 9 was not recalled expressed the belief that it could safely remain in space for much longer than the 110 days or so of Soyuz-T 5 and Soyuz-T 7.

The difficulty in controlling the station's attitude imposed "a substantial change" on the research. After adopting gravity-gradient stability, Lyakhov and Alexandrov concentrated on making semiconductors and alloys in the Kristall and the Magma-F, crystalline melts in the Pion-M, and electrophoresis of biological agents in the Gel. On 22 October, Progress 18 brought a revised Tavriya electrophoresis apparatus and the second pair of clip-on solar panels. Taking advantage of the visitor's propulsion system, they re-oriented the station to take MKF-6M pictures of the southern part of the Soviet Union for a geological survey. (Dense vegetation in the summer tended to mask subtle geological structures, but pictures taken later in the year could expose this detail.)

ADDING SOLAR PANELS

On 1 November, Lyakhov and Alexandrov ventured out to mount the first pair of solar panels. The size of the main panels was restricted by the shape of the stepped-cylinder compartment and by the aerodynamic shroud. Small clip-on units had been developed to overcome the degradation that had been noted in the case of Salyut 6's panels towards the end of that station's life, and attachment points had been bolted to the frames of the main panels to accommodate a strip on each edge. As Lyakhov and Alexandrov were not trained specifically for this task, they found the fifty steps in the attachment sequence daunting. Romanenko was in the control room to provide advice, and Titov and Strekalov were in the hydrotank to help resolve any problems. To extend the radio coverage, three communications ships had been deployed – the Cosmonaut Yuri Gagarin in the Mediterranean, and Cosmonauts Vladislav Volkov and Georgi Dobrovolsky in the Atlantic.

Alexandrov took up position on the foot restraint at the base of the dorsal panel, and Lyakhov handed him the folded panel and the toolkit. The panel was fitted to the lower attachment point and the handle inserted into its winch. As he unfolded the concertina-like panel up alongside the main panel, Alexandrov reported that the winch was difficult to use (this was the winch that Berezovoi and Lebedev had been eager to test). Telemetry indicated that he became most stressed during this phase of the operation. Once it was fully extended, the side strip automatically clipped into a fixture near the far end of the main panel. Finally, Lyakhov plugged in the electrical cables. The Sun-tracking motor had been deactivated so that the main panel could be locked in the best orientation for a cosmonaut on the anchor to access it. The second strip could not be fitted until the panel was rotated 180 degrees, and this could not be done until they re-entered the station – which is why installing the clip-on panels was to have been done while two crews were present. They returned two days later. This was the first time that a crew had made two spacewalks. Once the second strip was in place, Alexandrov took almost 10 minutes to tie a knot in the electrical cables to hold them in place. Their endeavour added 1.2 kW to the power available, which meant that full advantage could now be taken of the materials-processing facilities.

With the primary task of the flight extension over, the cosmonauts settled down to await the opening for the recovery window later in the month. Progress 18 left on 13 November, and Lyakhov and Alexandrov departed on 23 November. Whilst their 149-day flight had not set a human endurance record, it had considerably extended the demonstrated orbital life of the Soyuz-T ferry.

THE THIRD CREW

Salyut 7's orbit was high enough to enable it to survive until the next crew could be sent to repair its leaking propellant system. First, however, the engineers had to develop the procedure and build the tools for this ambitious task. Meanwhile, the station was repeatedly reported to be operating normally – no mention was made of the engine leak. By early 1984, with the orbit having decayed below 300 kilometres, it was decided to send up a full crew, and on 9 February Soyuz-T 10 docked at the front port. Leonid Kizim and Vladimir Solovyov had already trained extensively for spacewalks. If the complex repair task was completed successfully, they were to try to set a new endurance record. Oleg Atkov, a physician, had been added to the team to monitor his colleagues' health, and to perform his own investigation of the process of adaptation to weightlessness. He had the authority to end the

Vladimir Solovyov (left), Leonid Kizim and Oleg Atkov.

mission. It was announced that a medical specialist would form part of every record-breaking crew – although in the event this would prove impracticable. Kizim's first task was to assess the station's remaining attitude-control capability in the manual mode. By 23 February, when Progress 19 docked, the orbit was 282 by 286 kilometres, so the ferry boosted it to 306 by 327 kilometres. In a change to procedure to avoid the problem of cluttering up the station, items were to be unloaded only as needed, and Atkov looked after the cargo so as to enable Kizim and Solovyov to concentrate on their work, and then he filled it with the accumulated trash to enable it to depart on 31 March.

Soyuz-T 11's arrival on 4 April brought Yuri Malyschev, Gennadi Strekalov and Rakesh Sharma, an Indian cosmonaut. It was the first time that six people had been on a station. Nikolai Rukavishnikov had been assigned to this flight, but when he had fallen ill Strekalov had stepped in.[6] Whilst not a member of Intercosmos, India (like France) had collaborated with the Soviet Union on space research for several years. In this case, most of the research work was biomedical. Vektor used an Indian-built electrocardiograph to monitor the cardiovascular system by measuring chest motion; Anketa and Optokinez investigated relationships between the vestibular and visual systems; Membrane measured calcium loss in bone; and Ballisto monitored changes in the heart's shape and location as the body adapted to space. Sharma's most novel test was a Yoga experiment in which he adopted various standard positions while Atkov observed his muscular activity with the Miokomp and Briz apparatus (it was hoped that Yoga exercises might help to prevent muscular atrophy). MKF-6M and KATE-140 photographs were taken of India for the Terra project, to identify oil and gas resources, to assess the biological productivity of the Nicobar, Andaman and Laccadive Islands, to assess water resources in arid regions, to evaluate sites for hydroengineering projects, and to measure ice-melt run-off rates in the Himalaya and Karakorm regions. The Earth studies were to prepare for India's own remote-sensing satellite, IRS, which was under development for launch on a Soviet rocket.

Malyschev, Strekalov and Sharma

Yuri Malyschev (left), Rakesh Sharma and Gennadi Strekalov.

---

[6] Rukavishnikov had suffered a run of awful luck!

returned to Earth in Soyuz-T 10 on 11 April. The residents reported that their guests had worn them out. Two days later, they transferred Soyuz-T 11 to the front port, and four days after that the newly vacated port was taken by Progress 20, which boosted the orbit back above 300 kilometres. In addition to consumables, it delivered replacement nickel–cadmium batteries, two dozen specialised tools, and one-off components to facilitate the repair of the unified propulsion system.

ENGINE REPAIRS

Kizim and Solovyov ventured out on 23 April to start the complex operation. Atkov monitored their progress from the safety of Soyuz-T 11's descent module in case his colleagues were forced to abandon the station. A foot restraint had to be affixed to the periphery of the main compartment, alongside the engine. Doing so proved more difficult than expected, because it involved driving pins into the hull. Curving in a 120-degree arc around the engine bay, this was to provide a workstation for one cosmonaut during the repair operation. The other man was to stand on a restraint on an *ad hoc* platform on the orbital module of the Progress, which unfolded on radio command from Kaliningrad. Finally, the tools and other apparatus were attached to the main workstation so that they would be able to set straight to work on the next excursion. When they returned three days later, Solovyov stood on the platform in the gap between the rear of the station and the ferry and Kizim strapped himself to the ladder that was wrapped around the engine bay, in order to be able to access the work site from different angles.

Leonid Kizim outside.

The first task was to replace a valve associated with the oxidiser leak. Kizim opened the access panel, then used a pneumatic punch to open the thermal blanket that protected the bay, in order to access the plumbing. Nitrogen gas from a tank in the Progress was pumped through each stage of the replenishment system in turn to confirm that the length of pipe believed to be fractured did indeed leak. This section of pipe was to be isolated by replacing the nearest valves with ones with connectors for a bypass pipe. The disconnection of the valve was aggravated by the discovery

Leonid Kizim with a spacewalker's toolkit.

that some of the bolts had been sealed with glue to prevent their being loosened by vibrations, and one took an hour to release. When the station passed into the Earth's shadow they continued under the illumination of small portable spotlamps because, with only five hours of life-support, and with an hour reserved for making their way back and forth along the station, there was no time to waste. Finally, they managed to install the first valve. It had been hoped to add the pipe too, but they were so far behind schedule that they had to stuff the insulating blanket into the hole and retreat to the airlock. Returning three days later, they affixed the bypass pipe to the valve. On 4 May, they replaced the second valve and connected the pipe to provide an independent path for propellant paralleling the leaky pipe. Having done all that they could, they restored the insulation and closed the access panel. The next task was to 'close' the leaky pipe, but the tools to accomplish this task were still under development, so the final phase of the repair operation had to be left for a later date. Its function as a foot restraint over, Progress 20 left on 6 May, and the port was taken four days later by Progress 21, which delivered a third pair of clip-on solar arrays.

On 18 May, continuing their record-breaking series of external operations, Kizim and Solovyov fitted the second pair of clip-on strips to either side of the solar panel alongside the airlock hatch. This time Atkov remained in the main compartment in order to rotate the main panel through 180 degrees while his colleagues fetched the second clip-on from the airlock, to enable them to complete the task during a single excursion. The winch confirmed its reputation for being awkward to use – its handle broke! Finally, Kizim and Solovyov moved over to the far side to inspect the other lateral panel. They cut out a segment of its transducer array for return to Earth for its degradation to be assessed. A few days later, the station was turned to measure the efficiency of the new panels under a range of insolation angles.

Progress 21 left on 26 May, and Progress 22 arrived on 30 May. It delivered a great deal of medical apparatus for Atkov, more nickel–cadmium batteries, scientific equipment for the next visiting crew, and more air to replenish that being consumed in repressurising the airlock after spacewalks. During June and July, the crew spent most of their time on geophysical observations. They surveyed sites in Siberia that were being considered for hydroengineering construction, and the entire length of the Baikal–Amur railway, which was now almost complete. The Ukrainian harvest was starting, so they took multispectral imagery of this to estimate its yield. In oceanic studies, they confirmed earlier reports that, under certain lighting conditions, shallow water is transparent, but extended this by reporting that they could see submerged seamounts in open ocean. They made extensive use of the Niva video while making these observations, because it enabled them to provide a running commentary to the photographic work. On 14 July, Progress 22 raised the station's orbit to 365 by 383 kilometres, then undocked the following day. Having decided that the efflux from the thrusters of departing ferries degraded the efficiency of the transducers in the solar panels, this time the springs in the docking system were used to push it away, and it did not fire its thrusters to increase the rate of withdrawal until it was well clear (this would become standard procedure). As it withdrew, the Astra apparatus monitored the gaseous environment around the station. It did so again three days later when the next spacecraft approached.

Soyuz-T 12 carried three cosmonauts, even although the station's orbit was back at its normal operating altitude. It had been launched by a new form of the

Igor Volk.

Semyorka with more powerful engines. Vladimir Dzhanibekov and Svetlana Savitskaya were accompanied by Igor Volk. It had been rumoured that a woman would fly, and this had been expected to be Irina Pronina, who was Savitskaya's backup on Soyuz-T 7, but she had retired. Savitskaya therefore became the first female cosmonaut to make a second flight. Although it was not announced, Volk had piloted the Soviet equivalent of the US Air Force's 'DynaSoar' spaceplane in atmospheric tests in the mid-1960s, and was in training to fly the Buran space shuttle.

He was flying to gain experience of weightlessness, and immediately upon returning he was to assess how spaceflight degraded normal piloting skills.[7] Atkov paid particular attention to Volk, who tested a medication to ameliorate the most unpleasant symptoms of weightlessness, and to isolate its effects he was banned from participating in the exercise regime followed by the others. Savitskaya worked with an improved Tavriya electrophoresis system. The small quantities of material produced by these experiments were sufficient to keep specialists busy for months. In fact, trial production of drugs refined from the results of her previous experiments had already begun. This time, she made an anti-influenza preparation for a vaccine, interferon, a genetically engineered anti-infection preparation, and an antibiotic for agricultural use which (when added to animal and poultry fodder) would increase the body weight of the animals by up to 20 per cent. These electrophoresis experiments proved so successful that an industrial-scale unit was developed.[8] The Electrotopo-

---

[7] In effect, this question had already been answered, because pilots were manually landing NASA's Space Shuttle.

[8] When this eventually flew, it was named Svetlana.

graphy experiment was also rerun. Because the first results had been flawed by water vapour, Kizim suggested an improved procedure. At the request of the Byurakan Astrophysical Observatory in Armenia, Savitskaya used Chrétien's Piramig to photograph the Earth's L4 and L5 Lagrangian points (the gravitationally neutral points 60 degrees ahead of and behind the Earth in its orbit around the Sun) to help to investigate the distribution of interplanetary dust.

Vladimir Solovyov (left), Vladimir Dzhanibekov, Svetlana Savitskaya and Leonid Kizim.

The main event of the visiting programme was the spacewalk by Dzhanibekov and Savitskaya on 25 July. Kizim, Atkov and Volk remained in the station, on one side of the airlock, and Solovyov retreated to Soyuz-T 11's descent module on the other side. The purpose of the excursion was to test the Universal Manual Toolkit (URI). This was a development of the Isparitel electron-beam apparatus built by the Institute of Electrical Welding in Kiev.

Vladimir Dzhanibekov in the hatch.

Savitskaya stood on the anchor just outside the hatch and erected the workstation that carried a variety of tools to process metal in space. An electron beam was used to cut titanium and stainless steel plates. She reported this to be very easy, and said that she could see the beam. A small handheld non-contacting infrared thermometer was used to measure the tempera-

Svetlana Savitskaya tests a vacuum-welding tool.

ture of the material. Steel and titanium plates were then joined by tack-welding. She reported producing good seams. Metal plates were joined using solder composed of tin and lead. She reported that she did not think that the seam looked very good. Finally, a small silver granule was heated in a crucible and sprayed onto an anodised aluminium plate. She complained that this was more difficult than she had expected because she could not see the spray. This entire test was completed during a single daylight pass, and Savitskaya remained in position while the station passed through the shadow. Back in sunlight, she and Dzhanibekov exchanged places, and he repeated the test. Later, Academician Boris Paton, director of the Institute of Electrical Welding, said that these simple tests had demonstrated that in the future robotic systems would be able to assemble structures using such tools. Before closing the hatch, Dzhanibekov retrieved some of the exposure cassettes and deployed some new ones. The Astra spectrometer detected the increase in the stream of gas trailing the station resulting from venting the airlock.

The visitors departed in Soyuz-T 12 on 29 July. Their flight had lasted longer than the standard week-long international visit. The residents returned to mapping fault lines and ring structures – the latter are most readily detected from space, and they had already reported two possible candidates near the Syrdarya and Amudarya rivers, each 10 kilometres in diameter – in order to assist prospectors.

On 8 August, Kizim and Solovyov set off to complete the engine repairs using a miniature pneumatic press that was delivered by Soyuz-T 12. In fact, Dzhanibekov and Savitskaya had been assigned to use it during their extended visit, but their hosts had argued that the repair was their responsibility – after all, they were more familiar with the work site. A video was played on the uplink to show them how to seal the bypassed pipe. With Progress 20 gone, they had to work without the advantage of its anchor point. After opening the insulation blanket they used the pneumatic press to pinch the fractured pipe. Driven by compressed air at 250 atmospheres pressure, it delivered a force of 5 tonnes, which was sufficient to crush the pipe and block it. Once this seal was verified using nitrogen, the thermal insulation was reset. They left the curved anchor in place, but returned the toolkit to the airlock. Before retiring, they used another tool that fitted closely around the edge of a 20-square-centimetre transducer in one of the solar panels, to cut it out without risking contamination for later examination.

What Lyakhov had described as "a slight leak" had taken *24 hours* of external activity to repair. In contrast to Lyakhov and Alexandrov's spacewalks to attach the first set of clip-on solar panels, this impressive activity was hardly mentioned (*Tass* reported simply that the engine had been "serviced"). But it was later admitted that when making the station serviceable by its crew, replacing valves in the engine bay had been considered to be the "absolute worst case" scenario. Now that it had been achieved, the leak was admitted to have been "very serious". Cosmonauts had once again demonstrated the ability to overcome a problem so serious that it would have obliged an earlier station to be abandoned. In a very real sense, it was having a crew onboard that had enabled Salyut 6 and Salyut 7 to be sustained for so long. Although some observers criticised the amount of time that cosmonauts spent maintaining the systems – arguing that this was time lost to research – this narrow viewpoint failed to recognise the value of the *engineering knowledge* derived from sustaining ongoing operations. The learning curve was steep, but the crews were determined to climb it. Knowledge gained from overcoming adversity was empirical, but this was the only way to proceed.

Progress 23 arrived on 16 August, and the station's propellants replenished over the next few days to restore the engine to use. The cargo included the RS-17 X-ray telescope and French–Soviet Sirene X-ray spectrometer. It was unloaded in double-fast time because the Sirene apparatus, which was to be installed in the rear transfer compartment, could not be operated while the port was in use. Progress 23 departed on 26 August to facilitate the observations. In early September, Salyut 7 contributed to the Black Sea-84 and Gyunesh-84 projects for the Intercosmos organisation.

On 7 September, Kizim, Solovyov and Atkov exceeded the 211-day endurance record. Later in the month, they began preparations to return to Earth, and did so on 2 October, having set a new endurance record of 237 days and undertaken a record-setting six spacewalks.

A crew of three had proved to be more efficient during a long flight than a crew of two. Atkov had focused on the adaptation of the cardiovascular system. He used the Aelita to make vector cardiograms, the Reograf to make rheographs and used the French Echograph that showed ultrasonic images of internal organs. He had also used the Glucometer to monitor carbohydrate exchange in cells, and established that this process operated at a greatly reduced rate in space, suggesting that cellular functions might be disrupted on very long flights. In the Membrane experiment, he had studied the exchange of calcium in cell membranes. This had shown that if they took an anti-oxidant preparation, calcium metabolism was reduced. Blood samples were spun in a small centrifuge, to separate the components for storage and analysis on Earth. The wisdom of including a doctor in the crew was demonstrated by the fact that Atkov had performed a series of 30 different medical tests a total of 200 times, providing an unprecedented detail in monitoring the adaptation process. The 182 days endured by Soyuz-T 11 suggested that once in-space handovers became routine, six-month tours would be the norm, and the onerous task of pushing the endurance record would be pursued by medical specialists serving with a succession of resident crews.

CRISIS!

At the end of 1984, with Salyut 7 in a 365- by 370-kilometre orbit, Boris Belitski, speaking on *Radio Moscow*, said that its systems were "in good shape", and that it still had "plenty of life" left. At the end of January 1985 its systems were said to be "functioning normally" with its internal conditions "optimum" for the next crew. On 11 February, after telemetry indicated that there was a fault, contact was lost. With the station dead and drifting freely, it would be impossible for an automated ferry to dock to provide an independent manoeuvring capability. On 1 March, *Tass* pointed out that the station had been in space for 34 months, and had "completely fulfilled" its "planned programme". This statement had a greater sense of finality than the one that had announced that the stricken Salyut 6 had achieved its "primary" mission. It was hardly credible that Kizim and Solovyov would have made such a great effort to repair the engine if Salyut 7 was to be written off following their departure. Clearly, something had gone disastrously wrong. It was therefore expected that its successor would be launched. However, Vladimir Dzhanibekov and Viktor Savinykh set off for Salyut 7 in Soyuz-T 13 on 6 June. Dzhanibekov was making a record fifth flight. On approaching the station in Soyuz-T 6, he had made a manual approach from a range of 1 kilometre, and was the best choice to try to dock with an uncooperative station. Savinykh had entered training in September 1984 to launch early in the new year for a 10-month tour with Vladimir Vasyutin and Alexander Volkov, and had been selected to accompany Dzhanibekov because he had trained (with Vasyutin) for a spacewalk to affix the final pair of clip-on solar panels. The repair mission had been organised as soon as Salyut 7 had fallen silent. If they managed to board the station, they were to find out why it had fallen silent, and then, if possible, restore it to life.

Soyuz-T 13 flew a propellant-efficient two-day rendezvous. The transition from the final transfer orbit to the point at which a straight-in approach could be initiated had to be accomplished without the benefit of an Igla radar transponder. The way in which Vladimir Titov had flown Soyuz-T 8 in without his radar provided the *modus operandi*. Although the ground radars had been able to get Titov within 1 kilometre, when he closed in he had encountered difficulty determining range and closure rate. If this problem could be overcome, there was every chance that Dzhanibekov would be able to repeat this process. The slow rendezvous had ensured that the tracking radars had determined the relative orbits very precisely, so that accurate steering cues could be provided. When 10 kilometres out, Dzhanibekov sent Salyut 7 a command to turn on its Igla transponder and orient itself to face the approaching ferry, but there was no response – he had not really expected there to be. He closed to 3 kilometres, then activated the special optical sight that had been fitted to enable him to approach free of the difficulties that had dogged Titov. This incorporated both low-light optics and a laser rangefinder. The image intensifier would show him the station's position and orientation in the Earth's shadow. The rangefinder would measure the separation

Soyuz-T 13 lifts off.

and compute the rate of closure. At the usual pause point, 200 metres out, he studied the station. Although the station was in a very slow roll of about 0.3 degree per second, which was easily matched, it was not tumbling. He slowly flew around it in order to examine it from all angles, seeking any sign of catastrophic damage. It was apparent that the solar panels had ceased to track the Sun. The fact that two lateral ones were angled 75 degrees apart implied that the station had suffered a power supply failure, rather than a loss of communications. Dzhanibekov aligned Soyuz-T 13 to approach the front port. He controlled the ferry using the optical sight and Savinykh called out the range and rate. Once they had matched the station's roll, he moved in and docked at the first attempt. As he reported their success to the relieved flight controllers, the station entered the Earth's shadow.

Soyuz-T 13 approaches the 'dead' Salyut 7.

Standard procedure after docking was to establish the electrical and hydraulic links, test the hermetic seal of the tunnel, equalise pressure with the station, and then open the hatches. Unfortunately, because there was no power available they could not tell if the station was still pressurised. Opening the valve was, therefore, the moment of truth. There was air in the station. They sniffed it for toxic fumes. Satisfied, they swung back the probe and drogue assemblies and stared into the dark void (the porthole covers had been left closed). Using torches, they ventured into the station. Dzhanibekov said that the air was musty. He estimated that it was $-10°C$, but it was difficult to tell, as it was off the bottom of the scale of the thermometer. The water pipes had frozen, condensation had frosted the walls, and very peculiar microgravity icicles had formed. Even wearing fur hats, fur coats and fur boots, the cosmonauts had to return to their ferry every hour or so to warm up. They had to retreat there to eat and sleep. Time was not on their side. Unless they could restore Salyut 7 to life within 10 days, which was the limit of their powered-up ferry, they would have to abandon it and return to Earth.

It was obvious that the power supply had failed and the batteries had run flat, and it did not take long to find the fault. The switching mechanism that directed the feed from the solar panels to the batteries had failed and, ironically, the batteries had drained their remaining charge running the motors to keep the panels facing the Sun. Fortunately, the panels were still feeding power. If the batteries could be recharged, it should be feasible to restore the station to full operation. Only two of the batteries were ruined – the other six should be serviceable once recharged. The faulty circuit was isolated, jump leads run from the panel feeds to the battery recharger, and then Soyuz-T 13 was used to turn the station so that its irregularly oriented solar panels collected as much energy as possible. The first battery recharged in a few hours, and the telemetry link was reactivated to enable Kaliningrad to examine the state of the other systems. As it would take several days to restart the environmental systems, a ventilation tube was run in from Soyuz-T 13 to circulate the stagnant air, because it contained pockets of carbon dioxide that gave the cosmonauts headaches. There was soon sufficient power to run the lights and air heater. The thermal regulation system would not be able to be reactivated until the humidity (initially at about 90 per cent) had been reduced. The main fault had been overcome, but there was a long way to go before the station would be able to support scientific research. As the vital systems were assessed, the parts and tools needed to effect repairs were loaded into the next supply ferry. Once the attitude-control system was restored, the Igla would be able to orient the rear port towards the approaching automated ship.

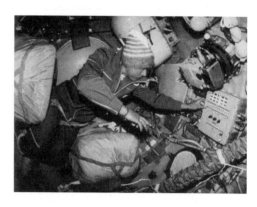

Viktor Savinykh inside chilly Salyut 7.

At the end of the week, the temperature finally rose above 0°C and the ice began to thaw. The humidity was still high, so it was crucial to ensure that the temperature did not melt the ice more rapidly than the condenser could extract the moisture, and it was several more days before the humidity fell sufficiently to enable the heaters to be safely activated. This was just in time, as Soyuz-T 13 was nearing its limit. *Tass* reported that Salyut 7 would soon be in "normal working mode". Dzhanibekov and Savinykh had successfully achieved the most important part of their mission. Once again, human intervention had saved a crippled station. The irony of the crisis was that if cosmonauts had been present when the recharger had failed, they would have pre-empted the draining of the batteries. Clearly, although an operational station was a prerequisite for continuous occupancy (which was the objective of the programme) *maintaining a crew aboard was likely to be crucial in keeping a station operational.*

Viktor Savinykh (left) and Vladimir Dzhanibekov inside chilly Salyut 7.

The situation improved rapidly, and on 19 June the MKF-6M and KATE-140 cameras were used to take pictures of the Kursk Oblast region for the Kursk-85 programme. It was something to do while waiting for Progress 24, which docked on 23 June to deliver propellant, water, air, clothes, three replacement nickel–cadmium batteries, and a new water heater (the original had been split by the ice). It departed on 15 July, with all of the damaged items. Immediately after the next resupply ferry attained orbit on 19 July, a fault led to it being assigned the anonymous designation of Cosmos 1669. However, its cargo was simply too valuable to be written off, and the flight controllers were able to overcome the fault. It then followed the standard rendezvous and docked without incident. Its cargo included two new spacesuits (the originals had been deemed unsafe, due to being frozen), the Mariya spectrometer to study high-energy particles in near-Earth space, the Rost plant chamber and several exposure cassettes for deployment outside. This spacewalk was made on 2 August. The new semi-rigid suits offered improved peripheral vision and increased mobility. The primary task was to install the third pair of clip-on solar panels (delivered long ago by Progress 21). This time, the lateral panel was commanded to rotate through 180 degrees by remote control. An experimental solar transducer was then deployed to enable its performance to be evaluated. Well ahead of schedule, they replaced the Medusa and several technology cassettes, and deployed the Comet micrometeoroid detector (a French package with several chambers that could be opened from within the station; the first was to be exposed for Comet Giacobini-Zinner, the second for Halley's comet). Although many satellites had studied meteoroids in the past, these had tended simply to measure the rate and size of the particles striking the detectors. This package was an advance because not only could it focus on specific cometary streams, the material was to be returned to Earth for analysis. In addition to being of intrinsic interest, it had been suggested that cometary dust might have played a key rôle in the onset of the Earth's ice ages.

Throughout August, Dzhanibekov and Savinykh concentrated on photography using the KATE-140 and MKF-6M, while flax grew in the Magnetobiostat, the Gel electrophoresis apparatus purified biological agents, and Mariya data was collected. Cosmos 1669 left on 29 August, redocked to test an alternative control system, and a few hours later left for good.

AN ORBITAL HANDOVER AT LAST

Soyuz-T 14 arrived on 18 September with Vladimir Vasyutin, Alexander Volkov and Georgi Grechko, who brought the Skif spectrometer and EFU-Robot electrophoresis apparatus. On 25 September, Dzhanibekov and Grechko left in Soyuz-T 13, leaving Savinykh, Vasyutin and Volkov onboard Salyut 7. This handover was notable not only for being the first time that one crew had relieved another, but also because it was the first time that a cosmonaut launched as part of one crew had returned as part of another. The reformed crew expected to remain onboard until March 1986, by which time although Savinykh would have been up for 10 months as planned, his colleagues would have served only six months. Given the earlier

Vladimir Vasyutin (left), Georgi Grechko, Viktor Savinykh, Alexander Volkov and Vladimir Dzhanibekov.

announcement that a doctor would join crews on endurance missions, it would turn out to be unfortunate that this crew did not include a medical specialist.

Soyuz-T 13 re-rendezvoused with the Igla inactive, in order to further assess the transition from the transfer orbit to the 1-kilometre point using the optical sight and laser rangefinder, but no attempt was made to redock, then it withdrew. It landed the next day. The day after that, a Proton put Cosmos 1686 into orbit, and this began a slow rendezvous with Salyut 7. It arrived on 2 October. This was the first time that such a module had docked with an inhabited station. *Tass* said that it was "similar in design" to Cosmos 1443, but neglected to explain how it differed. In fact, the capsule had been heavily modified, and was no longer capable of returning to Earth. A bank of instruments had been affixed to its nose (instead of the re-entry package) and the control panel mounted inside. Vasyutin's crew had trained to use the instruments. The main module had a "miscellaneous cargo" of 4,500 kilograms that included the Kristallisator semiconductor furnace and a deployable girder (a package that was too bulky to fit inside a Progress ferry). During October, unloading cargo was interleaved with gathering Aerosol and Mariya data, processing materials in the Pion-M furnace and evaluating the ability of the module's advanced control system to manoeuvre the complex (it proved to be capable of executing all the desired manoeuvres without the crew's intervention).

In its 25 October report, *Tass* made its last reference to the crew being in "good health and feeling well". Over the period of several days, Vasyutin succumbed to an illness which began with "a slight uneasiness" which was aggravated initially by his inability to sleep and then by losing his appetite. Although initially the cosmonauts kept this to themselves, as soon as it became evident that it was a serious ailment he reported his symptoms and transmitted biomedical telemetry to enable the medics at Kaliningrad to assess his condition. Although they were ordered to prepare to return to Earth, because orbital dynamics did not offer a favourable recovery opportunity until mid-November an emergency return would be in less than ideal conditions. On 29 October, *Tass* noted that the crew was implementing the flight programme, and that the systems on the complex were functioning normally, but did not include the stock phrase indicating that the cosmonauts were in good health. Similar reports followed. By 17 November, Vasyutin's condition had deteriorated to the extent that he had periods of acute pain. The doctors decided that he was too ill to continue, and ordered him home. Immediately after landing on 21 November, he was examined. Although his condition was "satisfactory", he "required hospital treatment" and was flown to Moscow. The report of their successful return said that the flight had

A magnificent view from Soyuz-T 13 of Salyut 7 with Soyuz-T 14 attached, and with a full set of clip-on solar panels.

been curtailed because of the commander's "sickness". This vague reference was the first official acknowledgement that there had been a medical problem. It was first suspected that he had appendicitis, but it proved to be an infection of his prostate that had manifested itself as an inflammation and a fever. In retrospect, considering that so many cosmonauts were spending long periods in space, it was remarkable that such an illness had not occurred earlier.

It had been intended to send two crews to visit the residents after Cosmos 1686 had departed. One of these was an international flight, with a Syrian cosmonaut. The other was to be an all-female crew with Svetlana Savitskaya commanding Yekaterina Ivanova and Yelena Dobrokvashina. These flights had been initially postponed when Salyut 7 fell silent, and rescheduled after the Soyuz-T 13 crew rescued it. The Syrian flight was further postponed following the recall of Vasyutin's crew. The all-female crew was disbanded early in 1986 when Savitskaya became pregnant. Although soon after it was launched Salyut 7 was said to have been designed for a four-year life, Soviet spokesmen at the recent congress of the International Astronautics Federation in Sweden indicated that it was soon to be superseded by a facility constructed from modular components. The Salyut 7–Cosmos 1686 complex, meanwhile, continued in its automated regime with recently delivered experiments awaiting attention.

Table 3.1 Salyut 7 docking operations

Spacecraft	Docking		Port	Undocking		Days
Soyuz-T 5	14 May 1982	1536	front	27 Aug 1982	–	105
Progress 13	25 May 1982	1157	rear	4 Jun 1982	1031	9.94
Soyuz-T 6	25 Jun 1982	2146	rear	2 Jul 1982	1501	6.72
Progress 14	12 Jul 1982	1541	rear	11 Aug 1982	0211	29.44
Soyuz-T 7	20 Aug 1982	2232	rear	29 Aug 1982	1847	8.84
Soyuz-T 7	29 Aug 1982	–	front	10 Dec 1982	–	103
Progress 15	20 Sep 1982	1012	rear	14 Oct 1982	1646	24.27
Progress 16	2 Nov 1982	1622	rear	16 Dec 1982	–	44
Cosmos 1443	10 Mar 1983	1220	front	14 Aug 1983	1804	157.24
Soyuz-T 9	28 Jun 1983	1346	rear	16 Aug 1983	1825	49.20
Soyuz-T 9	16 Aug 1983	1845	front	23 Nov 1983	–	99
Progress 17	19 Aug 1983	1747	rear	17 Sep 1983	1444	28.87
Progress 18	22 Oct 1983	1434	rear	13 Nov 1983	1808	22.15
Soyuz-T 10	9 Feb 1984	1743	front	11 Apr 1984	1427	60.86
Progress 19	23 Feb 1984	1121	rear	31 Mar 1984	1245	36.06
Soyuz-T 11	4 Apr 1984	1831	rear	13 Apr 1984	1427	8.83
Soyuz-T 11	13 Apr 1984	–	front	2 Oct 1984	1030	172
Progress 20	17 Apr 1984	1322	rear	6 May 1984	2146	19.35
Progress 21	10 May 1984	0410	rear	26 May 1984	1341	16.40
Progress 22	30 May 1984	1947	rear	15 Jul 1984	1736	45.91
Soyuz-T 12	18 Jul 1984	2317	rear	29 Jul 1984	1326	10.59
Progress 23	16 Aug 1984	1211	rear	26 Aug 1984	1913	10.29
Soyuz-T 13	8 Jun 1985	1250	front	25 Sep 1985	0758	108.80
Progress 24	23 Jun 1985	0634	rear	15 Jul 1985	1628	22.41
Cosmos 1669	21 Jul 1985	1905	rear	29 Aug 1985	0150	38.28
Cosmos 1669	29 Aug 1985	–	rear	29 Aug 1985	–	–
Soyuz-T 14	18 Sep 1985	1814	rear	21 Nov 1985	–	56
Cosmos 1686	2 Oct 1985	1316	front	permanently docked		–
Soyuz-T 15	6 May 1986	2058	rear	25 Jun 1986	1858	49.92

Dates are Moscow Time.

A step towards continuous occupancy

Table 3.2 Salyut 7 crewing

Cosmonaut	Arrival	Departure		Days
Anatoli Berezovoi	CDR	Soyuz-T 5	Soyuz-T 7	211.38
Valentin Lebedev	FE	Soyuz-T 5	Soyuz-T 7	211.38
Vladimir Dzhanibekov	CDR	Soyuz-T 6	Soyuz-T 6	7.91
Alexander Ivanchenkov	FE	Soyuz-T 6	Soyuz-T 6	7.91
Jean-Loup Chrétien	CR	Soyuz-T 6	Soyuz-T 6	7.91
Leonid Popov	CDR	Soyuz-T 7	Soyuz-T 5	7.91
Alexander Serebrov	FE	Soyuz-T 7	Soyuz-T 5	7.91
Svetlana Savitskaya	FE	Soyuz-T 7	Soyuz-T 5	7.91
Vladimir Lyakhov	CDR	Soyuz-T 9	Soyuz-T 9	149.41
Alexander Alexandrov	FE	Soyuz-T 9	Soyuz-T 9	149.41
Leonid Kizim	CDR	Soyuz-T 10	Soyuz-T 11	236.95
Vladimir Solovyov	FE	Soyuz-T 10	Soyuz-T 11	236.95
Oleg Atkov	CR	Soyuz-T 10	Soyuz-T 11	236.95
Yuri Malyschev	CDR	Soyuz-T 11	Soyuz-T 10	8.02
Gennadi Strekalov	FE	Soyuz-T 11	Soyuz-T 10	8.02
Rakesh Sharma	CR	Soyuz-T 11	Soyuz-T 10	8.02
Vladimir Dzhanibekov	CDR	Soyuz-T 12	Soyuz-T 12	11.80
Svetlana Savitskaya	FE	Soyuz-T 12	Soyuz-T 12	11.80
Igor Volk	CR	Soyuz-T 12	Soyuz-T 12	11.80
Vladimir Dzhanibekov	CDR	Soyuz-T 13	Soyuz-T 13	112.13
Viktor Savinykh	FE	Soyuz-T 13	Soyuz-T 14	168.12
Vladimir Vasyutin	CDR	Soyuz-T 14	Soyuz-T 14	64.87
Georgi Grechko	FE	Soyuz-T 14	Soyuz-T 13	8.88
Alexander Volkov	FE	Soyuz-T 14	Soyuz-T 14	64.87
Leonid Kizim	CDR	Soyuz-T 15	Soyuz-T 15	49.92
Vladimir Solovyov	FE	Soyuz-T 15	Soyuz-T 15	49.92

Table 3.3 Salyut 7 spacewalks

Date	Hours	Activity
30 Jul 1982	2.5	Berezovoi and Lebedev exited the airlock, retrieved several exposure cassettes from near the hatch, deployed others and recorded the state of the solar panels.
1 Nov 1983	2.8	Lyakhov and Alexandrov attached the first clip-on solar panel to the dorsal solar panel.
3 Nov 1983	2.9	Lyakhov and Alexandrov attached the second clip-on solar panel.
23 Apr 1984	4.25	Kizim and Solovyov installed a curved ladder in a 120-degree arc around the engine compartment.
26 Apr 1984	5.0	Kizim and Solovyov opened up the engine bay, cut the thermal insulation, and replaced a valve to support a bypass around the fractured propellant pipe.
29 Apr 1984	2.75	Kizim and Solovyov fitted the bypass pipe.
4 May 1984	2.75	Kizim and Solovyov replaced a second valve to complete the bypass.
18 May 1984	3.1	Kizim and Solovyov attached the second pair of clip-on solar panels, this time to the lateral panel alongside the airlock hatch.
25 Jul 1984	3.6	Dzhanibekov and Savitskaya tested the URI toolkit, retrieved some exposure cassettes, and deployed others.
8 Aug 1984	5.0	Kizim and Solovyov employed a pneumatic press to crimp the fractured pipe, leaving the bypass to carry the propellant flow. On their way back, they cut out another sample of solar panel.
2 Aug 1985	5.0	Dzhanibekov and Savinykh attached the third pair of clip-on solar panels and an experimental transducer, retrieved the Medusa and other materials cassettes and replaced them, then deployed the French Comet experiment.
28 May 1986	3.8	Kizim and Solovyov retrieved the Comet, Medusa, Spiral, Istok and Resurs cassettes, and then erected and retracted the URS girder.
31 May 1986	5.0	Kizim and Solovyov mounted experiments on top of the girder, re-extended it, then jettisoned it. Finally, they retrieved the transducer that Dzhanibekov had installed.

4

A base block for modular construction

MIR

Early in the morning of 20 February 1986, a Proton rocket, still the most powerful in the inventory, rose on a brilliant pillar of flame above the Kyzyl-Kum desert and then streaked northeast into the pre-dawn twilight. Within its payload shroud was the first element of what was to become the Mir space station. Even though rumours that a new vehicle was being prepared had been leaked, the actual launch was a surprise, as the Salyut 7–Cosmos 1686 complex was still operational. In fact, it was Salyut 7's troubled history that had prompted the decision – after the early return of the Soyuz-T 14 crew – to abandon it, and launch another station to mark the formal opening of the Congress of the Communist Party of the Soviet Union.

The word 'Mir' could be translated (depending on the context) as 'new world', 'peace' or 'community'. The assignment of a *name* for the new spacecraft, rather than Salyut 8, combined with the fact that the two orbits were coplanar, prompted initial speculation that Mir might dock with Salyut 7 and Cosmos 1686 to further expand that complex. It was soon announced, however, that the new vehicle was a third-generation design that was to serve as the core of an entirely new complex that was to be continuously occupied, and, to reinforce this point, it was reported that in addition to axial ports it had a ring of four radial docking ports; this feature had been widely predicted, so came as no surprise. Given that its overall mass and dimensions were dictated by the configuration of the Proton rocket, Mir's basic structure was necessarily similar to its predecessors: namely, a 20-tonne stepped cylinder some 13 metres long and 4.15 metres in diameter at its widest point. When its configuration was announced, Mir was criticised in the West for having little scientific equipment, but to argue this was to fail to appreciate that the Soviet space station programme had reached the point where it was feasible to construct an orbital complex from highly specialised modules. The first element, the 'base block', was a habitat, not a laboratory. It seemed a fair bet that it would soon be expanded by modules similar to Cosmos 1686.

On achieving orbit, Mir had deployed a pair of solar panels to provide electrical power. These had a larger area (each 38 square metres) than those of Salyut 7 (each 20 square metres), and were deployed by a new mechanism that initially unfolded a truss from each side of the narrower section of the stepped cylinder, as a backbone, after which the individual panels mounted along it unfolded fore and aft. Each panel produced 4.5 kW. Although this was sufficient to operate the life-support systems, more power would be required to run the equipment that was to be delivered in the specialised laboratory modules. The dorsal panel of the second-generation Salyuts had not been fitted prior to launch, because the new units were so bulky that, when folded, there was no room in the aerodynamic shroud for a third panel – this was to be mounted on the already installed motor by spacewalkers. Each of the main panels could be rotated in only one degree of freedom over a 180-degree arc (± 90 degrees). The power level decreased sinusoidally with deviation from face-on illumination. An all-sky sensor noted when the station was in sunlight, and it activated the automated system. Each mount had a Sun sensor, and a controller which compared the angle of the panel with the direction of the Sun and rotated it for the maximum output. With both panels mounted on the same axis, the vehicle clearly had to be oriented so that they *could* face the Sun. As the station entered the Earth's shadow, the panels were rotated to the predicted angle for reacquiring the Sun. On a daylight pass the power ran the systems and charged the twelve nickel–cadmium batteries, each of which had a storage capacity of 60 amp-hours. The electrical distribution system used 28 volts of direct current, with 5-amp, 10-amp, 20-amp and 50-amp taps. On emerging from shadow, it took some time for the transducers to reach their maximum output, so the batteries continued to supply power during this lag, prior to being switched over to recharge. Although the power system operated automatically, the cosmonauts, being engineers, were familiar with it, and could intervene in an emergency.

The Mir base block undergoing preparation for launch.

Mir had seven computers in an integrated complex known as Strela. The main unit was the Argon 16B, which was also used by the Soyuz-T ferry's flight control system. Serving the same purpose onboard Mir, it dealt with attitude control. Frames of reference were drawn from infrared Earth-horizon sensors, a Sun sensor, and a sophisticated sextant. Whereas the Delta navigation system of the second-generation Salyuts had required its state vector to be periodically updated, which had imposed a supervisory function on the crew, Mir could be told to establish a specific attitude, which it would then automatically hold. It could also be programmed to run through a series of manoeuvres, but its memory capacity meant that it could store sequences for only a few days at a time. It would automatically monitor when the vehicle was in communication with ground stations. Like the power generation and distribution system, therefore, the navigation system was highly automated to free the crew for productive work. In fact, the computers were able to manage resources in support of a wide range of research activities. Once the operating mode for a given situation had been specified, the computer complex eliminated the need for a cosmonaut to orient the station, to aim instruments, and then to track the target.

Although the Mir base block had six ports, only the axial ports were fitted with the antennas required to effect an automatic docking. It was not intended that more than two Soyuz spacecraft would be docked at any given time, so the environmental systems were capable of supporting six cosmonauts for brief periods, with indefinite residence by crews of two or three people. The standard crewing was nevertheless considered to be a commander and an engineer. Individual researchers would be able to make short visits during handovers, or a doctor could check out the residents and then accompany them on their return to Earth.

AN AMBITIOUS MISSION

Soyuz-T 15 was launched in the afternoon of 13 March. With refreshing openness, the event had been announced the previous day, and it was shown live on television. Furthermore, it was also announced that after docking with Mir to commission the new station, the crew would briefly revisit Salyut 7 in order to complete outstanding experiments. Having served as backups for Soyuz-T 14, Alexander Viktorenko and Alexander Alexandrov would normally have flown next, and if the aim had been just a brief visit to Salyut 7 they may well have done so. However, the decision to mount a *double-station* mission precluded this. Instead, "the most complex flight" likely to be attempted "for the next few years" (as chief cosmonaut Vladimir Shatalov put it) was assigned to Leonid Kizim and Vladimir Solovyov, who had not only trained for Mir but, two years previously, had spent 237 days on Salyut 7, and so were ideally suited for this unprecedented mission. Despite the workload, it had been decided not to fly a third cosmonaut, in order to accommodate additional propellant to perform the double rendezvous.

The *Semyorka* had put Soyuz-T 15 into a 193- by 238-kilometre orbit with

Vladimir Solovyov (left) and Leonid Kizim.

its perigee above the Soviet Union. Only after the many systems had been verified was the first manoeuvre executed. This was done at apogee half-way through the second orbit – that is, whilst over the South Pacific Ocean – and it raised the perigee by 100 kilometres to produce a 238- by 290-kilometre orbit with its apogee over the Soviet Union. The rotation of the Earth made the orbit migrate about 22 degrees westwards per orbit. Before the spacecraft's ground track took it south of the Soviet Union, the radars accurately determined the orbital parameters to assess the performance of the engine. At the end of the fourth pass, another burn raised the orbit to 290 by 330 kilometres, not so much to regulate the altitude as the period of the orbit in order to coordinate the motions so that at the end of the second day the ferry would be a few kilometres from the station (which was in a 332- by 354-kilometre orbit) and slowly closing on it. Given the ambitious plan for this mission, maintaining a high reserve of propellant was a major objective.

Mir had two rendezvous systems. The new Kurs was to supplement (and later supersede) the Igla that required the station to reorient itself during the final phase of the rendezvous to point the appropriate docking port at the approaching spacecraft. Although this reorientation was no big deal for the isolated base block, it would be impractical once a cluster of heavy modules were added. Using Kurs, the ferry could approach the station from any direction and then, when several hundred metres out, fly around for a straight-in approach to the required port. The rear port had both Igla and Kurs transponders, to enable Progress ferries to dock. Since Soyuz-T 15 was to dock with Salyut 7, it had only an Igla. In this case, it had been decided to leave the rear port clear to accommodate the first cargo ferry, which was already being readied. When the Igla paused at the 200-metre point, Kizim took command and flew around to the front port. Although their vehicle did not have a Kurs system, they had a laser rangefinder (similar to that used by Dzhanibekov and Savinykh in approaching the crippled Salyut 7) for range and closing rate during the manual straight-in approach, which was achieved without incident.[1] On pushing in Mir's Konus drogue assembly, Solovyov crossed the spheroidal multiple docking adapter to the internal hatch that opened into the main compartment. With the lights already on, the automatic video camera caught his entry. A few minutes later, having powered down the ferry, Kizim joined him to form Mir's first crew.

The Mir base block as seen from Soyuz-T 15.

---

[1] As events turned out, this would be the *only* Igla-equipped Soyuz ever to dock with Mir, because the new Soyuz-TM model was equipped with the Kurs system.

A NEW HOME

Even although cosmonauts can adopt arbitrarily orientations in weightlessness, the psychologists had required that they be provided with a clear sense of up-and-down, so in the main compartment, which, as in the case of the Salyuts, was a long narrow room, the 'floor' was carpeted, the 'walls' were green, and the 'ceiling' was white. A control panel with several video monitors spanned the floor immediately beyond the hatch, and the first thing that Kizim and Solovyov did was to check the status of the systems it displayed. The other end of the compartment had a slightly raised deck to accommodate embedded apparatus. From the viewpoint of a cosmonaut facing the controls, the multiple docking adapter was 'forward' and (appropriately) the engine was 'aft'. A set of cartesian axes had been devised to describe the layout of lockers, but this is so arbitrary that the more intuitively obvious reference frame will be used herein (a distinct sense of up-and-down may not be necessary for the cosmonauts in space, but it assists this Earthly writer). Thus, in addition to the front axial port, the multiple docking adapter had upper (dorsal), lower (ventral), and a pair of lateral left and right ports.

In contrast to the Salyuts, in which the conical case of the primary instrument had dominated the rear of the compartment, Mir was unobstructed. For the first time, two phonebox-sized cabins had been provided, located towards the rear, inset into the side walls. Each held a sleeping bag fastened to the wall, a small fold-out desk, and a porthole. These were for the residents. Visitors would continue the practice of fastening their sleeping bags to the ceiling. As Mir was to serve as the 'living room' for a modular complex, it had improved facilities for food preparation. The table had a water heater to reconstitute dehydrated food, and a stove to heat canned foods. Although the table folded against the wall, it could conveniently be left in position (just forward of the cabins) because cosmonauts could float over it rather than walk around it. There was a small fridge for fresh produce. Prepackaged food was kept in lockers near the hatch to the rear transfer tunnel. The toilet (ASU) was a tiny cubicle set immediately aft of the right-hand cabin. Unlike on the Salyuts, this had a door for privacy. In addition, this compartment included a 'wash basin' which incorporated a transparent hood that prevented water from escaping, and holes to accommodate the user's head and hands. Luxury indeed!

There was a comprehensive gymnasium which folded away when not in use. The veloergometer was stored under the floor, at the step in level between the sections of the compartment. Each cosmonaut had to cycle the equivalent of 10 kilometres a day. The treadmill was mounted behind the table, directly in front of the rear tunnel. Wearing a springy harness for traction, a cosmonaut would 'walk' the equivalent of 5 kilometres a day. The track provided a satisfying view along the axis of the station. The daily exercise regime was designed to minimise deterioration of the heart, bones and muscles. The exercise could be divided into short bursts, but each crew member had to total 2.5 hours per day (in order to burn off 450 calories) and they followed a four-day cycle in which the first day was to build up speed and strength, the second concentrated on exertion, the third emphasised endurance, and the fourth was a day of rest.

The thermal regulation system had two internal fluid loops (a moderately-warm and a cold loop) and one external loop through the radiators conformally mounted on the surface of the narrower part of the stepped-cylinder. Glycol was used as coolant due to its low freezing point. The heat exchanger was mounted in the unpressurised engine bay. The temperature and moisture content of the cabin were regulated by the internal coolant loops, which could maintain the temperature in the range 16–28°C. Every effort was made to recycle water, but this was categorised by its quality and utilised in different ways. Only the water vapour reclaimed from the station's cabin by the SRVK of the air conditioning system was potable. The water from the wash basin was pumped through a column of ion-exchange resins and activated charcoal, filtered, remineralised, purified and recycled through the hygiene system. The toilet incorporated a vapour-diffusion apparatus which heated the urine with the thermal regulation system's hot-loop to diffuse the water molecules through a membrane to condense on the cold-loop, so that the water could be recovered. The concentrated urine was then fed into another tank. The plan was to supply an electrolysis system which would release oxygen from the recovered water. Until that apparatus could be delivered, however, the filtration process would not be used, and the urine would be vented to space. Faecal matter was stored in a bag in a separate tank inside the toilet, and when this was full it was either ejected from the small airlock or dumped into a departing Progress ferry. Oxygen and nitrogen were drawn from high-pressure tanks and mixed to maintain a standard atmosphere at sea-level pressure. Progress tankers were to pump oxygen and nitrogen through pipes in the docking collar to replenish the tanks. A surge of oxygen could be released by heating a solid-fuel canister of a superoxide. Carbon dioxide was extracted by passing air through a lithium hydroxide canister.[2] Dangerous trace constituents (carbon monoxide, hydrogen, methane and ammonia) were extracted from the air using regenerated charcoal beds and catalytic oxidisers. These were interim solutions, however, because closed-cycle air cleansing and regeneration systems were to be incorporated in the add-on modules.

The Kaliningrad control room.

In general the cosmonauts were to follow a normal 0800 to 2300 Moscow day, five days per week. Unless there was a specific reason to disrupt their sleep cycle, their day began with two

---

[2] On earlier Salyuts, a similarly configured canister of potassium superoxide had been used, and this had both extracted carbon dioxide and liberated oxygen in a reaction which left a solid residue of potassium carbonate. In the lithium hydroxide scrubber, however, the surface of the crystal extracted the gas by adsorption, which did not yield oxygen.

hours of personal hygiene and breakfast. Work from 1000 to 1300 was followed by an hour's vigorous exercise, an hour off for lunch, another three hours of work, and another hour's exercise. The preparation of the evening meal began at about 1900. The meals were spaced out to achieve a high degree of assimilation. The diet was designed to yield about 100 grams of protein, 130 grams of fat, 330 grams of carbohydrates, and appropriate mineral and vitamin additives. In the evening, the cosmonauts were free to relax or to catch up on work as they pleased. Even in space there was paperwork to be done, and this was often processed late at night. Unless a task had to be performed at a specific time, the cosmonauts were free to define their own schedule and work at their own pace. On long missions, common sense prevailed – there was little reason to hurry to get started on a job one day and have to break off overnight, if it could reasonably be left to the next day. Ironically, on the Salyuts, crews had reported that working in their free time was more stimulating, because it was being done voluntarily.

During their first few days, Kizim and Solovyov adjusted the thermal regulation system to a comfortable 24°C, extensively tested the navigation and attitude-control computers, and one-by-one checked out the rest of the station's systems. Only then did they set about attending to the experiment packages that had been stowed for the launch. Progress 25 docked at the rear port on 21 March. Over the next few days, they unloaded its cargo. This was a time-consuming task, because it had been loaded to capacity and there was little room to move inside the cramped compartment. Each container had to be unbolted and unpacked, and the assorted items stored in the base block. This job was made difficult by the fact that the quick-release bolts holding the containers in the racks had been overtightened. In parallel with their manual labour, 200 kilograms of water was automatically pumped into the two Rodnik tanks in the unpressurised bay. As had the second-generation Salyuts, after being placed in a low (192- by 238-kilometre) orbit by its Proton launch vehicle, Mir had been required to climb using the twin-chamber engine of the unified propulsion system, which burned unsymmetrical dimethyl hydrazine with nitrogen tetroxide, fed by nitrogen-inflated metallic bladders within the tanks. As this had consumed a large fraction of the initial propellant load, Progress 25 replenished the depleted tanks. On 25 March, and again on 18 April, it fired its own engine to raise the orbit to 336 by 360 kilometres. Later resupply ships would progressively raise Mir to its circular 400-kilometre operating orbit. For attitude control, Mir had 13-kilogram-thrust engines arranged in clusters at 90-degree intervals around the periphery of the unpressurised bay. The 32 thrusters were split into two groups of 16, forming prime and backup systems, with separate propellant feeds. The nominal pointing accuracy was 1.5 degrees, but in its precision mode it could maintain a given orientation to within 0.25 degree. When instruments requiring finer pointing accuracy were installed, so too would be a better orientation system.

On a favourable pass, the integrated network of communications stations across the Soviet Union gave up to 20 minutes of coverage per orbit. With ships in both the Atlantic and the Pacific, this time could be doubled on a favourable pass. A series of *un*favourable passes, however, could result in the station being out of contact for up to nine hours. Geostationary relays were to overcome this limitation. Unfortunately, when Mir was launched, only the first such Luch satellite was in orbit. Launched the previous October, Cosmos 1700 had been located above the Indian Ocean. The plan was to install three such satellites around the equator at 95°E, 200°E, and 344°E as a Satellite Data Relay Network (SDRN) for continuous high-capacity communications with the scientific equipment to be installed on the expanded complex, to enable data to be relayed to researchers immediately instead of being stored on tape and returned later in descent capsules.[3] Mir's Ku-band system provided selectable channels across the 11–14 GHz range to relay voice, video and telemetry. As it had been tested prior to launch, Kizim and Solovyov just transmitted a television broadcast to verify that the steerable parabolic antenna mounted on a boom at the rear of the unpressurised bay could track the satellite properly. Although it could relay signals for 40 minutes per pass, on *every* pass, irrespective of whether the ground track passed over the Soviet Union, the link would nevertheless be of limited use until the network of satellites was completed because Mir's normal work cycle often made holding the antenna on a single satellite impracticable.

In early April, Kizim and Solovyov performed a Resonance test (which involved determining the station's natural vibration modes, in order to evaluate stresses on its structure) and continued installing apparatus, but they also made visual observations of the Earth and initiated long-term experiments in plant cultivators. Progress 25 left on 20 April. Progress 26 arrived on 26 April, with more propellant, food, water and apparatus, but the first item sought by the cosmonauts was the mailbag. The first of May was a holiday, and coverage of the Red Square parade was relayed up to them. Over the next few days, they loaded 500 kilograms of supplies into Soyuz-T 15's orbital module, recharged the ferry's battery, and verified its propulsion system, and Mir was returned to its autonomous operating regime – it was time to move house.

AN OLD HOME

Since late April, Mir had been some 3,000 kilometres behind the Salyut 7–Cosmos 1686 complex. On 5 May, Progress 26's engine was used to lower Mir's orbit slightly, to begin to reduce this range. Later that day, Kizim and Solovyov undocked Soyuz-T 15 and withdrew. All three control rooms at Kaliningrad were now in use, one dealing with the Salyut 7–Cosmos 1686 complex, one with Mir, and one with the Soyuz. The next day, after a series of propellant-efficient manoeuvres, the ferry activated its Igla and made an automated approach to Salyut 7. The orientation of the complex was being controlled by Cosmos 1686, in response to commands from the ground. When the Igla paused 200 metres out, Kizim once again took command, and manually docked at the vacant rear port. This was the first time that a spacecraft had transferred between two orbital stations – the ease with which it was achieved amply demonstrated how much the transportation system had matured. Only at this

---

[3] Luch was the equivalent of the Tracking and Data Relay System (TDRS) then being developed by NASA for the Space Shuttle and the Great Observatories that it was to deploy.

point was it announced that once the cosmonauts had finished Salyut 7's programme they were to load as much equipment as possible into Soyuz-T 15 and then, having already made history by making one transfer, they were to return to Mir.

On entering their old haunt, Kizim and Solovyov found it to be rather chilly, and so they turned up its heater and returned to their own vehicle for the night. The next day, they reactivated the station's systems and replaced components such as filters, fans, scrubbers, lamps and transmitters that had exceeded their recommended service life. To round off, they performed a thorough test of the Delta and Kaskad control systems. By mid-May, this maintenance had been completed, and they turned to the research programme and settled down to making Earth observations. At the end of the month, in preparation for a spacewalk, they cleaned and dried out the semi-rigid suits delivered by Cosmos 1669. On a previous visit, they had accumulated nearly 24 hours working outside Salyut 7. Within 15 minutes of venturing out on 28 May, they had retrieved the Comet dust trap, together with all the cassettes that had been mounted outside by previous crews to test the manner in which materials reacted to exposure to the space environment, including Medusa (biopolymers), Spiral (cables), Istok (connectors and bolts) and Resurs (structural metals). Once these cassettes had been secured in the airlock, they set about the main item on their agenda. The Ferma experiment involved erecting a girder which had been designed and supplied by the Institute of Electrical Welding in Kiev, and delivered by Cosmos 1686.

After affixing a small work platform immediately outside the airlock hatch, they mounted the 150-kilogram package on it, selected the automatic option to deploy the structure, and watched. When fully erect, the 15-metre-long girder comprised a set of aluminium-titanium alloy lattice-and-pin frames of a square cross-section about half a metre on each side. Having demonstrated that the deployment mechanism worked, they retracted it. When it was back in its container, they mounted a sensor on one of the portholes for use in a later experiment. By the time they closed the airlock hatch, they were 45 minutes late on the three-hour plan, but were jubilant because they had achieved all of their objectives. The television downlink had been broadcast live by domestic television and, for the first time, shown in part by the American networks. On 31 May, having refurbished their suits, they made a second spacewalk. The first task was to attach a flat plate to the top of the girder, which had been left retracted, mount a number of experiments on it, then extend the structure once again, this time manually. One such experiment was a low-power laser positioned to illuminate the sensor that had been deployed at the end of the first excursion. This measured the rigidity, stability and vibration modes of the structure. They used a welding kit – an improved version of the URI tested by Savitskaya – to electron-beam weld sample elements of the lattice to test this as a means of locking such joints into position. A similar girder was to be erected on Mir, so verifying this deployment mechanism and assessing its dynamic characteristics had been the principal reason for returning to Salyut 7. Another package on the girder was Fon; this sampled the environment near the station. Previous data from the Astra mass spectrometer had demonstrated that a station left a tenuous 'tail' of gas. The tests over, the girder was jettisoned. The final task was to retrieve the experimental solar transducer that Dzhanibekov had left, so that engineers could measure how it had degraded.

After a well-deserved rest, Kizim and Solovyov spent early June making Earth observations, paying particular attention to the area affected by the fallout from the explosion of a nuclear reactor of the Lenin Nuclear Power Plant at Chernobyl on 26 April. Their data – together with that from the Meteor weather satellites – was used to direct aircraft that were seeding clouds in an effort to prevent the normally heavy rains from flushing contaminated soil into the rivers and the reservoirs supplying the major population centres of the Ukraine. They also tended plants in the Biogravistat and Oasis cultivators, and processed semiconductor in the Kristallisator furnace and biological agents in the Gel electrophoresis apparatus. In the following week, though, they dismantled such apparatus as could be salvaged and loaded some 400 kilograms into the orbital module of Soyuz-T 15. This trove included the KATE-140 mapping camera, PCN low-light camera, Echograph biomedical apparatus, Kristallisator and Pion-M furnaces and EFU-Robot electrophoresis system, together with a number of spectrometers, television and video apparatus, film with 3,000 exposed images, the new spacesuits (because there were none on Mir) and the sample cassettes recovered from outside Salyut 7.

In mid-June it was announced that Salyut 7 was being mothballed. On 25 June, having restored the complex to its automated operating regime, Kizim and Solovyov closed the hatch for the last time and undocked. During their 50-day stay, they had successfully completed the work left outstanding by Vasyutin's crew.

ANOTHER NEW FERRY

Meanwhile, Soyuz-TM 1 was launched without a crew on 21 May 1986 to evaluate "the onboard systems and assemblies and the structural elements" of the upgraded version of the Soyuz-T, and conduct "joint tests" with Mir – the Kurs in particular. After flying a two-day rendezvous, the spacecraft manoeuvred to line up with Mir's front port, with the station remaining stable, then closed in and docked. After raising the station's orbit slightly, the ferry undocked on 29 May and returned to Earth the next day. The new spacecraft had the unified propulsion system of the Soyuz-T and Progress, an improved inertial system for navigating independently of ground radars, and it could relay via the Luch satellite whilst in line-of-sight with Mir. In addition, various modifications (most notably a lightweight parachute) had given mass savings that allowed an increase in payload into orbit of 250 kilograms, and down to Earth of 70 kilograms (an increase to 120 kg that had required uprating the retrorockets that were fired immediately prior to touching down). As would later become evident, the recovery criteria had been considerably relaxed, with the result that the new model could land at any time of the day.

THE WANDERERS RETURN, BRIEFLY

By mid-June, Mir was only 1,000 kilometres ahead of Salyut 7. Having replenished the station, Progress 26 left on 22 June – it had taken the full loads of two tankers to replace the propellant consumed in the initial climb. Soyuz-T 15 lowered its orbit on 26 June to rendezvous with Mir for the second time. As previously, after the Igla made an automatic approach, Kizim took over and flew around to redock at the front port. On this occasion, video downlink was released, and this revealed that the collar of the port had been rotated compared with that of the Salyuts, and the ferry docked with the plane of its solar panels at 45 degrees to those of the station, but the reason for this new configuration was not elaborated.

On their first day back, Kizim and Solovyov unloaded the salvaged apparatus. Early in July, during a television report, they showed off their information system (a computerised manual to assist in operating the station's equipment) which they had just added to the Strela computer complex. It offered the advantage that it could be updated to reflect changes in equipment and operating procedures, and was not only to supersede the usual load of paper manuals but also the Stroka teletype uplink. On 2 July Kizim exceeded the 362-day accumulated endurance record previously held by Valeri Ryumin. Most of early July was spent taking pictures for the Intercosmos organisation's Geoex-86 remote-sensing programme, which focused on the German Democratic Republic. Their results were correlated with data from other satellites, aircraft and ground teams.

By this point, Kizim and Solovyov were well into the increased exercise regime that added half an hour to their daily routine in order to strengthen their load-bearing muscles and generally develop stamina in preparation for returning to Earth, but the announcement on 14 July that they were to return in two days came as a surprise to observers who (accepting the statement that Mir was to be continuously inhabited) had expected this commissioning crew to hand over to their successors some time in September, but speculation that the 'emergency' return was prompted by a serious systems failure was unfounded. The truth was that delays in preparing the first of the new modules had made the intended handover impracticable. Put simply, Kizim and Solovyov had grown bored after completing the limited number of experiments available. They loaded Soyuz-T 15 with the exposed film and the sample cassettes that they had retrieved from Salyut 7, undocked and landed on 16 July. As they sat in reclining couches alongside the capsule awaiting medical tests, they were in hearty spirits. It had been a remarkable mission: after spending 50 days commissioning Mir, and 50 days finishing off research on the Salyut 7–Cosmos 1686 complex, they had occupied Mir for another 25 days. In the process, Kizim had taken the accumulated endurance record to 373 days.

Vladimir Solovyov (left) and Leonid Kizim immediately after landing.

INTERMISSION

In late August, the Salyut 7–Cosmos 1686 complex began to manoeuvre, prompting speculation that it might be about to rendezvous with Mir to enable Cosmos 1686 to jump ship, but over a four-day period each of the docked vehicles fired its engine to climb to a 480-kilometre circular 'storage' orbit. Cosmos 1686 then manoeuvred the complex into the gravity gradient for stability. Previous stations had been de-orbited, but this time it had been decided to monitor it to assess the degradation of its basic systems. Salyut 7 had suffered its problems, but it had hosted ten Soyuz ferries, 13 Progress cargo craft and two of the large modules.

Meanwhile, it was reported that Mir operations would not resume until early in the new year. The launch of Progress 27 on 16 January was shown live by domestic television. In the six months that Mir had been dormant, its orbit had decayed to 312 by 340 kilometres, so after docking on 18 January the ferry boosted it to 340 by 365 kilometres. A second Luch satellite was launched at the end of the month, but it was stranded in a useless orbit when its geostationary transfer stage malfunctioned. It had been meant to replace Cosmos 1700, which, upon running out of station-keeping propellant in October 1986, had drifted off its Indian Ocean relay station. Without a Luch relay, Mir was now in contact with Kaliningrad only for brief periods several times a day.

5

An astrophysical laboratory

SECOND START

In the spirit of *glasnost*, the launch of Soyuz-TM 2 was announced in advance, and the television coverage began with the suiting-up process. In fact, Yuri Romanenko and Alexander Laveikin had started out by backing up Vladimir Titov and Alexander Serebrov, but had switched when Serebrov fell ill. Ironically, because of the delay in preparing Mir's first expansion module, and the resultant slippage in this mission, Serebrov had recovered by the time it set off on 6 February 1987. It flew the now standard two-day rendezvous, and once the final transfer orbit had closed the range to 100 kilometres the Kurs system was activated. The ferry initially approached the rear of the station and then, because Progress 27 was at that port, automatically flew around to the front. *Tass* reaffirmed that Mir was to be continuously inhabited, and added that this crew hoped to extend the endurance record. This was Laveikin's first flight, but Romanenko had spent three months on Salyut 6 a decade earlier and had commanded one of the visiting Intercosmos missions. They spent the first few days readjusting to the Kaliningrad duty cycle, because orbital dynamics had obliged them to make the rendezvous during the night. The initial phase of adaptation was "rather painful" for Laveikin, so he "rested" for four days while Romanenko undertook the manual work of unloading Progress 27. A medical check-up at the end of the week indicated that Laveikin had physically recovered, but he later said that it

Alexander Laveikin.

was almost a month before he felt fully at home in space. During the second week, they carried out maintenance on expired components of the environmental and thermal regulation systems, reconfigured the electrical power distribution system in preparation for the arrival of the new module, and refurbished the spacesuits. Having replenished Mir's propellant and water tanks, Progress 27 undocked on 23 February.

When Romanenko and Laveikin were launched, the first laboratory module was expected to follow in "a few months". Until it arrived, the primary research activity would be Earth studies. One of their first observations was a striking oceanic feature involving a nine-ring concentric wave some 300 kilometres across, centred on an area of apparently "serene" surface. The Sever camera was "a fixed unit using a movable apparatus", and was to be used to take oblique imagery that would highlight surface relief. The Kolosok experiment investigated aerosol structure in microgravity. The new Pion-M experiment studied heat and matter transport in a fluid (it revealed that silica aerogel in suspension formed saucer-shaped structures, fluoroplastics formed tree-shapes, and glassy pellets formed arbitrary but very robust clumps) in order to assist in the design of an industrial-scale apparatus to process biological (particularly colloidal) materials in bulk. Progress 28 arrived on 5 March with a new KATE-140 mapping camera, several spectrometers, the Gamma-1 biomedical test kit, and the Korund-1M furnace (which was an improved version of the one tested on Salyut 7, although as it consumed 1 kW its use had to be restricted until the station's power supply could be augmented). They installed the KATE-140 and conducted a survey of water run-off in the Caucasus, coordinating the overhead KATE-140 views with oblique-imagery from the Sever camera. After Progress 28 left on 27 March, taking away the accumulated trash, they switched to general housekeeping duties while the materials experiments continued.

KVANT 1

Early on 31 March, a Proton placed a heavy satellite into an orbit that was coplanar with Mir's. Almost half of the 22-tonne mass was an 8-metre-long stripped-down TKS with a laboratory (named Kvant 1) to expand Mir. One year earlier, journalists had been shown the mock-up in Kaliningrad of a rear-mounted module. To preserve the two-ended docking capability, it was evident that this would have to incorporate a rear-facing port. None of the modules that had docked with the second-generation Salyuts had had such a configuration, so it was something new. This was evidently that module. It was mated with its tug by a conventional probe-and-drogue docking system, and once the module was safely in place the tug would depart.

As with Cosmos 1686's slow rendezvous with Salyut 7, the tug made a series of engine firings designed to close in on Mir at a low relative motion. On 5 April, when its final transfer orbit brought it within 20 kilometres of Mir, the Igla was activated. Kvant 1 extended its docking probe when it was 500 metres out. At 200 metres, as it rolled onto the proper orientation, it lost transponder lock, began to deviate from the straight-in path, and aborted its docking attempt. Watching the video downlink from its docking system, the flight controllers feared a collision. Romanenko and Laveikin had retreated to Soyuz-TM 2, but flight director Valeri Ryumin urgently asked them to re-enter Mir and look through the portholes to try to locate Kvant 1. Romanenko watched it drift by in a slow roll, at a distance of only 10 metres. Later in the day, having analysed Kvant 1's telemetry, the engineers identified the problem. Ryumin told reporters that they had been "unduly cautious" in defining the tolerances for the approach of the large module, and this was readily rectified. However, it would take several days to set up another rendezvous. On 9 April they again sealed themselves into Soyuz-TM 2, and Kvant 1 made its second approach. When 1 kilometre out, its closure rate was 2.5 metres per second. By the time it was 25 metres out, this had been reduced to 0.3 metres per second. This time, the final phase of the approach was nominal. Kvant 1's probe slid into Mir's rear port and achieved a soft docking. Unfortunately, when the command was sent to retract the probe to draw the two collars together, the main latches failed to engage. It halted several centimetres short of its fully retracted position. When the troubled station flew out of radio range, the cosmonauts re-entered the base block and inspected the newcomer through the small porthole in the rear transfer compartment, but could see nothing amiss. If the module could not be hard-docked, it would have to be jettisoned. When the telemetry failed to offer any clues, it was decided to send the cosmonauts out to inspect the docking mechanism.

Two days later, Romanenko and Laveikin opened one of the vacant radial ports in the multiple docking adapter at the front of the base block. Upon noticing that his suit was losing pressure, Laveikin had an anxious moment until he recycled a switch and rectified the fault. Then they cautiously made their way back along the 13-metre base block using strategically positioned hand-holds. They took a television camera with them to show the flight controllers what they found, but this failed to work. As soon as Laveikin reached the engine compartment, he peered over the rim and saw "a white object" in between the collars. Kvant 1 was commanded to extend its probe to open the gap. Laveikin then reached in to extract the obstacle, which turned out to be a cloth bag of hygienic towels that had somehow become caught in Mir's docking assembly when Progress 28 had been loaded with trash. Satisfied that nothing else was blocking the mechanism, they asked the flight controllers to command Kvant 1 to retract its probe, which it did, and achieved a hard docking. Having demonstrated yet again the ability of a crew to troubleshoot a problem, Romanenko and Laveikin made their way back inside. The first task once safely back inside was to reconfigure the attitude-control computer to recognise the altered centre of mass of the complex which, with Soyuz-TM 2 at the front and Kvant 1 (with the tug in place) at the rear, was 33 metres in length and almost 50 tonnes in mass. However, it took much longer than expected to fine-tune the attitude-control system, and a great deal of propellant was consumed. It was

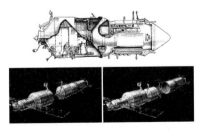

Graphics showing Kvant 1 and its TKS tug.

later revealed that in a medical check-up upon his return, an anomaly was found in Laveikin's heart rhythm, an irregularity which prompted the medics to ban him from spacewalks pending an investigation. The next day, 12 April, Cosmonaut Day, they removed the docking assemblies and entered Kvant 1 for an initial inspection. Later that evening, the tug undocked. As it withdrew, antennas for both the Igla and the Kurs rendezvous systems deployed on the exposed rear of the module. Unfortunately, the tug had used so much propellant in setting up the second rendezvous that it was unable to fully de-orbit itself, and it had to be left for its orbit to decay naturally.

Kvant 1 was a stubby cylinder 5.8 metres long and of the same diameter as the main body of the base block. It comprised three separate hermetic compartments. At the front, adjoining Mir, was the short forward transfer tunnel. A lightweight internal hatch led to the main laboratory compartment, in which the controls for the scientific apparatus were located. The hatch beyond accessed the longer tunnel that led to the rear-facing docking drogue. Equipment was wrapped around this tunnel, in an unpressurised compartment. Of the 1,600 kilograms of scientific apparatus built into Kvant 1's structure, fully 800 kilograms was Svetlana, the semi-industrial-scale electrophoresis processor for biological materials.[1] The rest of the compartment was taken up by a variety of high-energy instruments (collectively known as the Röntgen apparatus) for high-energy astrophysical observations, which had been supplied by a range of collaborating international partners.

The Max Planck Institute and the University of Tübingen in Germany supplied a scintillation spectrometer, known as the High-Energy X-ray Experiment (HEXE), that incorporated four identical Phoswich detectors sensitive to X-rays in the energy range 15–200 keV and had a 1.6 degree square field of view. The European Space Agency's Sirene-2 high-pressure gas-scintillation proportional-counter (an improved version of apparatus on its EXOSAT observatory) was sensitive to X-rays in the 2–100 keV energy range and had a 3 degree square field of view. It was to observe emissions from extremely high-temperature rarefied gas in space. The TTM wide-angle coded-mask imaging spectrometer supplied by the Netherlands Space Research Organisation in Utrecht and Birmingham University (an improved form of apparatus tested in 1985 on the Spacelab 2 Space Shuttle mission) was sensitive to X-rays in the 2–30 keV energy range and had a 7.8 degree square field of view. It was to refine the location of discrete X-ray sources, with an angular resolution of 2 minutes of arc. The Soviet-supplied Pulsar X-1 spectrometer had four identical Phoswich detectors for hard X-rays and soft gamma-rays in the 50–800 keV energy range and a 3 degree square field of view. For survey work, the Soviets also supplied a 180-degree wide-field scanner that was sensitive up to 1,300 keV and had a one millisecond temporal resolution. Most of these instruments were designed to be capable of being operated remotely by their research teams. A new Luch satellite (Cosmos 1897) was launched on 26 November to replace its now defunct predecessor

---

[1] This was named after Svetlana Savitskaya, who had tested an experimental form of the apparatus onboard Salyut 7 in 1984.

above the Indian Ocean, but a single relay satellite would not really be sufficient for real-time operations.

In addition, the Byurakan Astrophysical Observatory in Armenia supplied the Glasar telescope for spectrography in the wavelength range 1,150–1,350 Angstroms to conduct a survey of bright quasars, active galactic nuclei and stellar associations at ultraviolet wavelengths that are absorbed by the atmosphere, and so are unable to be studied from the ground. Although it used an electronic image-intensifier, exposures of up to ten minutes were required to record faint stars. There was a small airlock in the rear transfer tunnel of Kvant 1 to enable a cosmonaut to reload the film cassette.

A variety of other apparatus had been built into the module to assist in operating the expanded complex. The Elektron apparatus electrolysed water reclaimed from urine and liberated oxygen (the hydrogen it produced was vented into space, further adding to the station's gaseous tail) thereby reducing the amount of oxygen that had to be delivered by the Progress ferries. It used long flexible air tubes that were strung throughout the complex by the crew. The Vozdukh was a molecular sieve to scrub carbon dioxide from the air (which was also vented), thereby reducing the number of lithium hydroxide canisters that had to be ferried up. These regenerative systems (together with the moisture condenser in the air conditioner of the base block) were a step towards a closed-cycle environment. The Vika utilised an exothermic charge to thermally decompose sodium chlorate at 1,000°C, yielding a surge of oxygen to top up the station during crew handovers. However, these use-once canisters would have to be ferried up. Outside the pressurised shell were six 165-kilogram magnetically suspended flywheels. These gyrodynes were spun in pairs at 10,000 rpm, one pair on each of the three cartesian axes, and converted electricity from Mir's solar panels into torque to control the complex's orientation without consuming propellant. Once the required position had been attained using thrusters, the gyrodynes were to damp out perturbations due to active elements (such as the crew) and maintain it with an accuracy of 1.5 minutes of arc (an order of magnitude better than was possible using the base block's thrusters, and a prerequisite to effectively operating the telescopes). In addition, the main compartment had been packed for launch with 2,500 kilograms of assorted cargo. The bulkiest item was a 340-kilogram solar panel (actually in two segments) to provide the power required by the new equipment. There was pressure from astronomers to use the telescopes to study the supernova SN1987A that flared in the Large Magellanic Cloud two months previously, but this would be impractical until the solar panel could be erected. Unfortunately, this would require a spacewalk and the flight surgeons, having spotted Laveikin's heart irregularity, had banned him from going out while his condition was evaluated – for which his cardiac trace was transmitted by telemetry during strenuous exercise. The largest of the four portholes in Kvant 1 was in the floor. It was 43 centimetres in diameter and was high-fidelity optical glass. Although a mock-up displayed at the Paris Air Show had an MKF-6M multispectral camera mounted to use this porthole, this had not been included in the cargo. A high-precision star tracker was to be installed on the 23-centimetre porthole to calibrate the gyrodynes. The others were small (8-centimetre) observational ports. With Kvant 1 on the rear of Mir, it became impractical to fire the base block's main engines, and since the module had no propulsion of its own (which was why it had been delivered by a tug) the complex would now be dependant on docked Progress ferries to perform major orbital manoeuvres.

It was later disclosed that Kvant 1 had originally been designed for the front of Salyut 7, but that station's troubled history had imposed so many delays that it had been decided to modify the module to function at the rear of Mir. This had required that pipes be installed to enable a Progress tanker to pump fluids around it into the base block. It is noteworthy that if the module *had* been sent to Salyut 7, its double docking ports would have enabled it to become a *permanent* adjunct to that station, whereas modules like Cosmos 1443 could only be temporary because they had to be jettisoned to release the front port for the next ferry. The evolutionary nature of the development of the elements for the construction of a modular complex is thereby evident. Years later, it was revealed that the Korolev Bureau had intended that Mir be expanded using a number of modules of this type, each having a mass of about 8 tonnes, but had been ordered instead to use 20-tonne modules based on Chelomei's TKS. Development of small modules (most particularly a biotechnology laboratory, a power plant, and an X-ray telescope, initially to be delivered by tugs, and later by the Buran shuttle) continued, but Kvant 1 was both the first and the last of its type. As a hybrid, therefore, it seems with hindsight to have been something of a 'cuckoo'. It had been intended to add it during the second part of the Soyuz-T 15 crew's residency, but the modification work had fallen behind. As a result, this initial phase in the expansion of the Mir complex was far behind schedule.

After a few days spent unloading cargo from Kvant 1, Romanenko and Laveikin updated the attitude-control system's mass model to take account of the shift in the centre of gravity due to the redistribution of so much apparatus, and then powered up and tested each gyrodyne. Progress 29 docked at the rear of Kvant 1 on 23 April. It created a four-vehicle in-line complex 33 metres long. Although the 750 kilograms of propellant it delivered was much less than previously – because the station would now fire its engines only to adjust its orientation – it carried 170 kilograms of water, 250 kilograms of food, 140 kilograms of film, 275 kilograms of replacement parts, and 140 kilograms of miscellaneous items (some of which were for the long-delayed Syrian mission). This was unloaded during the next week and then, while fluids were pumped in through the pipes that ran around the intervening module, the crew took a well-earned rest on 1 May to watch the Red Square celebrations on the television uplink. After several days of testing the ability of the reprogrammed attitude-control system to handle the unprecedented four-vehicle stack, they turned their attention to the environmental support systems in Kvant 1, starting them up and verifying their functionality. After Progress 29 left on 11 May, they resumed fine-tuning the ability of the Argon to manoeuvre the complex using the gyrodynes. Progress 30 arrived on 21 May. They unloaded its cargo for two days, and then set up the Yantar electron-beam spray in the scientific airlock (in the floor, between the veloergometer and the treadmill). This extended work with the Isparitel apparatus on Salyuts 6 and 7, and was tested by vaporising a copper-silver alloy to create a thin layer over a polymer film. The newly delivered high-precision star tracker was then installed in Kvant 1's

23-centimetre porthole. With this rapid build-up of the complex, the early Western criticism that Mir lacked scientific equipment now seemed rather naïve.

A NEW LAUNCHER

On the evening of 15 May, the first Energiya was launched. Despite *glasnost*, there was no live coverage of the launch. This Saturn V-class heavy-lift launcher carried its payload in a 38-metre-long cylindrical container strapped to the side of the 60-metre core segment. Unfortunately, although the rocket functioned flawlessly, the payload did not. It was not properly oriented when its orbital insertion engine fired, and so it drove itself back into the atmosphere and burned up above the Pacific. The payload was reportedly an engineering mock-up of Polyus, a platform intended to carry an industrial-scale factory for materials processing and biotechnology. It was envisaged that such platforms would either be added to a future orbital complex or operate autonomously and be visited occasionally for maintenance. Nikolai Gerasimov, head of the Salyut Bureau (part of the former Chelomei Bureau), said that Polyus could be used to deliver 40 tonnes of cargo to a station in low orbit. Dr Vladimir Pallo, the platform's designer, said that it was essentially an enlarged version of the Mir base block, with a mass of 80 tonnes. Dr Boris Gubanov, chief designer of the Energiya Bureau (formerly the Korolev Bureau), which was responsible for the new rocket's development, said that it was to be used to launch the elements of Mir 2, which was to be an elaborate facility equivalent to NASA's Space Station Freedom concept.

THE DORSAL SOLAR PANEL

For ten days in late May, when the Sun passed through the plane of Mir's orbit, the time available for the two solar panels to generate electricity was so brief that they would be unable to recharge the batteries if any power-hungry apparatus was used during the shadow pass. During this time, therefore, although plants continued to grow in the Rost and Phyton cultivators and electrophotometric and spectrometric atmospheric data were collected, little other scientific work could be performed. Then, on 6 June, with the power crisis over, the Röntgen telescopes were powered up for 'first light' trials. At the top of the list of X-ray sources was, of course, the supernova. This trial revealed that when Mir was near the South Atlantic Anomaly the X-ray detectors suffered electron precipitation.

In early June, the medics declared Laveikin fit for a spacewalk to install the new solar panel. The timing was dictated by a requirement that the three or four orbits during which they would be outside offer daylight passes over the Soviet Union for maximum communications. The spacewalk had originally been scheduled for April, immediately after Kvant 1's arrival, but concern regarding Laveikin's heart had led to its postponement to early May, then to late May, and finally to mid-June. The solar panel had been delivered in two segments, so two spacewalks would be needed. The 2-metre-diameter spheroidal multiple docking adapter was too small for the men

160 An astrophysical laboratory

in spacesuits and the bulky apparatus. However, Soyuz-TM 2 was at the front port to provide an escape route in the event that they were unable to re-enter the base block, and its orbital module was able to be used as a vestibule. On 12 June the spacecraft's internal hatch was closed to protect its descent module, the apparatus was put in the orbital module, and the multiple docking adapter was depressurised. With the upper port open, Romanenko moved out, and Laveikin retrieved the apparatus and pushed it out. They fixed the stubby cylindrical container onto the motor mount in the roof of the base block, then aligned the panel so that its arrays projected 'horizontally' out to each side. The container held an extensible girder like that tested by Kizim and Solovyov on Salyut 7. After refurbishing their suits, they went out again on 16 June and affixed the second segment to the top of the first. They extended the upper segment to draw out the panels in concertina fashion, and then repeated the procedure for the lower one to complete the 10-metre-long structure. Finally, they plugged the cables into colour-coded sockets to feed the power into the electrical system of the base block. In case they found themselves ahead of schedule, they had prepositioned several exposure cassettes in the depressurised orbital module. These were retrieved and affixed to anchors on the base block. With an area of 22 square metres, the new solar panel added just 2.5 kW to the 9 kW from the lateral panels, but 11.5 kW could simultaneously run one of the furnaces, the X-ray telescopes and the gyrodynes for stability for such observations. Upon closely monitoring Laveikin's heart after these spacewalks, the medics realised that whilst the data from the Gamma-1 monitor was able to confirm extra-systolic activity, it was insufficient for a diagnosis, and so they decided to recall him as soon as possible. Since the problem appeared only when he exerted himself, he was told to take it easy.

July saw a very varied programme of astrophysical observations; taking pictures using the Glasar telescope; additional Yantar experiments; making semiconductors in the Kristallisator furnace; electrophoresis in the Gel apparatus; making polymers in the Biostoykost apparatus; growing plants the Phyton and Rost cultivators; taking photographs of the Urals, Moldavia, the Crimea, the Pamir mountains, the Caspian depression, and the Soviet Far East; and participating in the TeleGeo-87 programme for the Intercosmos organisation. (An Earth-study programme was run each summer; this one concentrated on Poland.)

THE LONG-DELAYED SYRIAN MISSION

Progress 30 departed on 19 July, and Soyuz-TM 3 lifted off three days later with Alexander Viktorenko, Alexander Alexandrov and the Syrian, Mohammed Faris. This was Viktorenko's first flight, but Alexandrov had served a 149-day tour on Salyut 7 in 1983. They had originally expected to visit Vasyutin's crew on Salyut 7 in 1985, but a series of problems had prompted their mission being reassigned to Mir, once the first module was in place.

After a flawless automatic docking at the rear, the visitors had to use a lever to crack their hatch's hermetic seal. Faris floated through the tunnel into the Kvant

The long-delayed Syrian mission 161

Soyuz-TM 3 lifts off.

module and proceeded to give his hosts the traditional Arab greeting. Romanenko observed that with three people onboard called Alexander, things could get a little confused. It was at this point that Vladimir Shatalov publically announced that Laveikin was to return home early, emphasising that although the medical specialists had "no serious misgivings", it was prudent to relieve him, which Alexandrov would do. Although he had been given just a month's warning that his mission would be more than a brief visit, Alexandrov was well prepared as a result of his tour on Salyut 7. Deputy flight director Viktor Blagov reported that it was intended to assign cosmonauts individual missions, and partial crew exchanges such as this would be commonplace.

In addition to the standard medical checks, Faris took pictures of Syria. The need for favourable daylight passes had been the primary factor in timing the visit. One objective was to use the KATE-140 to map ancient sites. The Euphrates experiment combined visual, photographic and spectrographic observations to search for water and mineral resources, and to assess local atmospheric pollution. Among the items delivered by Soyuz-TM 3 was the Ruchei electrophoresis unit. This was installed in Kvant 1 and utilised to purify interferon and an anti-influenza vaccine. The Svetlana separated active micro-organisms for agricultural antibiotics to assist in stock rearing. A range of other typical research was also conducted. Bosra secured data to improve mathematical models of the upper layers of the atmosphere and ionosphere. Kasyun employed the Kristallisator furnace to smelt an aluminium-nickel alloy, and Afamia utilised it to crystallise gallium-antimonide. Palmyra mixed two substances to make a crystalline structure resembling that of human dental and bone tissue.

On the evening of 29 July, Laveikin

Alexander Viktorenko (left), Alexander Alexandrov and Mohammed Faris.

joined Viktorenko and Faris in Soyuz-TM 2. No sooner had they settled themselves into the cramped descent module than they were told to delay undocking for two orbits, to avoid a rain storm at the landing site. Nevertheless, the wind caught the parachute and carried the capsule off course, and it nearly came down on a small settlement. Like most guest researchers, Faris had neglected to sleep, so he was exhausted. Laveikin, although said to be "pale" after 174 days in space, was flown to Moscow to be examined by heart specialists, who pronounced him fit to fly! Laveikin was no doubt keenly aware of the irony that if it had not been for the advanced biomedical apparatus on Mir, the doctors would never have become aware of his cardiac irregularity, and he would still have been in space with Romanenko on what was intended to be a record-breaking mission.

A LONG SIX MONTHS

On 31 July, Romanenko and Alexandrov undocked Soyuz-TM 3 from the rear port, pulled back a hundred metres, waited half an hour while the gyrodynes rotated the Mir–Kvant 1 complex end over end, and then redocked at the front port. Progress 31 arrived on 6 August. After three days unloading cargo, the cosmonauts started a run of atmospheric observations in the vicinity of a power station in Kazakhstan for an ongoing study of the propagation of pollution. At the end of the month, they held an emergency drill by retreating to the ferry to don the lightweight Sokol pressure suits – which is what they were to do in the event of a slow air leak. The first week of September was devoted to maintenance. One particularly tricky operation was the replacement of a pump in the thermal regulation system – they had to be careful not to spill the glycol coolant. Often, the X-ray telescopes in Kvant 1 were operated by remote control while the crew slept – some 300 observational runs had been made to date. As a treat, they were told that they could eat the radishes and onions that they had grown. The Biryuza apparatus was used to investigate the formation of "spatial structures during a chemical-oscillation reaction". Polycrylamide was made using the Gel electrophoresis unit to make the purification of biologically active substances on Earth more efficient. Progress 31 departed on 22 September. Progress 32 arrived four days later. As it made its final approach, the flight controllers monitoring its video downlink noted that Mir's new solar panel flexed and shook as the Igla commanded small manoeuvres to keep the rear port aligned.

In late September Romanenko took the 237-day single-mission endurance record that had been set in 1984 by Kizim, Solovyov and Atkov. He was in good physical and mental health, but because he was clearly growing tired his working day was cut to 5.5 hours and he was encouraged to exploit the full sleep period – once adapted to weightlessness, cosmonauts tended to need less sleep. Accordingly, Alexandrov did the heavy work of unloading the cargo while Romanenko continued experiments. By mid-October, however, he was irritable, and was suffering from fatigue and insomnia. Because he deemed the tasks monotonous, his work load was further reduced to 4.5 hours. In this regard, Vladimir Shatalov said that it was important for the morale of the cosmonauts on long missions that the results of their experiments be processed rapidly and the analysis relayed back, so that they could be confident that they were working effectively, otherwise they risked becoming disillusioned. He said that when the service life of the Soyuz had been just a few months, the frequent Intercosmos visits had meant that results could be returned to Earth quickly, but now that a ferry could remain in orbit for six months there was no need for such frequent flights, and unless a crew received visitors it might *never* receive such feedback. And, of course, specialists on the ground were frustrated by the delay in returning the results of their experiments. A small descent capsule was under development to enable results to be returned incrementally.[2] On 4 October, the cosmonauts took part in a televised link-up with the participants of an international forum in Moscow to celebrate the 30th anniversary of the launch of Sputnik. A stock of commemorative envelopes had been delivered by Progress 32 to be franked in space on the day. Ultraviolet photography with the Glasar telescope resumed. Observations by the Röntgen telescopes revealed a change in the X-ray spectrum of the supernova. Having increased for two months, its flux now began to decrease (this was correlated with a brightening of the envelope of ejecta surrounding the stellar remnant), but it began to increase again a few weeks later. On 7 November Romanenko and Alexandrov took the day off, and watched the Red Square celebrations of the 70th anniversary of the Revolution.

Having replenished the station's fluids, Progress 32 undocked on 10 November, withdrew several kilometres, then returned two hours later and redocked to evaluate a revised approach designed to minimise the manoeuvres imposed on the complex by the Igla, and this time the flight controllers monitoring its video downlink noted that the dorsal panel remained stable. The ship departed a week later. When Progress 33 approached on 23 November, it used this new procedure, which offered the bonus of using less fuel. Alexandrov set to unloading cargo. The Mirror furnace used a pair of lamps and a mirror system to smelt at temperatures up to 1,000°C. It was to assess procedures for a similar unit in the 'technology module' that was later to be docked with Mir. An improved Mariya spectrometer was also set up to study high-energy charged particles in near-Earth space, to resume a monitoring programme initiated by Salyut 7, in the hope of correlating flux variations with tectonically active surface areas.

On 2 December, Romanenko reached the 300-day mark. Although the muscles in his legs had atrophied by up to 15 per cent, this was within "foreseen limits". Viktor Blagov said that Romanenko was "very much" missing his home, his family and his friends. As he had late in his 96-day marathon on Salyut 6 in 1978, he had become increasingly grumpy. In fact, he was now so tired that almost all the work was being done by Alexandrov. Blagov explained that while it was planned for cosmonauts to generally serve six-month tours on Mir, it was also necessary to study adaptation to weightlessness over much longer periods. Romanenko later opined that the optimum length for a tour of duty would be three to four months.

---

[2] Recall that the Almaz reconnaissance platform was able to dispatch a small re-entry capsule.

HANDOVER

Progress 33 departed on 19 December, and four days later Soyuz-TM 4 docked with Vladimir Titov, Musa Manarov and Anatoli Levchenko. When Soyuz-T 8 had been launched in 1983, the jettisoned shroud had torn off the spacecraft's Igla antenna.

Vladimir Titov (left) and Musa Manarov.

Titov had performed a manual rendezvous with the Salyut 7–Cosmos 1443 complex, breaking off the final approach only when he entered the Earth's shadow and lost his bearings. Later in that year, during the final seconds of the countdown for his second attempt to reach that station, a fire broke out and he and Gennadi Strekalov survived only because the escape tower drew the capsule from the maelstrom of the exploding rocket. Titov had been scheduled to fly to Mir with Alexander Serebrov (with whom he had backed up Romanenko and Laveikin) and Valeri Poliakov, a physician who was to have assessed Romanenko's state of health prior to his return to Earth. This plan had been changed initially by the substitution of Manarov for Serebrov as flight engineer, and again at the last minute when Levchenko was assigned Poliakov's seat because the Ministry of Aviation had insisted that a Buran pilot trainee be flown as soon as possible in order to gain experience of weightlessness, as Igor Volk (leader of the Buran group) had done in 1984.[3] When Levchenko floated in through the hatch, Romanenko had to ask him who he was because he had never seen him before. Over the next week, Romanenko and Alexandrov briefed their successors on the status of Mir's systems. An orbital handover saved about a week that

Alexander Alexandrov (left), Anatoli Levchenko, Vladimir Titov and Musa Manarov.

---

[3] Unlike most military cosmonauts, the group selected to fly the Buran space shuttle were all highly experienced test-pilots. Unfortunately, Levchenko died in August 1988 following surgery to remove a brain tumour. Alexander Shchukin (his backup) was killed in an air accident a fortnight later, and Ural Sultanov was assigned as Volk's new co-pilot for the eagerly awaited Buran mission.

Musa Manarov watches as Anatoli Levchenko uses a straw to drink water from a spherical tank.

would otherwise have been wasted in powering down and then reactivating the complex. As more and more apparatus was delivered, it also became important for the retiring crew to show the next where everything was stored. And, as the complex was expanded and serviced, a handover enabled the newcomers to be briefed on the quirks of the systems as they *really* were – as no simulator was perfect. During the handover, a small aquarium and the Ainur electrophoresis apparatus for purifying proteins were installed, interferon was purified in the Ruchei electrophoresis apparatus, and some animal and vegetable tissue cultures were grown for return to Earth.

On 27 December, Romanenko and Alexandrov stowed their experimental results in Soyuz-TM 3's descent module and loaded trash into its orbital module. Their haul included 270 ultraviolet photographs by the Glasar telescope, samples processed in the Korund-1M furnace, Earth photographs and biological samples. Their last day in space was hectic, with the two crews sorting out their apparatus. At one point, one of them remarked that he would make faster progress if the flight controllers would just desist from trying to remind him of everything. When Soyuz-TM 3 departed on 29 December, it carried Romanenko, Alexandrov and Levchenko. The fact that they had never trained together was remarkable confirmation of Blagov's statement that cosmonauts would soon begin flying individual missions. A television camera in one of the recovery helicopters showed the capsule descending into the blizzard that was sweeping the frozen steppe. It was tipped onto its side when the deflating parachute caught in the wind. The wind prevented the erection of the inflatable medical tent, so it was decided to airlift Romanenko and Alexandrov to the nearby town of Arkalyk. Despite 160 days in space, Alexandrov walked unaided to the helicopter. Although Romanenko said that he would walk, the medical team insisted that he recline on a stretcher because he had been away 326 days, which was three months longer than anyone else. Nevertheless, the next day he was shown on television walking with his wife in the grounds of the Cosmonaut Hotel at the cosmodrome. In light of his rapid deterioration after about 8 months in space, it was concluded that 'routine' tours of duty should be scheduled for four to six months. Titov and Manarov however, were hoping to break Romanenko's new record. The fact that neither had spent time on an earlier station suggested that Romanenko's cumulative record of 431 days in space would be safe for some time to come.

166 **An astrophysical laboratory**

After 326 days in space, Yuri Romanenko walks with Alexander Alexandrov (left) and Anatoli Levchenko (right).

A YEAR IN SPACE?

Early on 31 December, as they became the first crew since Romanenko and Grechko a decade ago (on Salyut 6) to welcome in the New Year in orbit, Titov and Manarov transferred Soyuz-TM 4 to the front port. The following day, a two-way video link was set up with their families to celebrate Titov's birthday. Although they were not expecting any visitors until their ferry was replaced in the summer, they had a busy tour of duty ahead, with a number of spacewalks and the commissioning of the next expansion module. The first half of January was spent on astrophysical studies and smelting semiconductors. When Progress 34 docked on 23 January 1988 it brought propellants, more consumables, the usual replacement items for the environmental systems, a replacement segment for the upper solar panel, and specifically requested music cassettes. Once this cargo had been unloaded, Titov and Manarov had a varied programme for the remainder of the month: they studied thermocapillary processes with the Pion-M apparatus, performed the ERI experiment (to evaluate methods of depositing galvanic coatings) and took ultraviolet pictures with the Glasar telescope. The semi-automated Korund-1M furnace was in operation throughout. The remote-control astrophysical observations concentrated on the supernova. (Its flux had been increasing since November, and it was now 50 per cent greater than it had been at its peak in August, and it had been proposed that these variations might be due to the development of holes in the expanding ejecta which exposed the hot stellar remnant.) In mid-February, the video uplink was used to brief them on the procedures for their first spacewalk, and they refurbished the suits while Progress 34 replenished fluids.

On 26 February, Manarov, eager to start the spacewalk, floated out of one of the vacant radial docking ports prior to establishing contact with Kaliningrad. Once their new apparatus had been attached to the anchor point near the dorsal solar panel to prevent it floating away, they retracted the lower section and replaced one side of it with a new panel incorporating improved transducers, for a long-term assessment of the efficacy of different types of photoelectric semiconductor material and the rate at which they degraded in the space environment. In fact, only six of the new array's eight segments were to feed power for the station; the other two were instrumented to produce telemetry to enable their performance to be monitored by engineers. The new panel segments were smaller than those they replaced, but produced 20 per cent more power (the standard silicon-based transducer converted 11 per cent of sunlight into electricity, and this new gallium arsenide transducer had an efficiency of 14 per cent) and the power supply was actually increased slightly despite the fact that two of the segments were passive. The video downlink showed the two men working by the illumination of lamps while in the Earth's shadow. The main task complete, they collected more equipment and made their way along the length of the base block, deployed a handrail to form a bridge to Kvant 1, moved across and continued to the rear, where they set up several experiment packages and visually inspected Progress 34. On their way back, they photographed the surface of the complex to start long-term monitoring of its exposure to the space environment.

Progress 34 left on 2 March. During the next week, Titov and Manarov used the Mariya and Pion-M apparatus, and processed materials in the Mirror furnace. Earth studies were then resumed to assess the coverage of snow and glaciers in the Pamir and Tien-Shan mountain ranges for an ongoing assessment of irrigation in the Central Asian Republics. The X-ray telescopes continued to be used under remote control, most often for the supernova. They also installed a fax machine to permit sketches of their observations to be sent to scientists on the ground. The environment in the complex was itself a subject for study. A part of this was the Akustika experiment. This measured background noise at various points (the hum from the environmental systems was typically 80 decibels). The cosmonauts reported that there was a lot of dust adrift in the station (it made them sneeze), and they complained of an irritating smell from the environmental system. It was reported in mid-March that although they were following their programme, their working efficiency was having its "ups and downs". Progress 35 arrived on 26 March. In addition to the usual parts for the environmental and thermal regulation systems, it had 400 kilograms of food (much of it fresh fruit and vegetables, which were particularly welcome). After the cargo was unloaded, routine maintenance was interleaved with studying the atmosphere for the Climate experiment using the EFO-1 spectrometer and investigating the formation of crystals in the Kristallisator furnace.

On 12 April, marking Cosmonaut Day, Viktor Blagov reported that the first of the modules to be attached to the front of the complex was being prepared for launch "at the end of the year". This would incorporate a large airlock, in which would be an autonomous manoeuvring unit. The second module (devoted to astrophysics) was to be launched six months later. The third (materials processing) might be operated as a free-flyer in order to achieve ultra-pure microgravity conditions, and dock only to be serviced. The final module, he said, was to be for life sciences, and would have a sterile compartment. When finished, the complex would have a mass of more than 100 tonnes. For a variety of reasons, this plan to complete the complex at the end of the decade was to prove overly optimistic. In retrospect, considering the fast pace of this early schedule, and the fact that the construction of the modules had *begun*, it is astonishing that their eventual rôles would be so different from those outlined at this time. Later in April, Titov and Manarov complained of receiving conflicting requests for their time from different groups of researchers. Nevertheless, they worked with the Pion-M apparatus, made Earth-resources studies of Cuba, and took ultraviolet pictures using the Glasar telescope. Progress 35 left on 5 May, and Progress 36 took its place on 15 May. *Radio Moscow* reported that the cargo included a new detector for one of the X-ray telescopes, together with the specialised tools to install it. The TTM had been suffering intermittent electrical interference since late 1987. Because the apparatus in the unpressurised compartment at the rear of Kvant 1 had not been intended to be serviced, simply gaining access to the instrument would be tricky. A video had been sent up to show them how to replace the detector. The spacewalk to attempt this task was scheduled for later in the month. In the second half of May they finished unloading cargo, and installed "a considerable amount" of the recently delivered apparatus in preparation for the imminent Intercosmos mission, in order to ensure that their first visitors would be able to make an immediate start on their programme. Progress 36 replenished the station's fluids, refined its orbit and left. If Titov and Manarov had been on a six-month tour, by this point they would have been preparing to return, but they were in such good spirits that they had been given the go-ahead to try to set a new record.

BULGARIAN REFLIGHT

Soyuz-TM 5 was launched on 7 June with Anatoli Solovyov (not to be confused with Vladimir Solovyov), Viktor Savinykh, and Alexander Alexandrov (a Bulgarian research cosmonaut, not to be confused with the Soviet cosmonaut of that name). This was the second Bulgarian mission. In 1979 Alexandrov had backed up Georgi Ivanov, who had been prevented from reaching Salyut 6 when Soyuz 33's engine misfired. As that was the only Intercosmos mission to be aborted, this Mir visit had been laid on as a consolation. The primary purpose of the mission, however, was to exchange the expiring ferry. The flight had initially been scheduled a fortnight later to give the residents time to attempt to repair the TTM, but had been brought forward because the Rozhen astronomical experiment to be performed by the visitors had to be timed to avoid the illumination of a full Moon. The automatic Kurs system made a fly-around from the front to the rear of the complex, and initiated its approach, but when it deviated from the straight-in path Solovyov took over and withdrew to give the flight controllers time to analyse the telemetry. One orbit later, after the fault had been identified and overcome, the approach was resumed, this time successfully. If Soyuz-TM 5 had been unable to dock, Titov and Manarov would have been recalled, leaving Mir vacant.

An ambitious programme was planned with 46 experiments addressing remote-sensing, biomedical studies, space physics and materials processing for which over a tonne of new apparatus had been ferried up, including the Zora portable computer

Anatoli Solovyov (top), Viktor Savinykh and Bulgarian guest researcher Alexander Alexandrov.

to process the data in orbit to enable the cosmonauts to evaluate their results as they worked. The biomedical programme focused on physiology, psycho-physiology and radio-biology: Potential studied the interaction of the nervous and muscular systems; Prognoz studied changes in an individual's operational performance; Stratokinetika studied the body's movement; Pleven-87 used a battery of 15 psychological tests to investigate locomotor functions and volition processes, with the data processed by the Zora computer; Dosug assessed the influence of music, video and games on crew morale while off duty; Lyulin used a microprocessor to evaluate psychological and physiological analysis of the visitors' reactions, in particular of their reaction time; and a long-duration cassette tape gathered electrophysiological data while they slept for the Son experiment. In addition, Doza-B involved placing biological samples and radiation sensors throughout the complex to measure exposure levels.

The Earth-study programme involved surveying Bulgarian territory, but this was hindered by cloud cover. The Bulgarians had built the Spektr-256 (an improved form of the Spektr-15 spectrometer) to study the atmosphere, and its data was processed by the Zora computer. The Bulgarian-built Rozhen was a digitally processed electro-optical telescope employing the Parallax-Zagorka image intensifier and the Therma photometer. This was being evaluated for an autonomous telescope. Although it was primarily for studying celestial sources, in this case it was also to be used to observe aurorae and luminescence in the upper atmosphere. Its data was processed in orbit by the Zora computer and then downlinked. There were several materials-processing experiments: Voal produced an alloy of wolfram (a form of tungsten) and aluminium; Klimet-Rubidium made use of the Kristallisator to test making extremely lightweight batteries and condensers using a mixture of rubidium, silver and iodine (because this experiment consumed so much power, it was done while the cosmonauts slept); the Ruchei electrophoresis apparatus purified genetically engineered interferon; and an Australian experiment was set up to investigate the production of vaccines (of which more later). In addition, seeds of wheat, arabidopsis and ginseng were all planted in the Magnetobiostat.

On 16 June the visitors loaded 30 kilograms of material (including a large number of disks for the Zora computer) into Soyuz-TM 4, and left the next day. Savinykh later opined that the programme had been "a bit too much" for a week-long visit.

REPAIRING THE TELESCOPE

On 18 June, while the landing window was open, Titov and Manarov transferred Soyuz-TM 5 to ensure that it would be accessible from the multiple docking adapter in the event that they were unable to re-enter the base block after their spacewalk to repair the X-ray telescope, which had been set for 30 June.

The detector was a 40-kilogram cylinder, some 40 centimetres in diameter. It had been fitted with handholds. They manoeuvred it, together with the specialised tools they would need, through one of the vacant radial docking ports, along the length of base block, and across the bridge onto Kvant 1. A handrail on the side of the module led them to the unpressurised compartment, but thereafter their task became difficult because there were no restraints, with the result that they slipped behind schedule. Nevertheless, they managed to cut through the 20-layer thermal blanket protecting the telescope. Unfortunately, the 2.5-metre-long instrument had not been meant to be serviced by engineers wearing space gloves, and the small bolts proved extremely difficult to release. Although the procedure worked out during the hydrotank simulations had assigned 20 minutes to this task, it actually took an hour and a half. Hoping to catch up, they were shocked to discover that the apparatus before them incorporated several embellishments that had not been on the development unit used in training. Undaunted, they sawed through a number of bolts, wiped away an unexpected resin deposit, unfastened screws, and cut stainless steel clips to expose the detector, but when they inserted a special tool to release the brass clamp holding it in place, the tip sheared off the tool. Frustrated, they put a new blanket of thermal insulation over the hole to protect the apparatus and returned to the airlock with the new detector. On finally sealing the docking port, the two men had been out for over five hours and were exhausted. When back in communications range, they reported that they had broken the tool and abandoned the operation. The British and Dutch astronomers who had built the telescope were at Kaliningrad as advisers. Although "very disappointed" by this news, they were pleased that the cosmonauts had been able to gain access to the instrument. The machinists immediately set about making a stronger tool, to be sent up on the next cargo ferry.

BACK TO THE ROUTINE

The rear hatch and integrated life-support backpack of the Orlan extravehicular suit.

After cleaning their spacesuits and taking a few days off to rest, Titov and Manarov celebrated their 200th day in space on 8 July. Medical tests showed them to be in good health, so it seemed likely that they would be able to achieve their objective of a year in space. The next week was spent on Earth-resources observations to locate mineral deposits in the Soviet Union. Progress 37 docked at the rear port on 21 July. In addition to the usual consumables and replacement equipment, it delivered a pair of Orlan-DM extravehicular suits (these contained their own power and communications facilities, so did not require an umbilical link to the base block), a new colour video monitor, a new computer, apparatus for the next visitors, and new tools for repairing the telescope. Unloading was followed by several days of routine maintenance. The rest of the month was devoted to Earth studies. They reported on forest fires across the Soviet Far East, the Urals and southern Siberia, often being the first to spot such fires, and in many cases their information enabled the firefighting teams to reach the site before the fire became too serious. After replenishing Mir's fluids and refining its orbit, Progress 37 departed on 12 August. Three days later, Titov and Manarov passed the 237-day mark; Romanenko's 326-day single-mission record was in sight. The focus of the next few weeks was photographing Tadjikistan and Kirghizistan for the Tien-Shan-88 Intercosmos programme.

AN AFGHAN VISITOR

On 31 August, Soyuz-TM 6 docked at the rear with an *ad hoc* crew, hastily thrown together by bureaucratic and political factors. Vladimir Lyakhov had spent 175 days on Salyut 6 in 1979 and 149 days on Salyut 7 in 1983. Although Valeri Poliakov had been assigned to Soyuz-TM 4 to perform a complete medical check on Romanenko immediately prior to his to return to Earth from his record-breaking flight, his couch

172 An astrophysical laboratory

had been reassigned to a Buran trainee. Poliakov was the deputy head of Moscow's Institute of Medical Biology. It was at this point in his mission that Romanenko had begun to grow noticeably weaker. The insertion of this flight into the schedule gave Poliakov the chance to check on the current residents as had been intended in Romanenko's case, thereby satisfying the previously stated goal of having a doctor oversee an attempt to break the record. Rather than make only a brief 'house call', he was to remain onboard with Titov and Manarov and

Abdul Ahad Mohmand (left), Vladimir Lyakhov and Valeri Poliakov.

had the authority to curtail the flight if their physical or mental state deteriorated. As a bonus, he would be able to monitor his own adaptation to weightlessness. Furthermore, as the next handover was to be accompanied by a French visitor, Poliakov would not be able to return to Earth with his charges; he would have to remain onboard with their successors, and possibly try for the record himself. The reason for these hasty arrangements was the third crew member of the Soyuz. Abdul Ahad Mohmand was an Afghan. He had started out as backup to Mohammad Dauran, but Dauran had lost time in training through having his appendix removed. Earlier in the year, Glavcosmos chairman Alexander Dunayev had reported that because the mission had to be flown before Soviet military forces completed their withdrawal from Afghanistan, it would be inserted into the schedule prior to the handover later in the year. Although Mohmand received only six months of training (two years was more usual), this was greatly assisted by the fact that he could already speak Russian. In fact, Titov and Manarov were onboard Mir when he entered training, and the first time they met him was when he floated in through the hatch.

The decision to fly early had pre-empted the development of specifically Afghan apparatus, so 24 experiments had been devised using existing equipment. The main objective was to use the KATE-140 to photograph Afghanistan in order to map the mountainous regions, assess water and glacial run-off, and identify possible sources of oil and gas in lowland areas. This orbital survey would be significant because most of the country was remote and inaccessible (less than 30 per cent had been surveyed by conventional methods). The MKS-M and Spektr-256 spectrometers contributed to the ongoing Biosfera project and the EFO-1 provided data for the Climate project. In addition to the usual tests to monitor the initial phase of adaptation, Mohmand performed a vestibular experiment left over from the Bulgarian mission. While he and Poliakov slept, the Son apparatus recorded the electrical activity of their brains. Fish were delivered for the recently added aquarium, to evaluate a closed ecosystem. The Ruchei electrophoresis apparatus was used to make interferon (this was proving to be a very productive experiment because visitors could rush home with the product). Poliakov's programme had involved sampling his bone marrow (reportedly a painful process) prior to launch; this was to be compared with another sample immediately upon his return. On realising that he had readily adapted to weightlessness, he was reportedly "somewhat disappointed" not to have been able to experience for himself the unpleasant symptoms that upset some people.

On 5 September, Lyakhov prepared Soyuz-TM 5 for departure, even though it had been in space only a few months, on the principle that it was better to leave the residents with the most recent ferry. It was loaded with accumulated film, disks and other results from the main programme, including the ampoule from the Australian experiment which had crystallised an influenza virus antigen. (Such crystals grown in microgravity were more regular than could be produced on Earth, and better suited to analysis by X-ray diffraction to determine their structure – which was the first step towards developing a pharmaceutical product.) Compared to the hectic Bulgarian visit, Titov and Manarov had found this visit a welcome relief.

STRANDED IN ORBIT?

Lyakhov and Mohmand departed Mir in Soyuz-TM 5 on 6 September. The orbital module was jettisoned according to plan; then, as they ran through the final steps of the retrofire sequence, the navigational computer received conflicting signals from the primary and the backup infrared horizon sensors resulting from solar glare as the spacecraft passed over the terminator into sunlight. The automated control system required the sensors to confirm the appropriate alignment 30 seconds prior to firing the engine for the de-orbit burn, so the conflicting signals inhibited the manoeuvre. For years, recovery window constraints had required the spacecraft to cross into sunlight at least ten minutes beforehand in order to provide sufficient time for these sensors to verify the ship's orientation and stability. However, despite the increased flexibility of the latest model of the ferry, passing through orbital dawn at the critical moment had upset the system. Seven minutes later, while Lyakhov was interrogating the system to determine what was wrong, the sensor conflict cleared up and finally asserted the signal that confirmed that the spacecraft was correctly oriented, and the still-waiting computer fired the engine. However, the spacecraft was now far beyond the planned de-orbit point and if the manoeuvre was completed it would result in a descent thousands of kilometres downrange, which would make recovery difficult, so Lyakhov curtailed the manoeuvre. It was evidently a minor glitch – the engine was functional, and whatever had confused the sensors was unlikely to occur again. The manoeuvre was rescheduled for two orbits later. The engine ignited on time, but six seconds into the 230-second burn it shut off. Lyakhov immediately restarted it, but 50 seconds later the computer decided that the vehicle was out of alignment and shut it off again. The situation had suddenly deteriorated significantly. Lyakhov was later criticised for restarting it, but he had undoubtedly been aware that his ground track had drifted too far to try again on the next orbit. Flight director Valeri Ryumin decided that they would have to remain in space until the Earth had completed one revolution. Overnight analysis revealed that on the second attempt the computer had tried to execute the incorrect program – for some reason it had selected part of the rendezvous sequence that it had followed when taking the Bulgarian cosmonaut up to Mir. Having established that the engine had not really malfunctioned, the tension in Kaliningrad lifted.

In space, however, the situation was dire. Having jettisoned the orbital module, the capsule was very cramped for two fully-suited figures. It was little consolation knowing that this would have been much worse if the third couch had been occupied. There was no point trying to return to Mir, because the Kurs antenna and docking unit had been discarded with the orbital module. The descent module had not been designed for prolonged independent operation. Its temperature dropped to a chilly 10°C, which made the men glad they were wearing Sokol pressure suits. Lyakhov was not concerned at the lack of food (they had only dried rations for three days of basic survival following an emergency landing in a remote area) but he lamented the fact that the toilet had gone with the orbital module. (He recommended that in future the module should not be jettisoned until after the de-orbit burn had been completed, as in "the good old days", even although doing so would consume extra propellant.) The ship was oriented so that its solar panels faced the Sun to sustain its electrical systems, and placed into a slow rotation for stability.

Vladimir Dzhanibekov opined that this predicament was the result of a "blunting of vigilance". Viktor Blagov blamed it on "a combination of circumstances". Strictly speaking, however, the basic failure was the decision to make an early morning descent rather than the late afternoon descent that had been standard practice for so many years, as this necessitated the de-orbit burn at orbital dawn. The operational envelope had been stretched too far, and Lyakhov and Mohmand were paying the price. The Western media reported that they were "stranded" and running out of air, but the situation was manageable. As a result of the various manoeuvres (intended and unintentional) the ship had already descended from Mir's 338- by 364-kilometre orbit to one between 250 and 341 kilometres, and so it was well placed to complete the manoeuvre a day late and land at the planned recovery site. A revised computer program was read up to Lyakhov, and entered manually. Just to be certain however, the backup engine was to be employed. Television cameras recorded the parachute descent in pre-dawn twilight. In minutes, the two cosmonauts scrambled out of the capsule to greet the recovery team. Despite the apparent maturity of the technology, spaceflight was still a potentially lethal business.

BACK TO REPAIRING THE TELESCOPE

On 8 September, Titov, Manarov and Poliakov transferred Soyuz-TM 6 to the front port to make way for Progress 38, which arrived four days later. In addition to the usual stores, it delivered various apparatus for the impending French visit and a ham radio transceiver (this would serve as a backup in case the communications system failed). For the next several weeks, materials-processing was interleaved with visual observations of plankton and forestry. On 18 October Titov and Manarov began to prepare to finish the repair of the X-ray telescope by unpacking the new Orlan-DM spacesuits. These comprised an aluminium alloy body section with elasticated arms and legs for greater flexibility. The oxygen tanks and the lithium hydroxide scrubber that absorbed exhaled carbon dioxide could support external activity for seven hours in addition to several hours of pre/post-activity in the airlock (the previous suits had provided five hours, plus a reserve). Another enhancement was a set of biomedical sensors to provide continuous telemetry to enable the flight surgeon to monitor the cosmonaut's health. They did not require a communications and power umbilical, as the backpack had a power supply and radio transmitters for voice and telemetry but, as a safety measure, they were to remain tied to the complex by a nylon tether. The total freedom of action offered by these suits was to permit spacewalkers to use the autonomous manoeuvring unit that was to be delivered in the first of the large modules. (Titov and Manarov had hoped to commission this, but its launch had been postponed.)

Two days later, Poliakov retreated into Soyuz-TM 6 so that he would be safe in the event that it proved impossible to repressurise the multiple docking adapter after the spacewalk, and Titov and Manarov emerged from one of the vacant radial ports. As they made their way to the rear of the aft module, once again dragging the new detector, they ventured that the handholds were spaced too far apart. The downlink was shown live by domestic television. At the work site, they reopened the thermal blanket and readily unseated the clamp using the replacement tool. Working rapidly, they extracted the faulty detector, inserted the new one, and affixed a simple clamp to lock it into position. By the time they replaced the thermal insulation, they were an hour ahead of schedule, so on their way back they employed a fine brush to clean accumulated dust off the glass of two portholes and then erected the antenna for the ham radio and affixed an anchor point outside the docking adapter for an experiment to be set up during the French visit. Having achieved all their tasks, they closed the docking port with plenty of time to spare. The servicing of the X-ray telescope was a remarkable demonstration of real-time mission planning, engineering improvisation, and the prowess of the spacewalkers. To the delight of the astronomers, it was soon back in operation.

GOING FOR THE RECORD

During late October, the Mariya spectrometer collected data on charged particles in the near-Earth environment. Materials processing was resumed in early November, with aluminium and copper alloys and monocrystals of zinc being produced by the Mirror and Pion-M apparatus. By this time, Poliakov had Titov and Manarov on a "particularly rigorous" exercise regime. They were now regularly wearing the Tchibis for extended periods to increase cardiovascular capacity, and they began to consume salt water and other additives to build up body fluids. Dr Anatoli Grigoriev, director of the Institute of Medical Biology, noted that they seemed to have suffered none of the deterioration that had afflicted Romanenko in his final months. On 12 November, Titov and Manarov finally broke the 326-day record, and Poliakov said that he was content for them to attempt "a year and a day".

A FRENCH VISITOR

Progress 38 left on 23 November. Soyuz-TM 7 took its place on 28 November to deliver Alexander Volkov, Sergei Krikalev and Frenchman Jean-Loup Chrétien. The launch had been slipped by 5 days in order to enable President François Mitterrand to attend. Volkov had been a member of the Soyuz-T 14 crew that had been recalled from Salyut 7 when Vasyutin had fallen ill. This was Krikalev's first flight. Chrétien became the first guest researcher to be granted a second flight – he had spent a week on Salyut 7 in 1982. Titov and Manarov laid on a feast of jellied salmon, quail meat and candied fruit for the newcomers, who had in turn brought a range of specialities including vegetable soups, fish, ham, paté, and a range of cheeses and dressings. With six people onboard for the first time, the base block

Sergei Krikalev (left), Jean-Loup Chrétien and Alexander Volkov.

was a little cramped, especially since a month-long handover had been scheduled to permit Chrétien to undertake a much more ambitious programme than was usual for a visitor. He was to return with Titov and Manarov, which meant that Poliakov would have to stay on with Volkov and Krikalev.

Chrétien's researches included ten French experiments, involving both medical and technological topics. About 580 kilograms of equipment had been ferried up (in Progress 37 and Progress 38) specifically for this work. Of the medical experiments, Superpocket measured the neurosensory system and the reconditioning of postural reflexes; Physalie evaluated the coordination between the body's sensory and motor systems; Kinesigraph produced stereoscopic images to investigate the restitution of movement of the corporal segment; Circe monitored ionising radiation, gamma-ray and neutron dosages in the complex; a sophisticated X-ray scanner had been used prior to launch (and would be used again following his return) to measure calcium loss in his bones; and he was to return blood samples for analysis. In addition, an improved version of the Echograph

Jean-Loup Chrétien.

ultrasonic scanner was to measure blood flow in deep vessels (in particular, the main truncus and venous return), and to measure the capacity of the heart and other internal organs. The Echograph's monitor screen was to be used for the Viminal experiment in which a hand-operated indicator had to be keyed in response to visual cues to test visual acuity. The Ercos experiment was a cassette to be attached to the inside of the station's wall to test the extent to which high-density computer memory chips degraded by being exposed to cosmic rays (it was to be returned after six months). The main item on the technology programme, however, was the self-deploying Era truss, which was to be done outside, for which Chrétien was to become the first guest researcher to make a spacewalk. This had been scheduled for 9 December, then changed to 12 December by the slip in the launch. Having rapidly adapted to weightlessness, he decided to proceed as planned in order to exploit the favourable lighting conditions that had prompted the original date. (In doing so, he opened the possibility of making a second excursion if he was unable to complete his work.) He and Volkov devoted 8 December to preparing the Orlan-DM spacesuits.

After Chrétien exited one of the radial ports, Volkov pushed out the segmented Echantillon cassette – the Comes segment of which exposed small samples of paints, reflectors, adhesives, optical materials and filament-reinforced composites; MCAL was to determine how the absorptivity and emissivity of materials changed over time; and Mapol exposed polymeric materials under evaluation for creating inflatable structures. Echantillon also incorporated two dust traps (one active and one passive, called DIC and DMC respectively) to collect cosmic dust for later analysis. Chrétien mounted this 0.75-metre square box to a convenient anchor close by. (Intended to be left out for six months, in the event it was to be over a year before it was retrieved.) This done, Volkov passed out the Era experiment. The 240-kilogram main package (a 0.6-metre-diameter stack of 1-metre-long carbon fibre rods) was accompanied by a support platform, a video system, and a control panel. After the platform had been mounted on the anchor that had been installed by Titov and Manarov, the stack was affixed to an arm on the platform set at an angle of 45 degrees to keep it away from the surface of the base block. Once a 50-pin umbilical had been connected, Krikalev, inside,

How Jean-Loup Chrétien deployed the Era structure.

triggered the control system. The articulated pin-jointed rods were supposed to spring open automatically over a four-second period, to create 24 identical prisms to form a thick hexagon some 4 metres in diameter. Unfortunately, it did not budge. Although Chrétien nudged it in an attempt to shake it loose, it refused to deploy. At that point, the complex flew out of communications range with Kaliningrad. Rather than jettison the experiment simply in order to stick to the schedule, the cosmonauts decided to continue their efforts to deploy it. Several violent kicks by Volkov's boot did the trick. When the structure had unfolded, a set of accelerometers were attached to determine its vibration modes. Upon flying back into communications range, they replayed a video of the deployment. It was concluded that the structure had been locked in its folded position by water vapour that froze when exposed to vacuum. A hermetically sealed container would have precluded this. After the experiment, the structure was jettisoned. The objective of the experiment was to evaluate one of the options being considered by the French Space Agency for erecting antennas in space. The spacewalk had been scheduled to last for four hours, but the extra time required to complete the Era deployment stretched this to six hours, which was a new record. Despite improvements in the Orlan's environmental control system, by the time he was finished Chrétien noted that he was drenched in sweat, and the inner face of his visor was laced with condensation. Immediately afterwards, Poliakov gave Volkov and Chrétien a thorough medical examination.

Chrétien's next deployment experiment was conducted inside. Armadeus was to test the articulated mechanism for a solar panel using four motorised winding blades incorporating Carpentier joints to eliminate friction. He deployed it three times, with a pair of television cameras documenting it so that the kinematics could be analysed. The Physalie experiment involved filming himself perform a range of movements in order to provide biomedical data. After spending over two hours setting up the two cameras and a cluster of monitors, and attaching sensors on his neck, torso and legs, he vowed that if he got the chance to make a second spacewalk he would jettison the apparatus into space. It was a case of an experiment that seemed straightforward in the laboratory, but proved awkward in space.

In mid-December, the residents extensively photographed the region of Armenia that had recently suffered an earthquake in which 100,000 people lost their lives. Throughout the joint mission, the X-ray telescopes were operated remotely (the flux of hard X-rays from the supernova had decreased considerably of late). Titov and Manarov devoted an increasing amount of time to the exercise regime, and spent long periods in the Tchibis. On 15 December, they exceeded Romanenko's record by 10 per cent (the margin required by the International Astronautics Federation to make a new record official). It was therefore a happy pair who loaded experimental results into Soyuz-TM 6.

Titov, Manarov and Chrétien undocked on 21 December. As they waited for the de-orbit manoeuvre, the spacecraft's computer became overloaded and cancelled the burn. The problem was in a new software routine, so another attempt was scheduled for two orbits later utilising a backup program. In light of the hardships suffered by Lyakhov and Mohmand on Soyuz-TM 5, the practice of retaining the orbital module until after retrofire had been reinstated. This time the computer functioned perfectly.

About ten minutes after the completion of the de-orbit burn, the orbital module was jettisoned, followed by the service module a minute later. A television camera in a recovery helicopter caught the capsule's descent into a thick bank of low cloud. For once, it settled upright, and it was a struggle for the record-breakers to climb up to the hatch. The recovery team assisted the cosmonauts out, and then down the slide to the ground and over to the reclining couches. The frozen steppe was a shock after Mir's regulated environment. After 366 days in space, Titov and Manarov were in "good health". Titov had lost a little weight, but Manarov had put on 3 kilograms – gaining weight in space was unusual. Both were in better condition than Romanenko had been on his return. As there was an infectious outbreak at the Cosmonaut Hotel at the cosmodrome they were flown directly to Kaliningrad. Although the muscles in their lower legs had atrophied through lack of use, this was found to be due to loss of intramuscular fluid rather than deterioration of the tissue, and was easily restored; the bone calcium loss proved no worse than suffered by shorter-duration crews. A Tchibis configured as a lower-body positive pressure suit was used to prevent blood draining from the torso and, just as immersion helped in training for weightlessness, swimming assisted in adapting to its absence. Within days, Titov and Manarov were strolling around the space centre. With the endurance record now set at a year, it was clear that the task of breaking it could be assigned only to an extremely motivated individual serving with successive crews.

Musa Manarov after 366 days in space.

FRUSTRATED HOPES OF EXPANSION

Soyuz-TM 7 was moved to the front port on 22 December, and Progress 39 docked at the rear four days later. While Poliakov continued to observe his own adaptation, it was time to resume the assembly of the orbital complex. The schedule supplied by

Alexander Dunayev earlier in the year had called for the docking of the first module prior to Volkov and Krikalev taking over, but this had been delayed. When they took up residence, they hoped to receive the next *two* modules. Within weeks, however, Dunayev had to announce that the launch of the first module had been postponed again. Another change concerned foreign researchers. To date, these had been funded by the Soviet Union in the interests of fraternal cooperation, but there were to be no more 'free rides', foreigners would be accommodated only on payment of a *fee*, this being calculated to defray the cost of their participation in the programme (both the Soyuz and any Progress cargo shipments). In the years to come, this decision was to have far-reaching consequences.

To mark the start of 1989, Volkov, Krikalev and Poliakov opened the gifts which had been delivered by Progress 39, and then resumed a full research programme that included Earth-resources observations of Siberia and the Soviet Far East, using the Parallax-Zagorka image intensifier to study the vertical distribution of luminescence at polar, middle and equatorial latitudes, and of the luminescence generated by the complex itself as it passed through the ionosphere at orbital altitude; and using the Glasar ultraviolet telescope. Poliakov monitored the adaptation of the cardiovascular systems of his colleagues using the Reflotron apparatus supplied by West Germany, which, by providing instant blood analysis, enabled him to monitor changes in blood chemistry in real time – previously, samples had been collected and returned for later analysis. Given a drop of blood, it would measure a wide range of parameters, report on haemoglobin, cholesterol, uric acid and glucose, and then analyse and store the results, which Poliakov periodically downlinked to his colleagues at the Institute of Medical Biology. On 23 January, Volkov and Krikalev began several days of routine maintenance on the environmental and thermal regulation systems using components delivered by Progress 39. They then used a holographic imager to record damage to the portholes caused by cosmic dust. At the end of the month, they used the Yantar apparatus to apply coatings of alloys of silver-palladium and tungsten-aluminium to samples of polymer film. When Mir began to make favourable daylight passes of the Soviet Union in early February, observations were renewed to supplement data from automated satellites. Progress 39 left on 7 February. Progress 40 took its place five days later. Amongst the food, the cosmonauts found treats such as pickled cucumber and sweet-smelling honey. The Diagramma experiment was carried out to measure the physical characteristics of the gaseous environment at orbital altitude. Measuring the flow around the complex would help assess the aerodynamic drag that caused the orbit to decay. This was done using a magnetic discharge transducer on a short boom that was poked out from the scientific airlock.

In early February it had been reported that spacewalks planned to affix a pair of star trackers to assist in orienting Mir after it had been expanded had been cancelled due to a further delay in preparing the first module for launch. The plan for Volkov and Krikalev to receive two modules was now clearly impractical. As they awaited the first one, they worked on the experiments that were available. However, this had to take account of the deterioration of the storage batteries – which were in need of replacement. The real-time nature of the planning process was again made clear in mid-February, when Alexei Leonov reported that it had not been decided whether Poliakov would return to Earth with Volkov and Krikalev when they were relieved in April. (If he did stay on, Poliakov would almost certainly exceed the "year-and-a-day" record so recently set.) Leonov said that the next crew would "most probably" be commanded by Alexander Viktorenko. As Viktorenko had trained with Alexander Serebrov, it seemed likely that he would be the flight engineer, but a few days later it became apparent that the flight had been assigned to his backup, Alexander Balandin. The plan called for Viktorenko and Balandin to be launched on 19 April, and for Volkov and Krikalev (with or without Poliakov) to land on 29 April. Meanwhile, every effort would be made to complete the preparations to launch the first of the new modules.

The process of building up Mir would be complicated by the need to control the attitude of an asymmetric configuration. In a linear train, the centre of mass was at least on the major axis. Each new module would initially dock at the front axial port, then swing itself around to the appropriate radial port using its robotic Ljappa arm, which would grasp a fixture on the multiple docking adapter and use this as a pivot. Clearly, whenever such a module was imminent, the crew's Soyuz would require to be at the rear port. In this configuration, it would be impossible to receive any cargo ships. These operational limitations meant that sooner rather than later the first new module would have to be moved onto a radial port so that the crew ferry could swap ends to let a cargo ship dock. With a 20-tonne module mounted perpendicularly at the front, the centre of mass of the 'L'-shape would be 'outside' the structure, and the complex would be difficult to manoeuvre because an engine designed to deliver its impulse axially would tend to spin rather than push the complex. While the complex was in this asymmetric configuration, it would not be able to adjust its orbit and any experiment that required a given orientation to be set up would be impractical. Two factors were therefore key to the next phase of expansion. As soon as possible after the first module was swung off axis, the attitude-control computer would have to be fine-tuned to manage the asymmetric configuration. This would consume propellant, which would not be able to be replenished until a Progress could dock, which would be impossible if the complex could not control its orientation. Secondly, as soon as possible thereafter, the second module would have to be received and swung to the opposite side, to convert the 'L'-shape into a 'T'-shape and restore the centre of mass on the axis. Ideally, this phase of the expansion would be completed in a period of a few months. It was therefore vital that the first module should not be launched until the second was also nearing completion.

When Volkov and Krikalev were in training, they had expected that the first module would just have been docked. When they were launched, they had expected it to arrive within a matter of weeks, and the second to arrive towards the end of their tour of duty. Within weeks of their taking over, however, it had become evident that the second module would have to be left to their successors. Now, with continued delays in preparation, it looked as if they would be lucky to receive the first one. Yuri Semenov, the general manager of the Energiya Bureau, which was responsible for the preparation of the modules, later said that at this point it was decided not to send up another crew until both modules were actually finished. This decision offered two

182 An astrophysical laboratory

options: to maintain the current residents on Mir until both of the modules were certified ready for launch (so that Volkov and Krikalev would be able to commission the first module, as they had trained to do) or to recall them and leave Mir vacant until the modules were ready. The first option would be feasible if there was likely to be only a *slight* additional slippage, but if it looked like the delay would unreasonably extend the tour of duty, then there would be little choice but to give up (temporarily) the goal of maintaining a continuous human presence in space. As the Khrunichev factory worked on the new modules, the residents continued to pursue the available experiments and their successors trained in the expectation of taking over in April.

Earth observations continued throughout February, then Progress 40 replenished the complex's fluids, was loaded with trash, and left on 3 March. As it pulled back, a camera mounted in the rear docking unit recorded an experiment in which two large multi-link structures (each resembling an orange segment, but more sharply angled than an ellipse) unfolded from the sides of the ferry to assess the ability of form-remembering materials to re-establish a defined shape. A 'unique alloy' incorporated into the material 'recorded' a given shape and reformed it when a current was passed through embedded wires to provide the requisite thermal stimulus. It was hoped that this technology might facilitate the creation of a large reflector in space that could be used to collect solar energy. When the experiment was finished, the discarded ferry resumed its withdrawal. Soviet spacecraft had always been designed to be radio-controlled – Yuri Gagarin had been a mere passenger. As a result, they proved to be extremely capable as automated satellites. The use of the discarded ferry to carry out an experiment alongside the station permitted a test that would otherwise have been awkward to observe.

The research programme during March included astronomical photography using the Glasar camera; making measurements of the constituents of the atmosphere using the Spektr-256 spectrometer; a study of charged particles in near-Earth space by the Mariya spectrometer; and, of course, the continuance of X-ray observations while the crew slept. Progress 41 arrived on 18 March. Its cargo was unloaded intermittently over the following week. In addition to the usual consumables, it delivered several replacement storage batteries and a number of other components for the electrical system. On 26 March, the cosmonauts cast votes for the new Congress of People's Deputies – among the successful candidates were fellow cosmonauts Valeri Ryumin, Svetlana Savitskaya and Viktor Savinykh. Also, Krikalev was told (tongue-in-cheek) that he had been reprimanded for failing to report for service in the military reserve at the appointed place at the appointed time – the fact that he was in orbit had not been accepted by the military bureaucracy as a valid reason for absence.

By late March, it had been decided to bring the cosmonauts home on schedule at the end of April. A fault had been discovered in the service module of the next ferry. As there was not another flight-ready spacecraft available, Viktor Blagov reported that the launch of Soyuz-TM 8 had been slipped to August. Vitali Sevastyanov and Viktor Afanasayev had expected to supersede the Soyuz-TM 8 crew in September, so they were slipped in turn. They were to have been accompanied for the handover by

Rimantas Stankiavicus, who had taken over the Buran group when Igor Volk had been grounded.

In early April environmental studies were made of the northern Caucasus, the Black Sea and the Caspian Sea, and of the spectacular aurorae caused by an increase in solar activity. (The Circe experiment was regularly run to monitor the radiation level inside the complex; they were fairly safe, though, because with the exception of when they passed through the South Atlantic Anomaly, their orbit was considerably below the inner Van Allen belt.) On 10 April Progress 41 boosted the complex to a slightly higher-than-usual orbit as a preliminary to its evacuation. The following day, *Tass* announced that the cosmonauts had completed their research programme and had begun the process of "mothballing" Mir. In fact, this was a poor time to leave it vacant because, after three years of operation, the base block's electrical system was in need of a major overhaul. The last thing that flight controllers wanted was for it to suffer a power failure while unoccupied. During their final few weeks onboard, the cosmonauts serviced the converters and installed the new chemical batteries that had been delivered by Progress 41 (only a few of these 74-kilogram cells could be ferried up at a time, so it had not been possible to replace all twelve in the time available). On 21 April, Progress 41 departed. Because it had consumed so much propellant in raising the complex's orbit, it was unable to de-orbit itself, and had to be left for its orbit to decay. After two years of continuous occupation, Mir was returned to its autonomous flight mode on 27 April, and Volkov, Krikalev and Poliakov retreated to Soyuz-TM 7 and returned to Earth. This must have been a great disappointment to Poliakov, as if he had stayed on after the handover, and Stankiavicus had flown in the autumn, it would have been necessary for him to have further extended his flight to 18 months, which would have given him the endurance record by a wide margin. Although eager to examine their colleague, who had spent almost eight months in space, the medical team's first task was to attend to Krikalev's leg, which had struck the control panel in the final stage of the descent – it was sore, but not seriously injured.

Valeri Poliakov (left), Alexander Volkov and Sergei Krikalev.

Although Mir was largely dormant, the X-ray telescopes in the Kvant module could still be operated by remote control, so astrophysical observations continued.

ANOTHER INTERMISSION

At the Paris Air Show in June, Vladimir Shatalov noted that coordination between the various agencies supporting the space programme was at the root of the delayed construction of the Mir complex. He pointed out that without a central coordinating agency (such as NASA) it was becoming increasingly difficult to sustain continuous operations in orbit. He explained that because equipment had been ferried up to the complex as soon as it had become available, rather than when it was needed, much of the apparatus already in space was unusable without support equipment in the new modules. And with so much apparatus onboard, the base block was too cluttered for the cosmonauts to work effectively. He optimistically forecast that the first module (which would incorporate a large airlock and deliver the autonomous manoeuvring unit) would be launched "in either September or October", with the second (which would have an androgynous docking port for use by the Buran shuttle) following "at the end of the year". As regards Buran, he said that there was no particular urgency to initiate operations because this was meant to complement, rather than supersede the existing vehicles, and he added that improved versions of these would soon be introduced for greater manoeuvrability and a heavier payload. In addition, Progress would incorporate a small descent capsule by which the cosmonauts would be able to return the results of their experiment to Earth. But within weeks it was reported that although the Soyuz-TM 8 crew would be launched in early September (in order to receive the first module in October) the next module would not be launched until February 1990. The upgraded cargo ferry had the service module of the Soyuz-TM and the Kurs rendezvous system. Progress-M 1 was launched on 23 August, and two days later docked at the front. The power output from its solar panels was fed into the power grid, because every little helped. Despite the addition of solar panels, other savings had increased the overall cargo capacity of this revised configuration to 2,500 kilograms, so it was a significant step forward. Only now was it revealed that the front axial port incorporated plumbing to replenish the base block's fluids. This eliminated the need to transfer a newly arrived Soyuz ferry from the rear to the front port to make room for an incoming cargo ship.

6

A microgravity laboratory for hire

THIRD START

Soyuz-TM 8 was launched on schedule, on 6 September 1989. A new aspect of the television coverage was the advertisements erected around the pad and emblazoned on the rocket's shroud. As expected, Alexander Viktorenko was the commander, but he was, after all, accompanied by Alexander Serebrov. On their final approach two days later, they saw the complex start to oscillate when their ferry was only four metres off the rear port, causing the Kurs to abort. Viktorenko took over, withdrew twenty metres, inspected the docking port, and then closed in and docked without incident. An hour later, they opened the hatches, entered the complex and began the long process of returning it to life. Immediate maintenance involved the replacement of another three nickel–cadmium batteries that had been sent up on Progress-M 1 to further ameliorate the degraded power system. Then several days were spent loading and verifying new software that would enable the attitude-control computer to deal with the offset centre of gravity that would result when the first module was swung off axis. During this first week, the only active research was that using the remotely controlled X-ray telescopes, but by mid-month they had installed the Gallar furnace from Progress-M 1 in Kvant 1. This was an improved form of the Korund-1M, and on its first test it produced a monocrystal of cadmium selenide semiconductor.

In mid-September it was announced that the first module would be launched on 16 October, dock on 23 October, and immediately be swung onto the upper radial port. Soyuz-TM 8 would be relocated to the front on 25 October. Only then would the cosmonauts be cleared to open the new module. Once it had been checked, they were to make an 'internal spacewalk' by depressurising the multiple docking adapter in order to swap the docking drogue from the upper port to the lower port, to accept the next module. (Apart from the axial port, there was only one Konus assembly in the compartment, and it had to be placed on the port next to be used.) Progress-M 2 would dock at the rear port on 29 October. In December, it was said, Viktorenko and Serebrov would make one spacewalk to retrieve the Echantillon cassette. On another, they would put the two star trackers on Kvant 1 (this had not been done by Volkov and Krikalev). Finally, on two further spacewalks, they were to test the autonomous manoeuvring unit that would be carried by the first module. On 28 January, after the second cargo ship had gone, they would move Soyuz-TM 8 back to the rear port in order to free the front for the second module. This would be launched on 30 January, dock on 6 February, and immediately be swung onto the lower port to free the axial port for Soyuz-TM 9, which would dock on 13 February. Viktorenko and Serebrov – having received both modules during their tour of duty, commissioned the first one, and tested the autonomous manoeuvring unit – were to return to Earth in late February. General Kerim Kerimov, the chairman of the State Commission overseeing operations, said that the two cosmonauts had been assigned "a colossal amount of work". This was, in fact, essentially the mission that had been given to Volkov and Krikalev 18 months previously, and, as then, it would not be long before events on the ground rendered this ambitious schedule obsolete.

Alexander Serebrov (left) and Alexander Viktorenko.

In the second half of September, Viktorenko and Serebrov focused on preventive maintenance on the core systems of the base block, but additional materials were processed in the Gallar furnace and X-ray observations were undertaken on a regular basis. A major solar flare at the end of the month caused some concern. Its effects were detectable by the Circe apparatus (used to measure the level of radiation inside the complex twice a day) for about ten days – usually the effect of a flare lasted only a few days. They slept in Kvant 1 during this time, since it was more heavily shielded. It was later calculated that they had suffered exposure equivalent only to adding two extra weeks to their tour. In early October, while undergoing final checks, the first module was declared unfit for flight after the discovery of faults in computer chips drawn from the same batch as had been used in its Kurs circuitry. It was decided to replace the chips in the module, in case these were also faulty. *Tass* warned that the schedule had slipped by 40 days, but Viktor Blagov announced that the launch had been rescheduled for 26 November. This obliged Viktorenko and Serebrov to make the best use of apparatus already onboard Mir. Most of the next two months were spent taking pictures of the Ukraine, Krasnodar, Stravropol, Moldavia, the Caspian Basin and Turkmenia using the KATE-140, and using the MKS-M and Spektr-256 spectrometers to study the atmosphere above the tropics to measure the distribution of ozone for the Atlantika-89 project, carried out jointly with Cuba, to evaluate an apparent correlation between ozone levels and the formation of tropical hurricanes. As they waited, Viktorenko and Serebrov undoubtedly endured similar frustrations to those felt by their predecessors. When Blagov added that the launch of the second module had slipped to late March, they knew that they would not see its arrival, but the highlight of the mission, the spacewalk to test the 'flying-backpack', for which they had received special training, seemed certain.

KVANT 2

On 26 November a Proton rocket placed Kvant 2 into orbit. Unlike its predecessor, which had required a tug, this new module had its own propulsion system because it was a TKS augmented by apparatus, rather than a passive payload; as such, it was a refinement of the vehicles which had temporarily docked with the second-generation Salyuts. Unfortunately, it ran into trouble almost immediately. One of its two solar panels did not deploy properly. The outermost three segments unfolded as intended, but the innermost did not. In fact, because the innermost panel had not locked, the outer piece of the 10-metre long panel was free to swing, which made manoeuvring awkward. After analysing the telemetry, the flight controllers put the vehicle into a slow roll, and then commanded the motor that rotated the solar panel to cycle back and forth to cause the centrifugal force to draw the panel out. This worked. If it had proven impossible to deploy the panel by this means, the docking would have been attempted and the cosmonauts sent out to try to lock it with tools that would have to have been specifically made and sent up on the next cargo ship. On 1 December, Progress-M 1 pumped 'surplus' propellant into the base block's tanks and departed – this new capability of the unified propulsion system offered a significant bonus, as each ship carried more propellant than it required, in case it was obliged to make a second rendezvous; it retained only sufficient for the de-orbit burn. The next day, as Kvant 2 made its final approach, Viktorenko and Serebrov retreated to Soyuz-TM 8, at the rear of the complex. When the module was 20 kilometres out, its Kurs system inferred that it was closing too rapidly and aborted its approach. In the event, even if it had been able to continue, a docking would have been risky because at about the time it would have arrived the gyrodynes dropped off-line, leaving the complex in an uncontrolled orientation. Four days later, after a change of procedures, Kvant 2 made a second approach. This time Viktorenko locked the complex stable using the ferry's attitude-control thrusters. The chatter on the voice channel was aired live by *Radio Moscow* as Kvant 2 docked. The downlink from the camera in the front docking port showed the module's solar panels in an 'up-and-down' alignment, perpendicular to the main panels on the base block. The reason for the 45-degree offset in the docking collar was now revealed. The module's short twin-element Ljappa arm had a triple-petal androgynous grapple on its end which, when it engaged an attachment point on the multiple docking adapter, would enable the module to swing itself around onto a radial port. This twisting manoeuvre would rotate it by 45 degrees on its long axis. Since the front axial port was itself offset at 45 degrees, this supplied the 90-degree rotation needed to properly align the module on the upper port, with its solar panels parallel to those of the base block. The arm functioned flawlessly, and the transfer to the upper port took less than an hour.

A microgravity laboratory for hire

With Kvant 2 on its upper docking port, Mir was 'L'-shaped.

Just as Kvant 1 was also often referred to as the 'astrophysics module', Kvant 2 was the 'enhancement module' (or sometimes the 'D' module, from *dusnashcheniye*, meaning 'expansion', 'enhancement' or 're-equipment'). A great deal of apparatus was mounted on its outer surface, including six additional gyrodynes to assist in manoeuvring the expanded complex, and two tanks to increase the capacity of the Rodnik system. It comprised three hermetic compartments. The main compartment held apparatus to take the complex a step nearer a closed-cycle environment. There were two separate oxygen generators: Elektron produced oxygen on an ongoing basis by the electrolysis of water, and Vika repressurised the airlock by chemical reaction without imposing a demand on the main system. It also contained two cubicles, with a toilet and a shower, and had been loaded with miscellaneous cargo for launch. The middle compartment had built-in scientific apparatus. At the far end was an airlock with an outward-opening outer hatch that was a little wider than a docking port to accommodate spacewalkers wearing the autonomous manoeuvring unit, which was stowed in a large box on the wall of the airlock. Two Orlan-DMA spacesuits were delivered in the airlock. These incorporated changes reflecting experience with the Orlan-DM suits that had been delivered in 1988. The scientific apparatus comprised Volna-2 to study fluid flow in capillaries to evaluate a new propellant tank design, Inkubator-2 to hatch eggs for a study of embryonic growth, Epsilon to evaluate the efficacy of the module's thermal protection, the KAP-350 mapping camera, and the MKF-6MA – an improved form of the multispectral camera used extensively on the second-generation Salyuts. In addition, the ASPG-M

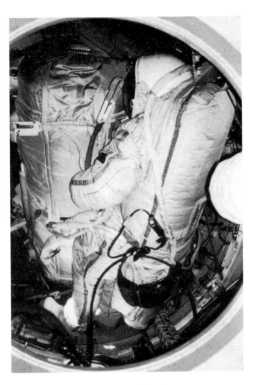

Kvant 2's airlock.

remote-control scan platform was mounted externally with a number of instruments, including the ITS-7D infrared spectrometer, the ARIZ X-ray spectrometer, the MKS-M2 spectrometer, and the Gamma-2 multi-spectral television camera cluster. Kvant 2 augmented the complex's power system with 7 kW and six nickel–cadmium batteries in pristine condition.

Soyuz-TM 8 was moved to the front port on 12 December. On re-entering the base block, Viktorenko and Serebrov opened Kvant 2's first compartment to inspect the cargo. Two days later, they took a portable television camera with them and gave the flight controllers a tour. They spent the next week activating and checking the module's systems. Because Kvant 2 projected out radially, whenever the orientation of the complex was changed this put stresses on the mating collar, and if this were to lose its hermetic integrity the complex would depressurise, so the structural stresses on the collar were measured using the acoustic probes of the Kontrol and Monitoring experiments; no unexpected stresses were revealed. When Progress-M 2 docked at the rear on 22 December, it made the first five-vehicle complex and brought the total mass to just over 72 tonnes. In addition to additional nickel–cadmium batteries for the base block and the 256-channel Spin-6000 X-ray and gamma-ray spectrometer, it delivered an experiment supplied by Payload Systems Incorporated in America as a commercial experiment to grow protein crystals – such packages were flown on the Space Shuttle, but that vehicle was unable to remain in space long enough for a

The scan platform on the exterior of Kvant 2.

slow-growing crystal. After the new batteries had been installed, the cosmonauts resumed their researches. One experiment grew chlorella and rice tissue cultures, and another was a long-term study of amphibians, crustacea and molluscs to observe the reaction of their nervous systems to weightlessness in order to identify functional changes to their vestibular apparatus. In addition, the Gallar furnace was operated for extended periods (often as long as six days) to process a wide variety of semiconductors. Towards the end of the year, Alexei Leonov said that there would be three launches in 1990. Vladimir Dzhanibekov said that for the foreseeable future resident crews were to be formed of two rather than three cosmonauts, who would continue to fly tours of five to six months. Vladimir Shatalov added that five two-man crews were in training, and that the first to fly would be Anatoli Solovyov and Alexander Balandin, followed by Gennadi Strekalov and Gennadi Manakov.

Cosmos 2054 was launched on 27 December to augment the SDRN network. In mid-1988, Cosmos 1897 had been repositioned above the Atlantic to relay telemetry from the Buran shuttle during the re-entry phase of its test flight, but it resumed its position above the Indian Ocean in early 1989. This second Luch satellite took up the Atlantic position to greatly enhance communications with Mir. To celebrate the New Year, Viktorenko and Serebrov opened the gifts delivered by Progress-M 2 and then feasted on crispy pickles, fresh lemon, canned sturgeon, blackcurrant juice and fresh fruit.

WORKING OUTSIDE

A major task for 1990 was external activity. On 8 January Viktorenko and Serebrov sealed off Kvant 2 and prepared to exit through a lateral port of the multiple docking adapter. Because they had inadvertently left open the pressure-equalisation valve to Soyuz-TM 8, the orbital module depressurised with the docking adapter, and fixing this problem put them behind schedule. Undaunted, they made their way to the rear of the complex with a pair of star trackers. These bulky 80-kilogram packages were mounted on each side of Kvant 1's unpressurised compartment in order to assist the gyrodynes in orienting the expanded complex. On their return, they retrieved sample cassettes from the exterior of the base block, and set up attachment points for their next spacewalk. Venturing out again three days later, they emplaced several cassettes to assess a variety of non-metallic test materials, and then made their way back to Kvant 1 to install the Arfa apparatus to study a correlation between seismic activity in the Earth's crust and charged particles in the ionosphere. On their way back, they retrieved the Echantillon cassette that Jean-Loup Chrétien had deployed (it had not been meant to remain outside for so long, but this did not matter) and dismantled the anchor set up for the Era experiment. Back in the docking adapter, they moved the Konus drogue from the upper to the lower port, in preparation for the next module. On 26 January, after a week spent unloading Kvant 2's cargo, they made their third spacewalk, this time from the airlock, immediately outside of which they installed an open framework 'dock' incorporating magnetic clamps to mate with the autonomous manoeuvring unit. They then used handholds to move down Kvant 2 to

Working outside 191

remove its now redundant Kurs antenna. Returning to the airlock, they deployed the Danko and Ferrit exposure cassettes and released the restraints that had held the ASPG-M scan platform alongside the airlock for launch.

The highlight of the mission was the test flight of the autonomous manoeuvring

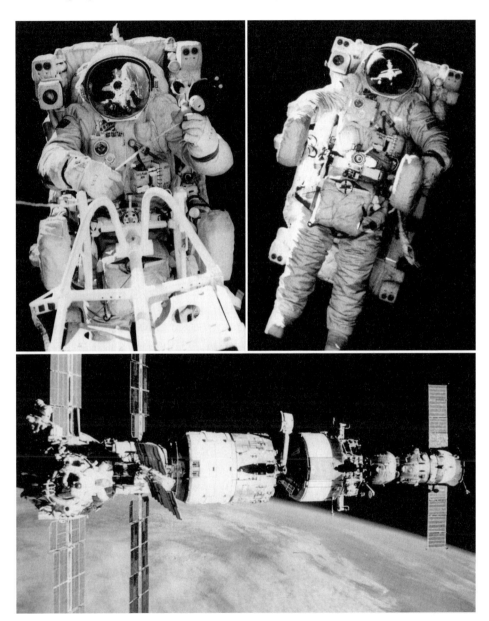

Safely tethered, Alexander Serebrov tests the YMK manoeuvring unit by moving in line with Kvant 2's axis.

unit. Developed by Zvezda (the manufacturer of the Orlan spacesuits), this flying backpack had been nicknamed 'Icarus' by its designer, Gai Severin. The YMK (the acronym for 'cosmonaut manoeuvring unit') was similar to the MMU backpack developed by NASA, but somewhat larger and covered by a thermal blanket. It had four 'T'-shaped thruster sets of eight pressurised-nitrogen jets (this obviated damage to apparatus on Mir's surface by corrosive efflux). Each armrest had a control panel of toggle switches and a joystick for specifying actions (the left one for translational and the right for rotational motions). It could be manoeuvred either manually or in one of two semi-automatic modes, one for optimum efficiency and the other for fast action. On 1 February, Serebrov suited up and donned the YMK, which fitted close around the integral life-support backpack and was fastened by a waist belt, and then the airlock was depressurised. He squeezed through the 1-metre-diameter hatch and 'docked' with the framework anchor that they had recently erected, so that he would remain in position while he deployed the armrests and checked out the systems of the YMK. For this test he was to remain linked to the anchor by a 60-metre nylon tether, just in case the backpack malfunctioned, whereupon he would be reeled back by Viktorenko using a winch. Satisfied, Serebrov disengaged and used the thrusters to move away. Viktorenko remained in the hatch and filmed his colleague's progress. There was a television camera built into the YMK, on Serebrov's shoulder. The first test involved slowly moving five metres out, halting, then returning. After repeating this three times, he moved out to about half the distance permitted by his tether for a deliberate series of slow rotations and translations. Domestic television interrupted its schedule to show the video downlink from Serebrov's camera. Once the test was over, Serebrov 'redocked', deactivated the YMK, folded up its arms, and re-entered the airlock. Four days later, Viktorenko repeated the evaluation. Before returning, he performed an experiment using the Spin-6000 apparatus to measure the X-ray and gamma-ray fluxes at distances out to about 45 metres from the complex to measure the radiation generated by its passage through the ionosphere. The YMK performed perfectly on both tests, but it was not for regular use by Mir residents – it was being tested for later use by spacewalking Buran cosmonauts, and so only a few bottles of nitrogen propellant had been supplied.[1]

To help to recover from their exertions, the cosmonauts used the new shower. A cubicle in Kvant 2, it was an improvement over the plastic curtain devices tested on the Salyuts that had taken several hours to set up, use, clean and pack away. As with its predecessors, however, it proved to be ineffective and the compartment was later converted into a steam room, and later still the plumbing was removed so that it could be used for another purpose.

---

[1] In fact, it would never be used again.

Soyuz-TM 9 lifts off.

A DAMAGED FERRY

As Progress M-2 departed on 9 February it tested the new Luch relay. Soyuz-TM 9 arrived four days later with Anatoli Solovyov and Alexander Balandin. As they had not received any visiting crews, Viktorenko and Serebrov were delighted to see their successors. Solovyov had visited Mir in 1988, but this was Balandin's first mission. The highlight of their tour was to be the commissioning of the 'technology' module – or at least that was the plan.

The late delivery of Kvant 2 and the postponement of the next module had obliged Viktorenko and Serebrov to devote more time to research than expected, and they had a large amount of material to return to Earth. The prize in the haul were monocrystals of various semiconductors produced by the Gallar furnace (one of the gallium arsenide crystals was almost 300 grams). It was later calculated that this was the first flight to 'break even' in terms of balancing its cost against the 'value' of its product, which raised expectations that the furnaces in the forthcoming 'technology' module would turn Mir into an orbital factory. Unfortunately, the assignment of a financial value to this semiconductor was hypothetical, because it did not really feed into a commercial market. Even so, it was genuinely hoped that the complex would eventually become self-financing. The Payload Systems protein crystal package was also returned. This had processed enzymes of hen egg white lysozyme and D-amino transferase by batch, vapour diffusion and boundary layer diffusion crystallisation. There had been considerable American corporate interest in sending experiments to Mir, but the US government's ban on exporting technology to the Eastern bloc had effectively precluded it – 'exporting' the proprietary apparatus was permitted only if the package was completely sealed. For this reason, the Payload Systems package had only an on/off switch. The contract was for six packages over a period of years, so it was not just a one-off novelty. Various experiments were undertaken during the handover, including growing plants in the Magnetobiostat, making semiconductors in the Gallar furnace, and the purification of biological agents in the Ruchei and Biokryst electrophoresis apparatus. In addition, a variation of the Resonance experiment was conduced to determine the vibration modes of the enlarged complex (this was done periodically to identify activities that placed undesirable stresses on the structure). Viktorenko and Serebrov later said that they had often spent as much as 80 per cent of a working day *setting up* apparatus, rather than using it. They had strung ropes to overcome difficulties moving around. They suggested that foot restraints be installed near fixed apparatus. They criticised Yuri Semenov for prohibiting the jettisoning of bags of waste from the scientific airlock, because this had resulted in 80 bags

Anatoli Solovyov (left) and Alexander Balandin.

being stored in the already cramped complex. Such criticism was welcome, however, as it was intended to improve operating procedures – operating an orbital complex on an ongoing basis with real-time planning was a learning curve.

On achieving orbit, the Soyuz-TM 9 crew had realised that something was stuck to the outside of their vehicle, because the field of view of the Vzor optical sight was partially obscured. As they completed their rendezvous, Kvant 1's aft-facing camera had shown that there were several loose objects projecting from one side of the craft. After they undocked in Soyuz-TM 8 on 19 February, Viktorenko and Serebrov flew around to inspect the other vehicle, and reported that three of the descent module's eight petal-shaped thermal blankets had been torn free when the aerodynamic shroud had separated. Unfastened at the base, the blankets had peeled back "like a flower" around the collar connecting the descent and orbital modules (one projected straight out at 90 degrees, and the others at 60 degrees). As they were preparing to de-orbit, Viktorenko and Serebrov were told that because conditions at the landing site had deteriorated they would have to postpone their return for 24 hours. Before they had powered down non-essential systems, however, they were told that a small clearing in the weather had been reported not far from the prime site, so they could return on their next pass. The half-yearly tours were pleasant for those crews that left in the winter and returned to glorious sunshine, but not much fun for those blasted by the chilly $-30°C$ north wind. The pictures taken on Soyuz-TM 8's flyby revealed that unless Soyuz-TM 9's loose blankets could be reattached they might easily block the infrared horizon sensors that were to orient the spacecraft prior to its de-orbit. Viktor Blagov announced that Solovyov and Balandin would have to conduct a spacewalk to either reattach or trim the blankets. Leonid Gorshkov of the Energiya Bureau said that the necessary tools would be fabricated and delivered in the next module (whose launch was due on 18 April), and the spacewalk would occur towards the end of the mission. Blagov explained that in the meantime the orientation of the complex would have to be controlled to ensure that the ferry's exposed surface was neither roasted nor frozen for prolonged periods – in space it was $+130°C$ in the Sun and $-130°C$ in the shade.

A NEW COMPUTER

Solovyov and Balandin flew Soyuz-TM 9 around to the front port on 21 February, and Progress-M 3 arrived on 3 March. The following day, the cosmonauts installed the batteries it had delivered. Most of the month was spent setting up the Salyut 5B computer to replace the Argon 16B that had been initially installed in the base block, a tricky operation involving integrating five processors, to control the asymmetrical configuration by coordinating the gyrodynes in the two Kvant modules. It had been delivered by Kvant 2. Then the mass distribution of the complex was modelled. This involved predicting the thruster firings required for a specific manoeuvre, performing that manoeuvre, then noting the discrepancy, further refining the mass distribution model, and trying again to iterate towards a solution. By mid-month, it had become evident that this was more difficult than expected, so it was decided to postpone the next module until the complex was able to be controlled using the gyrodynes. There was considerable excitement onboard Mir on 17 March when a quail egg hatched in Inkubator-2. This was the first time that an animal had been born in orbit. The chick flapped its tiny wings in a hopeless attempt to orientate itself in weightlessness. By early April, it was noted that the chick was not developing normally; it was frail and proved unable to feed from the dispenser. Half a dozen other chicks fared no better. Nevertheless, Dr Ganna Maleshko of the Institute of Medical Biology said that the hatchings had shown that the development of the embryo had proceeded normally. This was certainly a great step forward in the study of adaptation to weightlessness. These embryo studies were seen as a precursor for the development of a closed-environment orbital habitat in which fowl would provide a source of food for the crew. After maintenance on the environmental and thermal regulation systems, the cosmonauts resumed fine-tuning the ability of the Salyut 5B computer to control the asymmetric configuration using the thrusters. By early April they were able to move on to coordinating the gyrodynes in Kvant 1 with those in Kvant 2 – the latter being brought on-line the first time. The first all-up test was on 19 April. Although only 11 gyrodynes were available (one in Kvant 1 had failed), the computer performed flawlessly, marking a significant milestone in building the complex. On a pessimistic note, however, Viktor Blagov noted that the base block had been in space for over four years, and so was nearing the end of its notional operational life even before the complex had been completed – it had originally been intended to have all four radial modules in place by this time. Nevertheless, the longevity of the second-generation Salyuts suggested that with careful maintenance the base block should be able to be kept operational for at least twice this time.

The extended trials of the attitude-control system had used considerably more propellant than expected. Yuri Semenov announced that although the next module *could* now be launched, it would not be until the complex had been resupplied. It had been decided to leave Soyuz-TM 9 at the front port, dock a tanker at the rear, then move the ferry to the rear port after the tanker had departed, to leave the front port free for the module that was to be launched in late May. On 27 April Progress-M 3 departed. While they waited, Solovyov and Balandin started the Gallar and Mirror furnaces making semiconductors. The launch of Progress 42 on 6 May clarified the decision to retain the Soyuz at the front until the propellant had been replenished – the only tanker available at short notice was an Igla-equipped model that could dock only at the rear port; it turned out to be the last of the original model. It arrived on 8 May and left on 27 May, and Soyuz-TM 9 was transferred to the rear the next day. Three days after that, a Proton put the new module into orbit. The hastily arranged schedule had gone like clockwork.

KRISTALL

On 6 June, soon after the new module had activated its Kurs system and started its approach to Mir, one of the attitude-control thrusters malfunctioned by firing longer than required, and the automated system aborted and withdrew. A second attempt

Mounting Kristall opposite Kvant 2 restored Mir to a symmetrical configuration.

four days later (using only backup thrusters) ended with a successful docking. The next day, the new module extended its Ljappa arm and swung itself onto the lower port, transforming the complex into a 'T'-shape, restoring its balance, and completing this second phase of its expansion. The crew suggested that they be permitted to extend their tour by 10 days, to allow them time to complete the commissioning of the newcomer, and this was granted. Kvant 3 had been dubbed Kristall due to the furnaces it incorporated (Krater, Optizon-1, Zona-2, Zona-3 and Kristallisator). The internal hatch at the far end of its main compartment led to a spheroidal space derived from the Soyuz orbital module. This held the Priroda-5 Earth-observation apparatus, which incorporated a pair of KFA-1000 high-resolution cameras that (with Kristall on the lower port) faced forward. At the far end of the module was a docking port. This Androgynous Peripheral Assembly System (denoted APAS-89 because it was derived from the APAS-75 that had been built for Apollo–Soyuz) was to be used by Buran. A second such port was mounted in the side of the compartment that faced rearward. This was intended to take Pulsar X-2 (an X-ray telescope like Pulsar X-1 in Kvant 1) that Buran was to deliver and emplace using a robotic arm. Kristall also had Glasar-2 (to supplement Glasar-1 in

Kvant 1); the Ainur electrophoresis unit (to augment Svetlana in Kvant 1); the Marina, Buket and Granat cosmic-ray detectors; and the Svet cultivator. It had two Rodnik tanks and six nickel–cadmium batteries, but no gyrodynes (because no more were needed). The deployment mechanisms of its solar panels were designed to enable them to be retracted later on, and the mounts could be detached to facilitate their transfer to another part of the complex. The mast of each 15-metre-long panel drew out the 36 in-line rectangular segments (which did not have fore-and-after flaps that would inhibit retraction) in concertina fashion. The module augmented the complex's power supply by 8 kW. No sooner was this phase of the expansion complete than Boris Olesyuk, a worker at Kaliningrad, complained to *Moscow News* that the cosmonauts spent too much of their time maintaining the complex's systems. In addition to the need to unload the small Progress ferries, and the much heavier cargo delivered by the new modules, the installation and testing of apparatus took so much time that the scientific results were minimal. He observed that there was a great deal of equipment in the complex which the cosmonauts did not have time to use. He said that the Gamma-2 television package that they had set up six months previously had not been tested; neither, in fact, had the MKF-6MA Earth-resources camera. He claimed that the difficulties encountered in upgrading the attitude-control system derived from the fact that there were too many computers involved, and that the problem was making them cooperate. Further expansion was pointless, Olesyuk argued, without also increasing the crew to permit two shifts of three cosmonauts rather than single shifts by a team of only two. Unfortunately for this line of argument, the life support system could sustain a crew of six for only short periods. Nevertheless, it was soon announced that the funding to finish the last two modules was to be withheld until it was demonstrated that the apparatus in Kristall really did produce economically viable results. This effectively meant that the final phase of the expansion (which at that time called for the Spektr module to be added in late 1991 and Priroda in early 1992) would be postponed for a further three years. As a result, these modules, which were already under construction, were mothballed in the Khrunichev factory.

LOOSE INSULATION

Even although Viktor Blagov had explained in an interview on domestic television in February that Soyuz-TM 9's thermal blankets were loose, it was not until late May that the Western media picked up on this and reported that Solovyov and Balandin were stranded. Blagov insisted that reports that the crew had no "real possibility of returning to Earth" were "completely groundless", because, in an emergency, they could manually orient their ferry for the de-orbit burn. Nevertheless, despite efforts to minimise thermal stresses on the exposed descent module, there was still concern that the wrap-around heat shield might have deteriorated. On 4 July, Soyuz-TM 9 was returned to the front port to be better located for the spacewalk to attempt to repair its insulation. Over the next two weeks, the cosmonauts studied uplinked video of hydrotank tests of procedures for the operation, and on 17 July they set off to inspect the damage. Although exiting by a port in the docking adapter would have reduced the distance they had to travel to reach the work site, it would have meant making their way past the antennas mounted on the orbital module *en route* to the descent module. Because there were no handholds, their passage could easily damage apparatus mounted on the ferry. It had therefore been decided that they should exit through the airlock and make their way down to the base of Kvant 2. A kit of tools and a folded 7-metre-long metal ladder had been provided. At the base of the module, they were to unfold the ladder and angle it across the orbital module for direct access to the descent module beyond and to serve as a work platform. They were in such a hurry to begin that they released the hatch of the airlock before the compartment had fully depressurised. Other hatches opened inwards, but the airlock opened outwards so that spacewalkers could load the compartment with equipment without having to leave clearance for the hatch to swing aside. The hatch was designed to be opened in two stages. The first stage opened it a few millimetres to crack the hermetic seal, and a catch held it in position. The catch was to be released only when the airlock was fully depressurised, to enable the hatch to be opened. In this case, the catch released immediately and the pressure swung the hatch open beyond the designed limit of its hinge. The crew expressed surprise at the airlock suddenly being flooded by brilliant sunlight, but set about their spacewalk oblivious to the fact that the hinge had been damaged.

Free of umbilicals in their Orlan-DMA suits, Solovyov and Balandin each used a pair of short safety tethers ending with mountaineering clamps to switch between a succession of handholds. But dragging the bulky equipment slowed their progress, as did remaining in place while passing through the Earth's shadow, and it took almost two hours to reach the multiple docking adapter. Then the ladder proved to be more difficult to install than it had been in the hydrotank, and they fell even further behind schedule. Finally in position, they used a television camera to show the engineers in Kaliningrad the exposed surface of the ferry's descent module. After confirming that the thermal stress had not damaged the bonding of the heat shield, and that there was no sign of damage to the explosive bolts that linked the descent module to the service module, they turned their attention to the loose blankets. The preferred option was to clip these into place using spare pins, but the blankets had shrunk and would no longer reach the locking ring. The second option was to fold each segment and clamp it against the orbital module, where it would not obstruct the sensors further aft. Two of the blankets were secured in this way, but the third was so badly torn that they were unable to make as tidy a job as hoped, but fortunately this one was not so important, because it was far away from the sensors. By this point, they had been out for almost six hours, so they did not have time to dismantle the ladder prior to returning. Unhindered by tools, they made better progress. Having been on their life-support systems for so long they could not afford to stop, so they continued during orbital darkness and used the illumination of their small lamps to double-check each transfer of their safety lines.

Inside the airlock, Solovyov and Balandin found that no matter how hard they tugged on the hatch it would not close the final few millimetres. They left it ajar and passed into the middle compartment of Kvant 2, to use this as an emergency airlock, and by the time this was pressurised they had been on their suit systems for almost seven and a half hours, which was uncomfortably close to their absolute limit. The airlock was left exposed to vacuum. Despite puzzlement concerning the hatch, there was great relief in Kaliningrad that the issue of the ferry's thermal blankets had been successfully resolved. If the bonding of the heatshield had cracked, or the explosive bolts had eroded, then a replacement ship would have had to have been sent up. And although this could have been launched unmanned (as had Soyuz 34) in this case it had been decided to have Anatoli Berezovoi fly it to ensure that it would be able to dock if its Kurs malfunctioned.

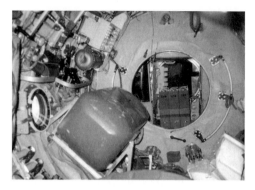

A view from Kvant 2's airlock through the inner hatch into its middle compartment, which served as an emergency airlock.

Another spacewalk was added to the schedule on 26 July. After depressurising the middle compartment, Solovyov and Balandin opened the internal hatch and used the television camera in the still-open airlock to show engineers the damaged hinges. A bolt had been pulled partly out, and an aluminium plate covering one of the hinges had been twisted up. This seemed to be what was preventing the hatch from closing. This done, they moved out and, unhindered, soon reached the ladder. This could not be left extended over the orbital module because it would swing back and block the port when the ferry undocked, but instead of bringing it back inside they were told to break it into two parts and affix it to the outside of Kristall on the other side of the docking adapter. They affixed one easily, but the other proved difficult and took longer than expected. When they returned to the airlock, Kaliningrad asked them to take another look at the hatch, and they discovered that part of the bent plate had snapped and become stuck in the hinge; once this was removed the hatch was able to be closed. Nevertheless, in the firm belief that no mechanical task was beyond the spacewalkers, it was decided that the damaged parts would have to be replaced as soon as possible.

The launch of the next crew had been postponed 10 days to allow Solovyov and Balandin time to finish commissioning Kristall's systems. They combined this with further smelting, and the mapping of a large expanse of forest in southern Siberia that had been swept by fire the previous year. After a flawless approach, Soyuz-TM 10 automatically docked at the rear on 3 August with Gennadi Manakov and Gennadi Strekalov. Having visited both Salyut 6 and Salyut 7, suffered the frustration of an aborted rendezvous with Salyut 7, and endured a launch pad abort, Strekalov was a veteran, but this was Manakov's first flight. There had been widespread expectation that a Soviet journalist would fly with them, but the spare seat had been loaded with cargo (including a flock of newly hatched quail chicks, being flown to test adaptation in the immediate post-embryonic phase to complement earlier studies of chicks that hatched in space). The residents brought their successors up to date on the state of the upgraded attitude-control system and showed them where things were stored – as apparatus accumulated, storing it so that it could be readily retrieved had become a major issue. Their flexible flight plan had resulted in a significant cargo for return to Earth that included gallium arsenide, zinc oxide, germanium oxide and samples of epitaxial silicon semiconductor, together with the radishes and lettuces grown in the Svet cultivator and 2,000 photographs of the Earth. Soyuz-TM 9 left on 9 August. It was now standard practice to jettison the orbital module prior to the service module, but in this case they were released simultaneously in order to preclude the loose blankets snagging at a critical moment. In the event, the re-entry was perfect. Unfortunately, Solovyov and Balandin had been so busy with spacewalks towards the end of their extended mission that they had neglected their exercise regime, and so suffered slight physical discomfort for a while after returning to Earth.

A MAINTENANCE MISSION

Manakov and Strekalov's main engineering job was to extensively rewire the base block's power supply, but they were also to try to repair the damaged airlock hatch before being relieved in December. Their research centred around keeping the Kristall furnaces making semiconductors (mainly gallium arsenide, zinc oxide and cadmium sulphide) in runs of up to ten days. On 17 August Progress-M 4 docked at the front. It brought the power cables (one disadvantage in using a 28-volt line was that the cable was rather heavy) and television equipment for a forthcoming visitor. A month later, just before the spacecraft departed, the cosmonauts attached an experiment to its docking assembly. After undocking, the ship withdrew 100 metres and paused so that the experiment, which created an artificial plasma, could be activated and filmed, and then it departed. Progress-M 5 took its place on 29 September. It brought more television equipment and, for the first time, a Raduga recoverable capsule. This could return up to 150 kilograms of compact cargo. Considering the experiment schedule, it had been decided to send a capsule up with every third or fourth resupply ferry – at 380 kilograms, however, these capsules would seriously diminish the cargo capacity. In early October, Manakov and Strekalov began to prepare to replace the bent hinge plate of the airlock hatch on 19 October, but Strekalov developed a head cold and the repair was postponed to 30 October, when they found that the hinge pin was badly deformed; it was clear that the entire assembly would have to be replaced. Although the next crew was assigned this task, the urgency to repair the hatch had evaporated once it had been established that it could be closed. The Raduga capsule was a 1.4-metre-long, bottle-nosed truncated cone slightly under 0.8 metres across at the wide end (in order to fit in the hatch of a Progress ship) and 0.6 metres at the narrow end. It was placed in Progress-M 5's collar instead of its docking probe, with the narrow end projecting inside the orbital module. Previous cargo ships had promptly de-orbit themselves over the Pacific, but when Progress-M 5 left on 28 November it flew a similar trajectory to that of a returning Soyuz. After it made the de-orbit manoeuvre, radar tracking stations determined the optimum moment to command the ejection of the descent capsule. This occurred at an altitude of about 120 kilometres, just before the ferry re-entered the atmosphere and burned up. The capsule descended into the recovery zone in Kazakhstan. As it descended through 15,000 metres altitude, an air pressure sensor deployed the parachute. At 4,000 metres, the radio beacon switched on to help the recovery team in locating it. Despite being used for the first time, the capsule was successfully retrieved. It delivered a 115-kilogram haul that included the results of the ongoing smelting operation.

A JAPANESE VISITOR

Toehiro Akiyama (left), Musa Manarov and Viktor Afanasayev.

When Soyuz-TM 11 docked at Mir's front port on 4 December, it delivered Viktor Afanasayev, Musa Manarov and Toehiro Akiyama, who was making a 'fee-paying' visit. In general, the backups for one main expedition flew the next. Afanasayev had backed up Soyuz-TM 10 with Vitali Sevastyanov. They had previously trained for the flight in 1989 that had been cancelled following the decision to leave the complex temporarily vacant. However, two months before Soyuz-TM 10 was due for launch Sevastyanov had been grounded by the doctors and Manarov (who had spent a year on Mir) had been assigned instead. This was Afanasayev's first flight. The fact that Akiyama was the chief foreign news editor for the Tokyo Broadcasting System gave rise to resentment in the Soviet media, which had hoped to see one of its own members flown first. The stereotypical Japanese tourist, Akiyama brought with him half a dozen cameras and a hundred rolls of film to supplement the television equipment that had already been ferried up. He was to make live television broadcasts whenever the complex was in communications range of Japan, which it was for up to ten minutes on favourable passes. Akiyama suffered motion sickness easily. During the two-day rendezvous he remained strapped in his couch, and on Mir he never felt comfortable. He had several experiments. Six Japanese tree frogs had been brought to record their adaptation to weightlessness – they had tiny suckers on their feet, and the objective was to determine whether the suckers enabled them to retain normal locomotion. A medical experiment for Tokyo University involved him wearing a cap incorporating sensors to measure the electrical activity of his brain, together with the state of his respiratory system. In addition, he regularly conducted a standard psychomotor test in which he moved his hands in a predetermined pattern while keeping his eyes shut. Meanwhile, experiments were set up involving wheat and barley seeds and a ginseng tissue culture, the Vita biotechnology experiment cultivated protein compounds, and Rekomb grew hybrid cells to be used later to make biologically active substances on Earth. On 10 December Manakov, Strekalov and Akiyama left in Soyuz-TM 10.

In addition to combining materials-processing and astrophysical and terrestrial observations, during their six-month tour Afanasayev and Manarov were to conduct a series of spacewalks to mount a crane on the base block, and then use this to start the process of transferring Kristall's solar panels to Kvant 1. The Gallar, Krater and Kristallisator furnaces were operated throughout December. In addition, the Pion-M investigated the melting and crystallisation of various materials

Toehiro Akiyama.

and the thermal exchange in liquids, and the Mariya and Granat spectrometers monitored the charged particles in the complex's orbital path.

SPACEWALKS AND TROUBLESHOOTING

After celebrating the New Year, Afanasayev and Manarov prepared their spacesuits. On 7 January 1991 they opened the airlock hatch and effortlessly unscrewed four bolts to dismantle the hinge, replaced it with a new one, and refastened the bolts to return the hatch to perfect working order. It was simply a matter of having the proper tools for the job. They then went out to affix a support bracket to a fitting on the conical skirt linking the two cylinders of the base block's main compartment (this had clamped the aerodynamic shroud in place) preparatory to installing a crane. On their way back, they retrieved the Danko cassette that had been deployed a year earlier, then extracted one of the television cameras from the Gamma-2 so that its lens could be adjusted (as the assembly was pressurised, the adjustment could not be done outside). Progress-M 6 docked on 16 January. Its cargo included the crane that had been designed by Vladimir Syromiatnikov of the Energiya Bureau. Afanasayev and Manarov went out to install it on 23 January. It took over an hour to haul the package down the side of Kvant 2 to the support structure that they had previously installed. The 45-kilogram telescopic boom was just 2 metres long when stowed, but could be extended to 12 metres. It was run out to its maximum length to assess its stability. Operated by a pair of hand-cranks, it could be elevated above, or lowered below the base block, and be rotated around the outside arc to reach as far back as Kvant 1. It could transfer a load of 750 kilograms between any two points on the left side of the complex, but care had to be taken to steer clear of all the solar panels. It was 'parked' against Kvant 2 to enable the cosmonauts to ascend it on their way back to the airlock and then slip down it at the start of the next spacewalk. Once back at the airlock, they retrieved the Ferrit cassette. Three days later, they went out to affix a framework mount to each side of Kvant 1 to accommodate Kristall's solar panels. They then installed a number of laser reflectors for a rangefinder to be carried by Buran on its first rendezvous (at that time, scheduled for 1992). Finally, they set up the Sprut-5 spectrometer on Kvant 2 (this measured the flux of charged particles around the complex and downloaded its data by the telemetry link).

The end of the month was devoted to routine maintenance on the environmental and thermal regulation systems, while the Gallar furnace produced gallium arsenide. Manarov exceeded Romanenko's 430-day record for total accumulated time in space on 6 February. Much of the following two months was devoted to Earth studies, but automated sensors continued to sample data on an ongoing basis, and the Optizon-1 furnace was operated intensively. The Pion-M (designed to investigate the processes of heat exchange in a liquid) had proved to be extremely sensitive to small vibrations from other apparatus in the complex, so it was used to *measure* the pollution of the microgravity environment. Progress-M 6 left on 16 March. When Progress-M 7 was 500 metres from the rear of the complex on 21 March, its Kurs system decided that it had drifted off course and aborted its approach (this was the first time that a cargo ferry had suffered a problem in the final approach). By a sad twist of fate, it carried the cake for Manarov's 40th birthday the next day. The ship made another approach on 23 March. At first everything went well, but when it was 20 metres out the flight controllers monitoring its downlink were alarmed to observe that it was misaligned, even although the Kurs thought that it was on course, and it passed within 5 metres of Kvant 1 and narrowly missed striking the base block's left solar panel. Although the diagnostics performed on the Kurs system did not report anything wrong, it was clear that there was a fault. However, it was not evident whether this was in the ferry or the Mir complex. There was only one sure way to find out. Afanasayev and Manarov undocked Soyuz-TM 11 on 26 March and manoeuvred around to the rear. When they let the Kurs system attempt an automatic approach it drifted off course, indicating that the fault was in the Kurs on Kvant 1. Having localised the fault, they then docked Soyuz-TM 11 manually at the otherwise unusable port. Analysis of the telemetry suggested that the Kurs antenna on Kvant 1 was misaligned. Progress-M 7 docked at the front port two days later, thereby confirming that there was nothing wrong with its Kurs. In addition to Manarov's cake, it carried a Raduga capsule and additional nickel–cadmium batteries for the base block. It was soon realised that the most likely cause of the problem was that Kvant 1's antenna had been disturbed by a cosmonaut during the spacewalk to affix the solar panel mountings. Consideration was given to replacing the damaged antenna with that at the rear of the base block, which had been redundant since Kvant 1 docked, but it seemed simpler to send up a

A view of Mir from Soyuz-TM 11 during a fly-around.

new one. However, there was no urgency because cargo ships could dock at the front port, and the replacement was assigned to the next crew – and at about this time, it was decided to postpone their launch by a week in order to give the residents time to finish their programme.

While Afanasayev and Manarov undertook routine maintenance, remote-control observations using the ASPG-M cameras were made to assess industrial pollution in the Ukraine. On 12 April (the 30th anniversary of Yuri Gagarin's pioneering flight) Boris Olesyuk reported in an article in *Trud* that although there were five furnaces in the Kristall module, there was insufficient power to operate more than two of them simultaneously. While this was true, it would be unreasonable to expect that every piece of apparatus on the complex should be operated on a continuous basis, so this was unfair criticism. On 26 April, while Afanasayev set up a thermomechanical joint near the airlock hatch, retrieved exposure cassettes, and replaced the camera in the Gamma-2, Manarov went to Kvant 1 to inspect its Kurs antenna and found that the

23-centimetre-diameter parabolic dish was missing! Before they shut the hatch, they retrieved the experimental joint. (This tested a material that reverted to a predefined shape on being heated, this energy being supplied by a filament to which an electrical current was applied. Such a joint was superior to a conventional mechanical hinge. The ability to test technologies on an ongoing basis assisted in the development of a mechanism for erecting structures in space.) While a cargo ferry was attached to the complex, it was responsible for orbital manoeuvres. In rendezvousing three times, Progress-M 7 had consumed more propellant than planned, and manoeuvring the complex had used most of what remained, and so Soyuz-TM 11 had to take on this rôle. The flexibility of the replenishment system of this latest version of the tanker was illustrated on 15 April when propellant was pumped from the base block *into* Progress-M 7's integrated propulsion system to enable it to de-orbit itself, in order to return a Raduga capsule. Unfortunately, when this was done on 7 May there was no sign of the capsule. If it failed to release, it would have burned up with the rest of the craft. If it was released and returned to Earth, then its radio beacon must have malfunctioned. This means of returning valuable cargo had been criticised by veteran cosmonauts, who felt that it would be better to send up an unmanned Soyuz to resupply the complex when there was material to be returned, even although this would preclude replenishment of the complex's fluids.

MIR'S FIRST WOMAN VISITOR

Soyuz-TM 12 lifted off on 18 May with Anatoli Artsebarski, Sergei Krikalev and Helen Sharman. As the Kurs system began its initial approach, Artsebarski realised that it was giving incorrect angular-separation data – he could see that Mir was not where the system believed it to be. He left the computer to close the separation to a few kilometres and then took over, closed in to 500 metres, paused to line up for a straight-in approach and docked at the front port. Even after the latches had engaged, he wryly pointed out that the Kurs believed it still had 100 metres to go. Given that the port had presented no problem for Progress-M 7, it was concluded that the fault was in the ferry. This was Artsebarski's first flight, but Krikalev

Anatoli Artsebarski (top), Helen Sharman and Sergei Krikalev.

had already served a tour on Mir. Sharman was a British food scientist. Sponsorship had been sought to pay the fee and provide experiments, but this had not been forthcoming and the fee had been waived in return for Sharman's participation in the ongoing investigation of adaptation – unlike Akiyama, she adapted well. As the first woman to visit Mir, she was presented with a tiny bonsai tree. With no special experiments, she spent many hours admiring the Earth – as indeed did most cosmonauts in their spare moments. On a personal level, therefore, she probably benefited more from her visit than had her often overworked predecessors. Her physical adaptation was regularly tested by a cardiac monitor and the Reflotron blood analyser, and her psychological adaptation by the Prognoz and Pleven-87 tests. The Electrotopograph-7K apparatus was put in the scientific airlock to expose samples to space. Several biological experiments were conducted, including Vazon to grow ginseng, onion and chlorella, Vita to study cells producing luciferase (a biologically active albumen) and Seeds (which required a bag of tomato seeds to be stored in Kvant 2's airlock during the handover so that genetic irregularities resulting from their exposure to ambient radiation could be studied on their return to Earth). A recent computer failure had prevented the solar panels from continuously tracking the Sun, so the batteries were not always sufficiently charged to run all of the systems through the shadow pass, and apparatus shut down as the power diminished, the comforting continuous hum of the life-support system faded and often the lights went out for a few minutes. This situation was aggravated by the fact that the Sun was passing through the plane of the complex's orbit, maximising its time in shadow. Afanasayev, Manarov and Sharman departed in Soyuz-TM 11 on 26 May. Although they had accomplished most of their programme, the retiring residents had not had time to attend to Kristall's solar panels.

Recovering Helen Sharman.

CONSTRUCTION WORK

Artsebarski and Krikalev had drawn a heavily construction-based programme calling for as many as eight spacewalks during their planned five-month tour. The first task was to transfer Soyuz-TM 12 to the rear on 28 May, in order to leave the front port free for Progress-M 8, which arrived on 1 June. The next week was divided between unloading its cargo (which included a replacement Kurs antenna and the tools needed to install it) and the routine maintenance which followed the arrival of parts for the environmental system. Then they set the semi-automated furnaces running and made a pre-harvest survey of Kazakhstan. The MAK-1 sub-satellite was ejected from the scientific airlock on 17 June. Carrying equipment to study the upper atmosphere, it was designed to radio its data directly to Earth, but its parabolic antenna failed to unfurl because its battery had gone flat (their predecessors had intended to release it a month earlier, but had been too busy); it was decided to send up a replacement as soon as possible. Artsebarski and Krikalev ventured out on 25 June to install a new parabolic dish on the existing mounting of Kvant 1's Kurs antenna. The mechanism was so delicate that they had to employ a dentist's mirror to see some of the smaller components. The only way to test their work would be for a ship to approach, but Soyuz-TM 12 was occupying the rear port. On returning to the airlock, they set up another experimental thermomechanical joint.

On 28 June Artsebarski and Krikalev ventured out again, this time to deploy the University of California's Trek cosmic-ray detector on Kvant 2. This was to be left outside for two years (but actually remained outside for much longer). Delivered by Progress-M 8, it comprised layers of phosphate glass designed to track the passage of the super-heavy nuclei component of the cosmic-ray flux. It was only the second American experiment to be delivered to the complex. Using the crane to move about, they also installed a number of small detectors to measure the spatial and energy distribution of charged particles near the complex. On the way back, they retrieved the experimental joint. On 15 July, they swung a work platform across to Kvant 1, and affixed it to the 'roof' of that module. Four days later, they returned with a large box containing the Sofora girder. Its tubular rods were connected by sleeve joints made from a 'smart' titanium–nickel alloy. The first segment had been erected prior to leaving the airlock in order to assist in the process. Once this had been affixed to the base plate, two additional segments were erected and mounted. The cosmonauts opined that it was confusing performing such assembly work in the ever-changing illumination of the daylight pass, but had no trouble working by the fixed lighting of their lamps in orbital darkness. More truss segments were added on 23 July, but the final (twentieth) was not attached until 27 July. By this time the girder projected 14 metres from the module. To celebrate their achievement, they put a *Hammer and Sickle* of the Soviet Union on the far end of the truss. During 24 hours outside, they had shown that orbital 'construction work' was feasible. Throughout this period, the automatic sensors continued to gather data for studies of space physics in the near-Earth environment (furthering the study of the correlation between tectonic activity and the charged-particle flux at orbital altitude); observations were made by the X-ray telescopes under remote control; EFO-1 data was secured for the Climate experiment; and the Vibroseismograph measured micro-accelerations in the expanded complex to assess its potential for microgravity research. In August, a post-harvest survey was made of Kazakhstan, and observations were made to assess the ecology of the Aral Sea.

As Progress-M 8 withdrew on 16 August, the cosmonauts monitored an attempt to deploy an aluminium foil balloon to act as a reflector. They had attached this package to the ferry's docking assembly. Unfortunately, although the cover released properly, the balloon inflated irregularly and its fabric ripped. It had been intended to track it from the ground in order to measure the density of the upper atmosphere. (A lightweight structure such as a balloon would have been extremely susceptible to perturbations by the rarefied gaseous environment at orbital altitude.) Progress-M 9 docked at the front on 23 August. Its cargo included a Raduga capsule and a special dispenser for Coca Cola that was flown for advertising purposes under the terms of a commercial contract with the American manufacturer.

In early September, in the aftermath of a failed coup to oust Mikhail Gorbachev as President of the Soviet Union, there was speculation in the West that Mir might be sold to NASA. When they heard this, Artsebarski and Krikalev enquired whether they were to be included in the deal. This amusing anecdote exposed a real debate on the future of the orbital complex. In the harsh economic realism that accompanied the political chaos, it seemed certain that the Buran shuttle would be cancelled; that the Energiya heavy-lift launcher (denied both Buran and the Polyus orbital factory module) would find itself redundant; and that with little prospect of generating a real financial return, Mir would simply be abandoned in orbit. After jokes that they seek to establish communication with NASA's Mission Control in Houston, Artsebarski and Krikalev were particularly productive in the second half of the month, using the Priroda-5 and KAP-350 cameras to assess soil conditions and crop growth in the territory adjoining the Kara-Bogaz-Gol Bay in Turkmenia, the Golodnaya Steppe in Uzbekistan and the Aral Sea in Kazakhstan. Much of this imagery was returned to Earth in the Raduga capsule released by Progress-M 9 on 30 September.

A KAZAKH AND AN AUSTRIAN VISITOR

On 4 October, following what flight director Vladimir Solovyov described as one of the smoothest approaches he had seen, Soyuz TM-13 docked at the newly vacated front port. Commanded by Alexander Volkov, for the first time a crew included two guests. Franz Viehböck was Austrian, and flying by commercial agreement. Takhtar Aubakirov, a Kazakh, was flying by invitation. Whilst earlier cosmonauts had been born in Kazakhstan, this was the first time that the Intercosmos scheme had been explicitly made available to one of the internal Republics. Aubakirov had entered training only at the start of the year in the expectation of flying on Soyuz-TM 14 in November, but in July budget constraints led to the decision to fly the two visitors together, to save the second rocket for a later mission. Because Aubakirov displaced the assigned engineer (Alexander Kaleri), Krikalev was asked to remain onboard for another six months. Having previously served with Volkov, Krikalev readily agreed, and the fact that he was familiar with the current state of the complex considerably eased the handover.

The visitors had a busy time. In addition to the standard medical tests

Takhtar Aubakirov (left), Alexander Volkov and Franz Viehböck.

to monitor adaptation, Aubakirov took part in several ongoing experiments, including Prognoz, Son-B and Batyr (to evaluate the effect of breathing exercises as a means of easing the initial phase of adaptation). He also had two biotechnology experiments: Altyn was a genetic study of wheat, and Maskat tested the ability of certain compounds to enrich genetic material. But his main task was to take pictures of Kazakhstan for the Aral-91 project to monitor the movement, concentration, composition, temperature and speed of dust and aerosols blowing off the recently exposed bed of the Aral Sea.

Viehböck's programme had been devised and funded by Joanneum, one of the large Austrian research organisations. About 150 kilograms of apparatus had been delivered by Progress-M 9. He had ten biomedical, three materials-processing and one Earth-observation experiments. Cogimir analysed his cognitive functions during adaptation; Lungmon tested a new electrical heart-and-lung monitoring unit; Dosimir tested a dosimeter; Pulstrans analysed pulse transmission and heart frequency during changes of body position, and during strain; Bodyfluids studied the composition and distribution of blood and bodily fluid in weightlessness by measuring the speed of sound in blood, to help determine the dynamics of transient fluid motions; Optovert measured eye movement in response to optokinetic stimulation; Mirgen used blood analysis to evaluate the effect of space radiation on genetic material; Motomir used a four-element ergometer to measure the force and velocity characteristics of the limbs to yield a neurophysiological analysis of body motorics; Mikrovib investigated skin sensitivity by analysing spontaneous and stimulated microvibrations of the body's surface in microgravity; Monimir used the ergometer to investigate the movement of the subject's head and arms to analyse postural reflexes to assist in the development of a computerised neurological analyser; and Audimir investigated changes to the auditory system. Many of these experiments would be added to the growing range of tests of adaptation to weightlessness made available to successive visitors, so that over time a substantial database of results would be accumulated for each study. The Fem experiment involved using the MKF-6MA camera for a multispectral survey of Austrian territory; Logion was to determine whether an ion-emitter could cancel the electrostatic charge that builds up on spacecraft in orbit (to test a method for making a magnetospheric research satellite more sensitive by discharging itself, to prevent its own charge interfering with extremely fine measurements); Brillomir was to measure critical fluctuations during the decomposition of binary liquids in microgravity; and Migmas tested a mass spectrometer built for the European Space Agency which was to be used on NASA's Space Station Freedom.

Franz Viehböck (left) and Anatoli Artsebarski.

Towards the end of the handover, the departing crew took air and water samples and collected smears from various points in the complex, just to check that there was no accumulation of toxins or microflora which would pose a health risk. Artsebarski, Viehböck and Aubakirov left in Soyuz-TM 12 on 10 October. It had been decided to use Soyuz-TM 13 to test the repaired Kurs antenna, because the crew would be able to complete the docking manually if it failed. On 15 October, after several successful approaches, the automated system was permitted to dock. But there was frustration four days later, when Progress-M 10 aborted its approach 150 metres from the *front* port, although it docked successfully on 21 October. For the rest of the year, Volkov and Krikalev combined materials-processing in the Gallar, Optizon-1, Kristallisator and Krater furnaces with terrestrial observations using the Priroda-5 and KAP-350, and an ultraviolet sky survey using the Glasar-2. In addition, the X-ray telescopes in Kvant 1 were remotely operated for astrophysical observations. The tranquillity on Mir contrasted with the shockwaves that were tearing apart the Soviet Union.

7

Expansion or abandonment?

UPHEAVAL

Mikhail Gorbachev resigned as President of the Soviet Union on 25 December 1991, and a few days later the *Hammer and Sickle* on the Kremlin was pulled down and replaced by the Russian flag. The Soviet Union formally ceased to exist at the end of the year. It was superseded by the Commonwealth of Independent States (CIS). The flag on Mir's Sofora girder was now providing 'top cover' for a State that no longer existed. The Western media dubbed Sergei Krikalev, who had been launched prior to the start of his country's demise, "the last Soviet citizen". Work in space continued as if nothing had happened. By this point the X-ray telescopes were being used on a five-day cycle. Accumulated results were put into Progress-M 10's Raduga capsule, but its departure was delayed while a fault in the gyrodyne system was investigated – the ferry was retained so that its thrusters would be able to be used to stabilise the complex – but it left on 20 January and returned its capsule to Earth. Progress-M 11 arrived a week later, with tools to enable the cosmonauts to gain access to the failed gyrodyne, the second Payload Systems experiment, and miscellaneous apparatus for forthcoming international visits. Although Volkov and Krikalev had been promised a jar of honey, none had been available. This seemingly trivial procurement problem in the post-Soviet chaos would soon turn into a nightmare. Boris Yeltsin, the Russian President, promptly established the Russian Space Agency under the chairmanship of Yuri Koptev to manage civilian space operations. It assumed control of existing launchers, spacecraft and Mir. Counterpart organisations were established by newly independent Ukraine and Kazakhstan, and these expropriated ground stations and tracking facilities. Kazakhstan even demanded that Russia pay a fee for each launch from the Baikonur Cosmodrome. In fact, the dispute over ownership of the facilities in Kazakhstan led to a suggestion that Mir-related launches be moved to the Plesetsk Cosmodrome, north of Moscow. Berthing facilities in the newly-independent Baltics were withdrawn from the ships of the tracking and communications fleet, preventing their use. Financial limitations that closed ground stations "temporarily" denied Mir use of the Luch network. For the foreseeable future, therefore, Mir would be able to communicate with Kaliningrad only while over *Russian* territory and its diminished network of tracking stations. In Kaliningrad, the flight controllers displayed placards to express their dissatisfaction with the inflation that was eroding their salaries, and many of them sought secondary jobs. It had originally been intended to complete the buildup of the Mir complex in 1990, and initiate construction of the more elaborate Mir 2 complex in 1992. Now, however, Mir 2 seemed to be a pipe dream. If Russia was to sustain orbital operations, it was evident that Mir would require to be kept in service until the end of the century, which was so far beyond its expected service life as to present a major engineering challenge. In fact, the first components would be technically obsolete even before the final module was launched. The base block had been built to operate for five years, but had already been in space for six years. And although the Kvants had been intended to be used for three years, Kvant 1 had been in place for five years and Kvant 2 for three years. Nevertheless, in a very real sense, even if the crews' time over the next several years was devoted entirely to maintenance, the engineering expertise that this endeavour would provide would itself represent a significant *result*, and would be a great step towards constructing and maintaining an industrial-scale orbital factory which might turn in the long-sought profit. But would the funds be provided to complete Mir?

In early February, Valeri Poliakov denied rumours that Krikalev was seriously ill, and pointed out that on his most recent medical checkup (made only a few days earlier) he had been confirmed to be in excellent health. Some reports of Krikalev's supposed illness also claimed that he had not been returned for immediate treatment because there were insufficient funds for a rocket to bring him back, which assertion demonstrated a misunderstanding of how crews rotated. (There was always a ferry *in situ*, and Soyuz-TM 13, which was in space, could return at any time.) Once this rumour was off, it fed on itself, and it was speculated that he would be replaced by a cosmonaut on the next visiting flight. Meanwhile, Volkov and Krikalev prepared for external operations. This spacewalk took place on 20 February. It ran into trouble almost immediately. The heat exchanger in Volkov's suit failed. The sublimator that chilled water for circulation through tubes in the inner garment malfunctioned, and he overheated and his visor fogged over. Rather than abandon the excursion, he plugged himself back into the airlock's environmental support system using an umbilical, and retrieved exposure cassettes from near the hatch while Krikalev made his way down Kvant 2, then along the base block, taking much longer than planned because Volkov could not swing him on the crane. On Kvant 1, he dismantled some of the auxiliary structures that he and Artsebarski had erected as preliminaries to the construction of the Sofora girder. Then he used an ion-emitter to electrostatically sweep dust off the lens of the television camera adjacent to the rear docking port. On his way back, he retrieved the experimental transducers that had been put on the base block's dorsal solar panel by Titov and Manarov many years earlier. Throughout the first quarter of 1992, the Kristallisator furnace was used for a series of experiments to investigate the process of crystallisation in silver-germanium and lead chloride-silver chloride, which are eutectic alloys, meaning that they have extremely low freezing points. The Gallar was also used on a continuous basis to produce a variety of semiconductors. During routine maintenance in early March, the cabling that had been run throughout the complex to distribute power was replaced. Progress-M 11 left on 13 March, and the next day Soyuz-TM 13 was moved so that the front port would be vacant when it left. (Despite it seemingly having been repaired, the Kurs antenna at the rear had not yet been entrusted with an automated cargo ferry.)

A VISITOR FROM UNIFIED GERMANY

Soyuz-TM 14's automated docking at the rear on 19 March confirmed the repaired Kurs. It brought Alexander Viktorenko, Alexander Kaleri and Klaus-Dietrich Flade. If Kaleri had not lost his place on Soyuz-TM 13, he would have been looking forward to returning to Earth now. Flade was making a fee-laying visit for Germany, which had unified in 1989.

Klaus-Dietrich Flade (left), Alexander Viktorenko and Alexander Kaleri.

For the TES experiment, Flade utilised the Kristallisator furnace to measure the heat capacity of a supercooled metallic melt (in this case antimony, two samples of sapphire and an alloy of silver and germanium) and how it varied with temperature. This would enable other thermophysical properties of these materials to be derived. A study of the specific heats of supercooling fusions was a basic science experiment that exploited the fact that direct contact between the material and the walls of the furnace can be eliminated in microgravity. The main focus of Flade's programme was to establish baseline biomedical data in preparation for operations on the European Space Agency's Columbus laboratory, which was to be mounted on NASA's Space Station Freedom. The biomedical tests concentrated on cardiovascular measurements and the investigation of hormones, plasma proteins, and redistributed body fluids in microgravity. The CHR experiment (supplied by the Genetics Institute of Essen University) involved sampling prior to, and after the flight to identify chromosomal aberration in the lymphocytes; ROK (Max Planck Institute) investigated responses to different orientations in weightlessness; OVI (University of Mainz) involved a special pair of goggles equipped with stimulators and sensors to study the influence on the eyes of vestibular disturbances in the absence of gravity; TON (University of Hamburg) used a specially built sensor to determine the interior pressure of the eye; HPM investigated hormonal changes by taking blood, saliva and urine samples; SUR studied whether the body's circadian rhythm changed in the process of adaptation to microgravity; HSD (University of Berlin) used the Tchibis suit to determine tissue layer thickness and compliance; VOG investigated eye movement by video-oculography for vestibular studies; PSY used a portable computer to test perception, speech and psychomotor coordination; ISX involved wearing knee restraints while doing calf exercises to evaluate the effect of isometric exercises on muscles, blood pressure and heart rate; KFV used applied potential tomography (APT) apparatus to investigate changes in the distribution and flow of body fluids (it was rather ironic that the APT apparatus had first been proposed for Sharman's flight); and DOM tested different dosimeters for measuring radiation inside the complex. After assessing the results of Flade's endeavours, Germany booked a second visit.

While Flade had a very busy week, Volkov and Krikalev briefed their successors on the status of the complex, and Viktorenko and Kaleri explained what life was like in independent Russia. With the Russian Space Agency's financial crisis deepening, serious thought was being given to "temporarily" vacating Mir, so the new residents began their tour knowing that they might be recalled at any time. Flade returned to Earth with Volkov and Krikalev in Soyuz-TM 13 on 25 March. This was the second time that three cosmonauts who had flown up separately returned in the same ferry. Although Krikalev had served a 312-day double tour, he had not even remotely threatened the year-and-a-day single-mission record held by Titov and Manarov. Nevertheless, he had set a new record for spacewalking by extending his accumulated time outside to 36 hours. Even so, the spacewalk to attend to Kristall's solar panels had been postponed yet again. Kristall had been attached to the lower port as an expedient to balance Kvant 2 until the module configured for that position in the complex arrived. Kristall was meant for a *lateral* port. It had been fitted with retractable solar panels so that it would not interfere with the base block's panels after it was relocated. Transferring its panels to Kvant 1 had been a high priority when it had appeared feasible to finish the construction of the complex over the period of several years, but when financial constraints ruled this out and it became apparent that Kristall would have to remain in place for the foreseeable future, the preparatory construction work had become a task that was handed on from one crew to the next and achieved only as opportunity arose in the spacewalk schedule.

A QUIET TOUR

As March gave way to April, Viktorenko and Kaleri performed photographic and spectrometric observations of agriculture around the Sea of Azov for a commercial contract with the Terra-K project that offered overhead imagery to local agricultural cooperatives. Meanwhile, the semi-automated furnaces churned out semiconductors and remotely controlled astrophysical observations were conducted using the X-ray telescopes in Kvant 1. Progress-M 12 arrived on 22 April. After unloading the cargo, and the standard maintenance on the environmental and thermal regulation systems, they stripped down and overhauled the ageing base block's communications system. In May, the Vibrogal experiment was run to characterise the vibrations polluting the complex's microgravity environment. On favourable ground passes later that month, the Gamma-2 cameras on the remotely controlled platform on Kvant 2 were used to monitor the ecology of water basins, heavily forested areas and agricultural land for the Terra-K contract. Other ongoing work monitored the spread of pollution from industrial sites, with Krasnodar Kray, the industrial part of Kazakhstan and the area around Chernobyl in the Ukraine being carefully monitored. Then, in June, the Gel apparatus was used to make more polycrylamide for use on Earth in pharmaceutical processes. Progress-M 12 replenished the complex's fluids, refined its orbit and then left on 28 June. Four days later, Progress-M 13 aborted its straight-in approach at a range of 150 metres, when its Kurs decided that its roll-rate was outside permissible tolerance. Analysis of its telemetry revealed that there was a fault in a new software routine, so it was rewritten and uplinked, and the manoeuvre was completed without incident two days later. Its cargo included a pair of gyrodynes and equipment for the forthcoming international visit. Although Kvant 1 had long exceeded its service life, five of its six gyrodynes were still functional. The gyrodynes in Kvant 2 had proved less reliable – four of its six had failed. To continue to use the gyroscopic system to orient the complex, some of the units were to be replaced. In contrast to most of the critical systems, these bulky packages had not been intended to be serviced in orbit, so rather then try to fix the failed units, it had been decided to install new ones. Each was a pressurised spherical casing about half a metre in diameter, having stubby axial projections that supported a magnetically controlled flywheel. Installing them would not be easy. On 8 July Viktorenko and Kaleri left the airlock, moved half-way down Kvant 2, and then utilised a heavy-duty cutter to slice through the thermal blanket to gain access to the control system in order to mount the two new units. On returning to the airlock, they evaluated a pair of binoculars especially designed to be used by a spacesuited cosmonaut. Over the next few weeks, they combined unloading cargo with ongoing studies of the ecological state of rivers and lakes using the MKF-6MA. Progress-M 13 vacated the front port on 26 July.

Alexander Kaleri and Alexander Viktorenko work on a gyrodyne.

A FRENCH VISITOR

Three days later, Anatoli Solovyov, Sergei Avdeyev and Michel Tognini arrived in Soyuz-TM 15. When its Kurs system abandoned the final approach, Solovyov took control and docked manually. Although this was France's third flight, it was the first to be undertaken on a commercial basis (in fact, since their earlier visits had been so productive, the French marked this docking by announcing that they wished to visit at roughly two-year intervals). In the post-Soviet financial climate, these fee-paying missions not only gave an opportunity to recoup some of the cost of essential crew exchanges, they were cited as international recognition of the orbital complex's merit, and thereby played a significant part in countering those who argued that Mir was a costly irrelevance that should simply be abandoned. But the crucial issue for those in favour of continuing operations (and indeed for resuming assembly) was whether the base block could be sustained for long enough to fulfil such advance bookings.

Full opportunity had been taken of the 12-day handover to provide Tognini with a varied programme of ten experiments involving medical and technological studies. Some 380 kilograms of apparatus had been delivered by Progress-M 13. The Alice materials experiment

Michel Tognini.

observed the phase-change phenomena at the 'critical point' of a gas-liquid. Superconductor in the Krater furnace studied the crystallisation of a high-temperature superconductor. The medical programme consisted of the Orthostatism experiment to measure orthostatic resistance by utilising the Echograph to monitor changes in the cardiovascular capacity and venous circulation, and invasive sampling to record hormonal changes; Illusion used the Physalie-M apparatus to expand on an earlier study of sensory and motor adaptation; Vinimal-92 studied sensory-motoric relationships using perception and orientation tests; Nausicca-1 measured radiation inside the station, which was correlated with its movement around its orbit; Biodose observed the biological effects of cosmic rays; Eceq measured the flux of the heavy-ion component of cosmic rays; and Immunologie-92 monitored characteristics of the immune system in space. The newcomers had a number of their own experiments, including the Altyn genetic experiment to observe the transformation of plant cells; Rekomb to produce cells and micro-organisms with given properties; and Reservoir to investigate the process of filling and emptying a capillary-tension reservoir. The residents continued to utilise the MKF-6MA and Priroda-5 cameras. Even when the inhabitants slept, the Mir complex remained active because the multispectral Gamma-2 camera package and the X-ray telescopes were operated by remote control and space physics sensors sent data by telemetry. Viktorenko, Kaleri and Tognini departed in Soyuz-TM 14 on 10 August.

THE VDU THRUSTER

This was Avdeyev's first mission, but Solovyov had served a 180-day tour in 1990. They set up the Krater furnace to produce epitaxial silicon, and then reran Tognini's Alice experiment. In effect, they were waiting for Progress-M 14, which docked on 18 August. It was the first cargo ferry to use the rear port since the repair of the Kurs antenna, and it was fortunate that it managed to dock because the most bulky item of its cargo could be unloaded only at the rear of the complex. This was because the compartment normally used for the 'wet' cargo had been modified to deliver an "external propulsion unit" (referred to as the VDU, as this was the acronym for the Russian phrase) to be mounted on the Sofora girder. This package could be retrieved only by spacewalking. On 2 September, a command was sent to an angled mounting in that section of the ferry to open a cover to expose this thruster block. Rather than lug the 700-kilogram package up the truss, a pivot had been built into a segment one-third of the way along the truss's length so that its upper part could be folded down to position its tip directly over the hatch from which the package would emerge. The next day, the cosmonauts made the first of a series of excursions to install it. They folded the Sofora girder down, clamped it in position, then retrieved the now useless *Hammer and Sickle*. Next, they used a special ratchet in the ferry to slide the boxy unit out of its storage compartment. Four days later, they ran an umbilical along the length of the girder and mated one end to the thruster block and the other to a plug on Kvant 1. Next, they attached several metal braces to the block so that it could be connected to the platform on top of the girder. After another four-day rest, they attached the thruster block and swung the girder up to a position 11 degrees beyond 'vertical', which placed the side-mounted thrusters of the block in the same plane as the peripherally located roll-control thrusters at the rear of the base block. By being so much further from the complex's axis, these thrusters would yield an 85 per cent saving in propellant in rolling the complex, which would in turn reduce the frequency with which the base block's tanks had to be topped up. However, because the VDU was self-contained (the umbilical was to issue firing commands and monitor the rate of consumption of the cold gas that it squirted) it would have to be replaced once its tank was empty. A fourth excursion had been planned in case the operation proved trickier than expected, and it was decided to use this time to undertake tasks that had been scheduled for later, so on 15 September they went

The Sofora girder with the VDU roll-control thruster block mounted on Kvant 1.

to the far end of the Kristall module to install a Kurs antenna for the axial androgynous docking port, and on their way back retrieved a micrometeoroid collector and several exposure cassettes.

MORE ADVANCE BOOKINGS

In early October, extending its international cooperation, the Russian Space Agency signed an agreement with NASA for a cosmonaut to fly on the Shuttle in 1994 and for an astronaut to make a lengthy visit Mir in 1995. NASA's rationale was to study adaptation to weightlessness in advance of the construction of its own Space Station Freedom. The Russian Space Agency hoped that the American participation would release the funds required to complete the complex, as for the foreseeable future Mir would be the only inhabited orbital facility for microgravity research available to the international community. And as if to confirm this, a month later the German Space Agency signed an agreement to use Mir as a base for evaluating a manoeuvrable free-flying robotic vehicle (called Inspector) that would be delivered by a Progress ship in 1997. Like the French, the Germans were clearly hoping that the Russians would be able to operate Mir for at least another five years. Additionally, the European Space Agency contracted in November for two extended visits in 1994 and 1995 to provide a total of six months onboard Mir. These flights were to yield insight into adaptation to weightlessness and test equipment and procedures to be applied on the Columbus laboratory. (With little prospect of NASA starting to assemble its space station any time soon, and the discouraging fact that its laboratory was to be the last module attached, the European agency was keen to exploit the opportunity offered by the Russians to carry out long-term research on Mir.)

END OF A TOUR

Their orbital construction successfully completed, Solovyov and Avdeyev resumed ongoing research. High on their list of priorities was monitoring the ecological state of water basins, forested areas and agricultural land for the Terra-K project, and the spread of pollution in the industrial Krasnodar Kray and Novosibirsk Oblast areas. The various furnaces were operated throughout, to create semiconductors and exotic materials (one six-day Krater run produced an alloy of barium oxide, yttrium oxide and copper oxide). Progress-M 14 left on 21 October and returned a Raduga capsule of Earth imagery, and Progress-M 15 arrived the following week with more storage batteries, assorted scientific apparatus, and several quail eggs which were hatched in Inkubator-2 while Solovyov and Avdeyev performed maintenance. On 20 November the MAK-2 satellite was ejected from the scientific airlock to study the physical characteristics of the ionosphere. December was devoted entirely to research. They reran the Alice experiment, continued the Glasar-2 ultraviolet sky-survey, took data for the Climate experiment using the EFO-1 electrophotometer and, at the request of meteorologists, reported on the track of a tropical cyclone in the Indian Ocean over a four-day period. They celebrated the New Year with an all-too-brief two-way video link with their families, and then set straight back to work. The next few weeks saw routine maintenance combined with MKF-6MA photography. As their tour drew to a close, they at least had the satisfaction of knowing that they had completed their programme.

SPACEPORT MIR!

Soyuz-TM 16 appeared on 26 January 1993 with Gennadi Manakov and Alexander Poleshchuk. Like the vehicle that flew the Apollo–Soyuz mission, it was fitted with an androgynous docking system. This was to test the port on the Kristall module – the Kurs antenna to facilitate this had been recently installed during a spacewalk. The automated system made the initial approach, but when it paused at 200 metres Manakov took over and flew around to line up with Kristall. Although this was the first time that a ferry had docked off the longitudinal axis, by now the Mir complex was so massive that there was little prospect of the impulse of contact inducing it to rotate. With Soyuz-TM 15 at the front and Progress-M 15 at the rear, this increased to *seven* the number of independent vehicles, which established a new construction record. The Resonance experiment determined the stresses on this configuration. Nevertheless, this did not mark an expansion of operational capability, because the androgynous port was not to be used regularly. An Israeli researcher had originally been assigned to accompany this handover, but this was cancelled, so Soyuz-TM 16 was notable for *not* delivering a fee-paying guest. Manakov had already served a tour on Mir, but this was Poleshchuk's first flight. Most of the brief handover was spent familiarising the newcomers with the state of the complex's systems, but there was time for several experiments. In addition to installing the Electrotopograph-7K in the scientific airlock to assess samples of construction materials, an experiment supplied by NASA was undertaken to monitor fluid motion in a granular material in order to provide data for the development of a nutrient-delivery system in a plant cultivator. Solovyov and Avdeyev returned to Earth on 1 February in Soyuz-TM 15. Manakov and Poleshchuk had a busy six-month programme ahead, starting with an experiment involving Progress-M 15.

Continuing the use of cargo ships for free-flying experiments, on 4 February the Znamya package was affixed to the docking assembly of Progress-M 15, which then withdrew 150 metres and set up a rapid roll-manoeuvre so that the centrifugal force would deploy eight triangular petals to create a reflector 20 metres in diameter. The attitude of the ferry had been chosen so that this mirror beamed sunlight towards the Earth, to trace a 'spot' of light some 4 kilometres wide across across Europe. To people fortunate enough to catch sight of it, this appeared like "a sparkling diamond" in the sky. This experiment was to evaluate the feasibility of employing orbital mirrors to illuminate polar regions enduring extended darkness. It had initially been planned for October 1992, but Progress-M 14 had been given a Raduga capsule. The experiment over, Progress-M 15 jettisoned the reflector and withdrew. It returned the next day and halted 200 metres out. A pair of controls – identical to those used in the Soyuz – for translational and rotational motions had recently been set up in the base block, with a video monitor to display the view from the ferry's docking camera to enable a ship to be flown under remote control. After commanding a series of manoeuvres to test this system, Manakov released the ferry, which left to de-orbit itself. As the probe had been removed for the Znamya experiment, it was not possible to redock. In addition to enabling the approach of a Progress to be manually completed in the event of a Kurs failure (three recent cargo ships had suffered such difficulties) Viktor Blagov ventured that this TORU system (the Russian acronym for 'remote-control flight system') would enable a cosmonaut to manoeuvre an unmanned vehicle alongside a complex that was festooned with projecting antennas and solar panels, as such manoeuvres could not be programmed into the automated system.

The TORU system on Mir replicated the functionality of the control system of the Soyuz spacecraft (shown here).

The Mir base block completed its seventh year in space on 20 February. It was hoped that continuing preventive maintenance would enable it to host the final two modules for a full programme of research. Progress-M 16 docked on 23 February. It delivered additional gyrodynes and the frames on which to fit them (this time *inside*, rather than outside) together with new electronics for the computerised flight-control system. The next few weeks were given over to routine maintenance and rewiring the Kvant 2 and Kristall modules in order to pool their power output, to enable it to be used more efficiently. On 26 March, Progress-M 16 undocked under TORU control, withdrew 75 metres, and then redocked. It left the next day. Progress-M 17 took its place on 2 April. When this had been unloaded, Manakov and Poleshchuk prepared for two spacewalks to mount electric motors on the frames on Kvant 1 that had been erected two years previously, in preparation for relocating Kristall's solar panels.

THE DISABLED CRANE

On 19 April, Poleshchuk made his way hand-over-hand down the side of Kvant 2 to the base block, and then used the crane to swing Manakov and a motor to Kvant 1. As it took three hours to attach the motor to the frame, by the time they managed to plug it into the power socket they were behind schedule. Worse, Poleshchuk's suit's ventilator had malfunctioned, and he had overheated. As they made their way back, they discovered that one of the crank handles of the crane was missing. It appeared to have worked free and drifted away! The installation of the second motor would be awkward without the use of the crane. It was decided to send up another handle on the next cargo ferry, which meant that the external activities had to be postponed. In the interim, attention switched to operating the furnaces. On 24 May Progress-M 18 docked at the front port. As Soyuz-TM 16 was on Kristall and Progress-M 17 was still at the rear, this was the first time that two cargo ferries had been in place. In addition to a new crank handle, the cargo included a Raduga capsule and apparatus for the next visiting researcher. It was not until 18 June that Manakov and Poleshchuk ventured out. On repairing the crane, they used it to swing the second motor across to Kvant 1, and installed it readily. Before returning, they used a television camera to show Kaliningrad how cluttered the surface of Kvant 1 had become as a result of all the construction work.

ANOTHER FRENCH VISIT!

As all of the docking ports were occupied when Soyuz-TM 17 arrived on 3 July, it paused 200 metres out and filmed the departure of Progress-M 18 before taking its place. A few hours later the cargo ship released its Raduga capsule, which returned near Orsk in the Urals in order not to involve the Kazakh authorities in its retrieval (this was to become standard procedure). As no other international participant was ready, Frenchman Jean-Pierre Haigneré, who had served as Michel Tognini's

With Soyuz-TM 16 on Kristall and Progress-M 17 on the Kvant 1, Soyuz-TM 17 had to wait while Progress-M 18 made available Mir's front port.

backup and was therefore fully trained, was invited to advance his visit by six months and accompany Vasili Tsibliev and Alexander Serebrov. This had both an up and a down side. While he would benefit from the longer than usual (three-week) handover, there had not been time to develop a new research programme, so most of his work had to use the apparatus sent up for Tognini's visit. However, 100 kilograms of additional apparatus had been delivered by Progress-M 18 for new experiments (Synergies, Tissue and Teleassistance) and to enable him to adapt existing equipment to vary the research objectives. The programme comprised the Orthostatism experiment, which innovatively combined Echograph, Diuresis and Tissue, and also the Haut Schicht Dicke apparatus that had been sent up for Flade; Viminal used a miniature flight simulator to study the process of adaptation to weightlessness; Illusions studied the adaptation of the sensory-motor systems; Biodose investigated the long-term effects of cosmic radiation on the body; Immunology studied the adaptation of the immune system; Synergies studied the rôle of the vestibular system in controlling dynamic equilibrium; Microaccelerometer made use of a video camera to measure microscopic accelerations on the complex; and Teleassistance was a repeat of the Orthostatism experiment using a link-up with experts on the ground to assess how easy it would be to provide on-line technical support during complex tasks. In addition, the output from the long-running Nausicca and Eceq experiments (both of which monitored radiation) was stored away.

Jean-Pierre Haigneré.

Manakov, Poleshchuk and Haigneré departed in Soyuz-TM 16 on 22 July. The delay imposed by the need to replace the crank handle had meant that the transfer of Kristall's solar panels had not been attempted – it was passed on to the newcomers, who already had several spacewalks on their schedule. Serebrov had visited Salyut 7 and had served a tour on Mir, but this was Tsibliev's first flight. Progress-M 17 left on 11 August. Progress-M 19 took its place two days later. Having Soyuz-TM 16 on Kristall had given unprecedented operational freedom, and this had been exploited to retain Progress-M 17 for 132 days. (NASA had approached the Russian Space Agency late in 1992 with a view to buying a Soyuz-TM to serve as the 'lifeboat' for its own much-delayed Space Station Freedom, but it would require this to remain in space for at least a year, which was twice the accepted limit of the Soyuz-TM. Since the unified propulsion system was common, it had been decided to demonstrate this using Progress-M 17. After withdrawing from Mir, it was manoeuvred into a lower orbit and powered down. On being powered up again on 2 March 1994 it performed several manoeuvres flawlessly. Knowing that the service module could survive an extended mission was a good start towards recertifying the Soyuz for NASA, but by that time the situation had changed.)

DAMAGE ASSESSMENT

Tsibliev and Serebrov made their first spacewalk on 16 September. They employed the crane to swing a small platform to Kvant 1, which they mounted on the 'roof' of the module, aft of the Sofora girder, and four days later they erected the 5-metre-long Rapana girder on the platform. This was a scaled-down test of a structure intended to hold the parabolic dishes of a solar-dynamics power system away from the body of the proposed Mir 2 module cluster. In this case, however, two exposure cassettes with construction materials were mounted on the girder.

In early August, the complex had been heavily bombarded by micrometeroids when the Earth crossed the orbit of Comet Swift-Tuttle, which had recently made its first return since 1862 and was associated with the annual Perseid meteor shower, so it was decided to add a spacewalk to the schedule to inspect the solar panels. After deploying additional sample cassettes on 28 September, Tsibliev and Serebrov made a video to document the surface of the complex, and, to their surprise, discovered a 10-centimetre-diameter hole punched straight through one of the solar panels. Upon closer inspection, they located 65 much smaller impact pits. At this point, Tsibliev reported that he was overheating. His suit's telemetry showed that coolant was not circulating properly, so the spacewalk was terminated. Viktor Blagov later reported that the complex had been reoriented to face each solar panel in turn directly towards the Sun to measure its peak output, and this showed that if they had been struck by micrometeoroids the damage had not significantly reduced their power output.

Radio Moscow reported on 8 October that the crew had been asked to extend its tour until early in January. The Energomash factory in Samara that manufactured the uprated engines used by the crew-rated rocket would not release any engines until it was paid, and the cash was not available. This kind of production bottleneck was to become a regular feature in the new corporate-orientated rather than state-orientated economy, and it would strain the patience of programme managers and crews alike. Soyuz-TM 18 (which was to have been launched on 17 November) would be able to lift off as soon as its rocket was ready, but it had been decided to postpone it so that the residents could finish a six-month tour. Progress-M 19 departed on 12 October, and returned its Raduga capsule. Although the crew handover could be postponed, replenishing the complex could not, so a rocket reserved for a military satellite was requisitioned to launch Progress-M 20, which arrived on 15 October. In addition to a Raduga (the third in a row, despite the announcement that these would be used only on each third or fourth mission), it delivered another two American biotechnology packages (both protein crystallisation experiments, one for Boeing and the other for Payload Systems; the company's third such experiment) and apparatus supplied by Germany. An impromptu spacewalk was mounted on 22 October to deploy a new micrometeoroid detector, and another a week later completed the documentation of the external surface. Apart from the large hole that they had spotted earlier, the solar arrays were in good condition, considering their age. However, the thermal insulation material was extensively coated by soot, which was believed to be a residue from the attitude-control thrusters. After inspecting the base of the Sofora girder to check that firing the thruster block on its end had not disturbed its mounting, they headed back, pausing only to retrieve a sample cassette which had been outside for several years.

The external video was used by engineers to assess whether the complex would support another three years of operations. Key to their assessment was the state of the solar panels. Analysis of the transducers that had been retrieved periodically had shown their efficiency to be deteriorating at a rate of 5 per cent per year. Nominally, the base block had 100 square metres rated at 10 kW, Kvant 2 had 50 square metres at 7 kW, and Kristall had 72 square metres at 8 kW, but progressive degradation had reduced this ideal 25 kW by 20 per cent, and with mutual shading and losses due to off-normal insolation this could easily be reduced by 50 per cent. Furthermore, much of that power was lost in the cabling. Despite the panels on the expansion modules, even under ideal conditions the power supply was little better than it had been at the base block's launch. The only way to overcome this power crisis was to install more solar panels. In early December 1993, following the announcement in September that America and Russia had agreed to integrate their plans and develop a *joint* space station, a project was launched in which an androgynous docking system would be installed in the Space Shuttle Atlantis to enable this to make a series of visits to Mir, starting in 1995. This was a commercial arrangement. For a fee, NASA would be able to maintain an astronaut onboard Mir on a continuous basis for two years, to carry out research and evaluate apparatus for the International Space Station. Given the Shuttle's capacity to carry cargo, it was not long before it was decided to develop a new solar panel combining the most efficient American transducer cell and the proven Russian deployment mechanism, and have Atlantis deliver it. Until this could be done, however, the complex would be chronically short of power.

Progress-M 20's Raduga capsule was returned on 21 November. It contained the contents of the sample cassettes, a small piece of the base block's thermal blanket retrieved on the latest spacewalk, and the Boeing package – even though the capsule used the new recovery area in the Urals, the package was handed over to a company representative in Moscow only eight hours later. (The experiment for microgravity pharmaceutical applications incorporated a dozen sample containers. It was a 'black box' that required only to be switched on, but the cosmonauts had provided a daily log of the ambient temperature and radiation level to assist the experimenters. The company was so impressed with the results that a month later it signed a contract to fly another experiment in 1994.) While they prepared themselves to return to Earth, Tsibliev and Serebrov continued to use the Glasar-2 telescope and the Kristallisator and Optizon-1 furnaces. To assist their successors, a video was downlinked showing where things had been stowed. Mir's cramped interior had become a storeroom, with frequently-used apparatus strapped to accessible surfaces and large bundles of other items tied in netting and stashed in little-used spaces.

As 1993 gave way to 1994, Vladimir Titov and Sergei Krikalev were training for a Shuttle mission, Norman Thagard and Bonnie Dunbar were training to launch on a Soyuz for a three-month visit to Mir, the Shuttle–Mir deal was in place, and talks were underway to define the configuration of the International Space Station. The continuing production problems that ruined flight schedules notwithstanding, a sense of optimism pervaded the programme. There was no more talk of abandoning Mir as a costly irrelevance – as the only inhabitable orbital complex in existence, it was a vital international resource. Riding the wave of optimism, the Khrunichev factory started to refurbish the partly built modules that would complete the complex.

POLIAKOV TRIES AGAIN

Soyuz-TM 18 docked at the rear port on 10 January with Viktor Afanasayev, Yuri Usachev and Valeri Poliakov, who had hoped to set an 18-month endurance record, but this would not be possible because his return date was fixed and the postponed launch had eaten into his assigned time. On 14 January, Tsibliev and Serebrov put their results (including the latest Payload Systems package) into Soyuz-TM 17 and undocked. Instead of pulling straight back, they were to manoeuvre close by Kristall and take pictures of the apparatus near the androgynous docking system in order to assist the Shuttle pilots in familiarising themselves with it. Instead of withdrawing to perform a distant fly-around, it was decided to pull back several metres and translate down along the length of the module. Unfortunately, Tsibliev did not realise that his translational controller was in standby mode, and so he was unable to prevent a slow drift that resulted in the ferry striking Kristall a glancing blow about a metre from the docking mechanism. The gyrodynes immediately restored the station's stability, and the cosmonauts onboard Mir did not feel the impact. On 24 January Soyuz-TM 18 undocked, withdrew 150 metres, and made a slow fly-around to look for any damage (there was nothing dramatic) and then redocked at the front. Progress-M 21 slipped into the rear port on 30 January.

Given their delayed launch, Afanasayev and Usachev had been assigned a short tour in order to restore the year's schedule. It was important to be able to draw up a long-term strategy, as the schedule would soon be integrated with visits by Atlantis. Difficulties in manufacturing spacecraft and rockets, however, made achieving this coordination increasingly difficult, and it was not long before the planned handover in April became impractical. The research programme included medical and technical experiments sponsored by Germany, using hardware supplied by Kaiser-Threde of Munich. (Some of this had been left by Flade in 1992, more had been ferried up by Progress-M 20, and the last-minute items had been brought by Poliakov as personal luggage.) When working on these experiments, the cosmonauts were to be linked by video with the researchers who had devised the equipment, both to provide instant feedback to the scientists and to receive their technical support.

History was made on 3 February 1994, when Sergei Krikalev was launched on Discovery as mission STS-60. It was fitting that "the last Soviet citizen" had more experience living and working in space than all of his astronaut colleagues combined.

Progress-M 21 left on 23 March and Progress-M 22 arrived the next day. In late March, the cosmonauts took part in an experiment that involved aiming an electron beam at the Swedish Freya satellite while it passed by, 600 kilometres further up the Earth's magnetic field lines. (This satellite had been orbited in 1992 on a Chinese rocket to study the magnetosphere.) The objective was to determine how a beam of charged particles was dispersed by the magnetic field, but the test prompted a report that Mir had tested a 'Star Wars' weapon. The long-running dispute between Russia and Kazakhstan over the Baikonur Cosmodrome was finally resolved on 28 March, when Kazakhstan agreed to lease it to Russia in return for the equivalent of $100 million a year in trade credits. Progress-M 22 undocked on 23 May. Progress-M 23 took its place the following day. Upon unloading it, Afanasayev and Usachev found that some of the food containers had been raided by opportunistic staff struggling to eke out a living in the chaotic economy. Financial restrictions had also prompted the Russian Space Agency to reduce the cosmonaut corps to 50 members, with 17 pilots at its core. Headed by Alexander Volkov, this group included veterans Vladimir Lyakhov, Vladimir Titov, Alexander Viktorenko, Anatoli Solovyov and Gennadi Manakov. Many other veterans had retired due to age. Others now resigned because they saw little prospect of being given a mission. In contrast, NASA was recruiting astronauts to gear up for the International Space Station.

The optimistic schedule was ruined when it became evident that the aerodynamic shroud for Soyuz-TM 19 could not be delivered on time, resulting in a further delay. On 2 July, Progress-M 23 returned its Raduga capsule to Earth. Soyuz-TM 19 took its place the next day to deliver Yuri Malenchenko and Talget Musabayev, neither of whom had flown in space before. Some twenty years previously, it had been decided not to fly 'rookie' crews. The Kurs automated rendezvous and docking system was sufficiently reliable to lift this rule. Musabayev had served as Aubakirov's backup for Soyuz-TM 13, and, in an effort to improve relations with Kazakhstan, had been assigned in place of Alexander Kaleri. His status on Mir became a matter of dispute: as a career cosmonaut, he was to serve a full tour as Mir's flight engineer, but did his Kazakh citizenship make it an international visit? Afanasayev and Usachev departed on 9 July in Soyuz-TM 18, leaving Poliakov with his new colleagues, who, in light of their late launch, had been assigned a short four-month tour in order to reinstate the schedule. Much of July was spent photographing Kazakhstan, particularly the area around the Aral Sea.

CRISIS AVERTED

Progress-M 24 aborted its final approach on 27 August. It was only 10 metres from the front port when it detected a misalignment, and it nearly hit one of the solar panels as it drifted by. It was not clear why it had failed to orient itself properly, but the Kurs was reprogrammed to accept increased tolerances. Another approach was made a few days later. The ship was perfectly aligned as it flew straight in, then pitched over at the last minute. This time the cosmonauts reported that they could hear it nudging the docking collar. The video from the docking camera confirmed that it was misaligned, and so it was ordered to withdraw. It had propellant for only one more rendezvous. On 2 September Malenchenko docked the troubled ferry using the TORU system. This was fortunate, as the financial restrictions that had delayed its launch had also necessitated the postponement of the next cargo ferry to November. Without Progress-M 24's consumables, the cosmonauts could well have been forced to vacate the complex in early October rather than hand it over to their successors, which would have ruined Poliakov's attempt at the endurance record. Furthermore, if the 275 kilograms of apparatus it carried for the European Space Agency had been lost, this would have meant the cancellation of that mission. After a detailed analysis of the telemetry, it was decided that the fault must be in the ferry's Kurs system – a software fault was suspected. On 9 September Malenchenko and Musabayev went out to examine the apparatus surrounding the docking collar, to verify that it had not been damaged. Then they used the crane to swing across to Kristall to inspect the point where Soyuz-TM 17 had struck a glancing blow – although there was a tear in the thermal blanket, there was no evidence of any damage, so they patched up the insulation. An anchor was attached to the shroud-fitting on the righthand side of the base block, in preparation for installing a second crane. On their way back to the airlock they installed the REM experiment sent by the European Space Agency. Four days later, having inspected the mounts of Kristall's solar panels and the motors fitted on Kvant 1 that would drive the panels once they had been redeployed, they retrieved the cassettes from the Rapana truss.

THE EUROPEAN SPACE AGENCY'S FIRST VISIT

Soyuz-TM 20, carrying Alexander Viktorenko, Yelena Kondakova and Ulf Merbold, lifted off on 4 October, thereby re-establishing the schedule intended to synchronise with the first American visitor. As a result of his three previous visits, Viktorenko had accumulated nearly a year onboard Mir. This was the first flight for Kondakova (the wife of former cosmonaut and now flight director Valeri Ryumin). Merbold had already flown on two Spacelab missions on the Shuttle as a European Space Agency astronaut, and was the first guest cosmonaut to have prior flight experience. Later in the day, Progress-M 24 undocked and performed a series of manoeuvres designed to shed light on the Kurs failure, but these revealed nothing. As Soyuz-TM 20 made its final approach the next day, its Kurs suddenly performed a yaw that placed it out of alignment. As this was what had happened to Progress-M 24, it demonstrated that, contrary to what had been thought, the problem was on Mir. Viktorenko took command and docked. Engineers immediately started reanalysing the telemetry in an effort to find out why a docking system that had worked flawlessly for years should suddenly malfunction.

The handover was to last almost a month to enable Merbold to make progress on the programme for the European Space Agency's two commercial visits. His focus was medical, but he also had four materials-science and two technology experiments. The

230 Expansion or abandonment?

Alexander Viktorenko (left), Yelena Kondakova and Ulf Merbold.

medical tests monitored the adaptation of his cardiovascular, neurosensory, and muscular systems to weightlessness, the long-term objective being to develop a way to counter atrophy in muscle fibres and reduce loss of calcium from bones on long missions. Most of the experiments reused apparatus sent up for the French, German and Austrian visits, demonstrating that Mir had indeed become a veritable laboratory in space. In fact, Poliakov was also using some of this apparatus on a regular basis.

Fortunately, Progress-M 24 had delivered a compact freezer to store the biological samples, a centrifuge to separate biological fluids, an IBM computer to process data, a variety of cameras and video equipment, and a passive cooler for the return of the refrigerated samples to Earth, and Merbold had brought 10 kilograms of apparatus as luggage. Merbold had given samples of blood, urine and saliva prior to launch, and his musculature had been recorded by a nuclear magnetic resonance body scanner to give baseline data. The aim was to enable his adaptation to be recorded in fine detail. In addition, he was to monitor his fluid and electrolyte balance, fluid transfer into and out of superficial tissue, chromosomal aberrations in the peripheral lymphocytes

Talget Musabayev (left), Ulf Merbold and Yelena Kondakova.

(that is, those in capillaries furthest from the veins into which they drain) and how changes in the central venous pressure affected the erythropoietic system (that is, the rate of production of red blood cells). In contrast to his Spacelab flights, Merbold's experimental programme on Mir was pursued at a relaxed pace; he worked autonomously, with much less contact with his support team, which had one 20-minute video conference per day. He also noted the different training processes. When he had been in training for a Shuttle flight, he had been issued voluminous technical manuals which explained each piece of apparatus about which he needed to know, but cosmonauts relied on word-of-mouth, mutual training, and calling in the engineer that designed a piece of apparatus if they needed additional technical information.

One week into the handover, a combination of activities inadvertently drained some of the storage batteries. Even with only essential equipment operating, the fact that there were six people onboard meant that the environmental system had to have priority, especially when in the Earth's shadow, so the gyrodynes were deactivated

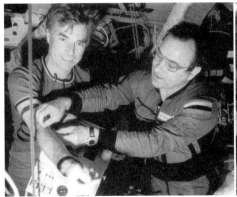 

Valeri Poliakov assists Ulf Merbold with an experiment. Yuri Malenchenko (wearing headphones) and Yelena Kondakova.

and for three days the complex was controlled manually using the thrusters to keep the solar panels face-on to the Sun to recharge the batteries. Recovery activities were intensive for two days and then, as the power level built up, the automated systems were reactivated. Three of the batteries were subsequently isolated from the power supply. No sooner had the cosmonauts recovered from this crisis than an small fire broke out in one of the Vika oxygen-producing systems, but Poliakov smothered it with a cloth before it became serious. Although Merbold had to reschedule his work as a result of the power shortage, this was not really a problem because a fault had disabled the Kristallisator furnace (in which he was to have undertaken experiments to study *in situ* metal matrix composites, undercooled melts, and exotic glasses to get data of benefit to terrestrial manufacturing) and this would be done by the residents once spare parts had been delivered to repair the furnace. Merbold packed his results on 1 November. The frozen physiological samples (34 of blood, 85 of urine and 125 of saliva) were returned in the passive cooler. Everything else was to be returned by Atlantis. Malenchenko, Musabayev and Merbold undocked Soyuz-TM 19 the next day, withdrew 200 metres, and then let its Kurs redock to verify the rear system for the next cargo ferry. They undocked again the following day and returned to Earth. Progress-M 25 docked without incident on 13 November and its cargo included the parts to fix the Kristallisator furnace.

Viktorenko and Kondakova had been assigned a four-month tour. They were to return with Poliakov in March 1995. The launch of the Spektr module had been scheduled for December, so Viktorenko and Poliakov inherited the task of relocating Kristall's solar panels. They were to do this in late November, but before they could unpack the spacesuits the task was cancelled because the module's launch had been postponed to February (and it would slip again, firstly to April and finally to May). A contributory cause of this delay was the late delivery of NASA equipment that was to be included in the module – this delay being a combination of late shipment and customs procedures upon entry to Russia.

A new geostationary relay satellite (this time openly called Luch 1) was sent up on

16 December to replace Cosmos 1897 over the Indian Ocean (the old satellite had exhausted its propellant in early 1993, and begun to drift). Cosmos 2054 was still in position over the Atlantic, but its use by Mir had been restricted since early 1992 – it was made available to improve communications with Mir at critical moments, but was otherwise leased to relay commercial television transmissions. Poliakov claimed the single-mission endurance record on 9 January 1995, having spent more than a year-and-a-day in space. With a total of just over 600 days in space, he also held the cumulative record by a wide margin. Given uncertainties of safe exposure levels to ambient radiation, he slept alongside the batteries in Kristall; his colleagues used the personal cabins in the base block. By this time, the engineers had discovered that the computerised flight control systems of both Progress-M 24 and Soyuz-TM 20 had been improperly programmed with the centre of gravity of their vehicle, which had induced pitching and yawing deviations as the system performed the minor attitude adjustments immediately prior to docking. This condition would not have occurred when the vehicles were subsequently manoeuvred manually. Now that the reason for the difficulties of docking at the front port were understood, it was a simple matter to revise the software. It was vital that the Kurs on the front port be functional for renewed expansion of the complex. To verify it, on 11 January 1995 the cosmonauts undocked Soyuz-TM 20, pulled back 160 metres, and let its reprogrammed Kurs redock; another problem fixed.

8

Shuttle–Mir

CLOSE ENCOUNTER

A step towards joint operations began with the launch of Shuttle mission STS-63 on 3 February 1995, which was to rendezvous with Mir. If the launch missed the five-minute slot, the additional propellant that it would need to complete the rendezvous would jeopardise the deployment and retrieval of a satellite, in which case Discovery was to proceed with its primary mission. Based on prior experience, it had a one-in-three chance of making it. Nevertheless, the launch took place at the opening of the window. Immediately on attaining orbit, two of Discovery's reaction control system thrusters developed a leak. This led to concern that the nitrogen tetroxide might coat instrumentation mounted on the Mir complex. It was decided that if the leaks could not be stemmed, Discovery would not be allowed to approach within 125 metres of the complex. The problem was solved during the three days of manoeuvres leading up to the final approach, and permission was granted to close in to 10 metres of the androgynous port at the end of the Kristall module. Flown by Jim Wetherbee and Eileen Collins, Discovery approached from below, then, with its nose high and its payload bay facing back, stationed itself directly in Mir's path. Mir was oriented to align Kristall's axis along the velocity vector, in order to aim its docking port at the newcomer. On Discovery, Vladimir Titov was responsible for communications with the Mir crew, and had been talking by VHF radio virtually continuously since they established line-of-sight contact. Discovery paused at 300 metres for permission to proceed. From this point, the upward-firing thrusters were not to be used, in order not to blast Mir with efflux. (This was a flight mode referred to as 'low-Z', as only those thrusters aimed obliquely 'up' could be used). Mir's solar panels had been set edge-on for further protection of their transducers. With Mir holding its orientation, the Shuttle was given permission to close in. (If Mir drifted, Discovery was to pause and wait for the complex to realign itself.) During the final phase, Collins maintained a commentary, providing range and closing-rate data using a laser rangefinder, and Wetherbee, at the aft station of the flight deck, viewed the androgynous port through

The APAS docking system as seen by STS-63.

a camera mounted in the upper window of a module in the payload bay that gave a view similar to that of a camera on the Russian-built docking system which Atlantis would have in its bay when it attempted to dock.

The US and Russian television networks relayed the video downlinks from the two vehicles in split-screen fashion as Discovery closed in. The rate was so low that it took about 40 minutes – almost half an orbit, much of it during darkness – to reach the 10-metre limit. Throughout, the video showed members of the two crews waving from their windows. After 10 minutes of staton-keeping, Wetherbee announced that he hoped their successors would be able to shake hands, then eased Discovery back to 125 metres. At this point, the Shuttle made a slow fly-around to photograph the complex using the large-format IMAX motion picture camera located in the payload bay. The joint exercise over, Discovery withdrew to complete the rest of its mission. From closing within the 125-metre point on the approach, to finally departing, it had spent about three hours in company with Mir. This "Near-Mir" flight, as it soon became known, set the scene for the first of the docking missions, scheduled for the summer. It had demonstrated that a Shuttle could approach Kristall's axial port within an 8-degree cone, with a 2-degree tolerance in orientation, then maintain its position at a point 10 metres out; and it had exercised communications and coordination between the crews and between Houston and Kaliningrad. In fact, it had been accomplished on-the-run, because the rendezvous had been added to a long-delayed and already very busy flight plan involving a laboratory, a free-flying satellite and a spacewalk. "When all was said and done", Wetherbee reflected, "it turned out to be easy". The way was now clear for Atlantis to attempt to close that final 10 metres.

Mir as viewed by STS-63.

AN AMERICAN CREW MEMBER

Progress-M 25 left on 16 February. Progress-M 26 took its place the next day, was unloaded, and departed on 15 March to clear the rear port for Soyuz-TM 21, whose arrival the following day was shown live by television networks in both Russia and America because it delivered Norman Thagard, who was to make NASA's first visit, having flown up with Vladimir Dezhurov and Gennadi Strekalov. Although this was Dezhurov's first flight, Strekalov had been in space four times previously and served a tour on Mir in 1990. Likewise, Thagard, a physician, was a veteran of four Shuttle missions. The two crews displayed the Russian and American flags in the base block to symbolise the new spirit of cooperation between the former rivals in space-faring. Unlike previous foreign researchers, Thagard was not limited to the time of the crew handover, even an extended one; he was a full member of the crew and was to serve a full tour. His programme was centred on 28 experiments, mostly biomedical. He was to study all the standard aspects of adaptation (the redistribution of body fluids, changes in red blood cell production and composition, muscle atrophy, and calcium loss) but was to take blood and urine samples regularly in order to track the changes on a continuous basis, in order to achieve an understanding of the processes as well as the outcomes. When NASA first devised Thagard's programme, it had expected that the first of the two remaining modules (Spektr) would have been commissioned, and it had planned to send up his apparatus in this. As soon as it became clear that Spektr would not be ready in time, it was decided to adapt existing apparatus (such as the European Space Agency's freezer) and send up just sufficient equipment on cargo ships to ensure that Thagard would be able to perform a core programme even if Spektr was late. Accordingly, Progress-M 24 and Progress-M 26 had each brought 100 kilograms of kit. The Space Acceleration Measurement System (SAMS) was a three-axis accelerometer to make continuous measurements of ambient vibrations in order to characterise the sites where NASA's microgravity experiments were later to be installed. It had been delivered on Progress-M 24, along with the Mir Interface to Payloads System (MIPS), which was a laptop computer configured to enable NASA apparatus to download data using Mir's telemetry link.[1]

On 22 March, Viktorenko, Kondakova and Poliakov left. Having set a 438-day single-mission record, Poliakov insisted on walking unaided to the medical tent. The plan was for Dezhurov and Strekalov to be succeeded by cosmonauts who would fly up in Atlantis, and then to return with Thagard on the Shuttle. The timing of the two missions was such that the two resident crews would each serve a three-month tour, so that the ferry that had launched one crew could be used to return the next at the conclusion of its six-month orbital life. This schedule was further complicated by the facts that Dezhurov and Strekalov had trained to commission Spektr, and Thagard could not complete his programme until it delivered the rest of his equipment. The final planning had to be flexible. If it appeared that Spektr's launch would be

---

[1] The days when American apparatus could be sent up only if it was encased in a 'black box' were clearly over.

delayed by just a few weeks, Atlantis's mission would be delayed by that amount to allow Thagard time to finish his work, but if the module was going to be significantly late they were to extend their tour and return in their own spacecraft when relieved in the normal manner. Unfortunately, this schedule (which the Russians had recently done so much to establish) promptly started to slip. Meanwhile, Dezhurov and Strekalov finished routine maintenance on the environmental and thermal regulation systems, and started to upgrade the base block's core systems in order to further extend its operational life. (This was their primary task, and it was to consume up to 40 per cent of their working time.) Progress-M 27 docked at the front port without incident on 12 April. Its cargo included the GFZ-1 sub-satellite (a small sphere incorporating laser retroreflectors to provide geodetic data, which was supplied by Germany) and 48 quail eggs – Thagard was to develop each egg for a specific time and then freeze it for return on Atlantis. On 18 April, Dezhurov and Strekalov cut the plumbing out of the shower cubicle in Kvant 2 – having been deemed a failure as a shower, it had been converted into a steam room, which had proved rather more successful; but the compartment was now needed for additional gyrodynes. The GFZ was ejected from the scientific airlock the following day.

With the launch of Spektr imminent, Kristall had to be moved, but this could not be done until its solar panels had been retracted; having been put off for so long, this acquired a sense of urgency. The first week of May was given over to the preparations. When Dezhurov and Strekalov went out on 12 May, their task was to retract one of the panels into its container, disconnect this from its drive motor, and use the crane to transfer it to a motor that had been mounted on the side of Kvant 1. However, the retraction was trickier than expected, and Thagard had to command the mechanism to cycle several times to complete the process. They had expected to accomplish the transfer on a single spacewalk, but had to stop after dismounting the retracted panel, and the power supply would be even more restricted until this could be redeployed. They returned five days later, and swung the bulky 500-kilogram box on the crane over to Kvant 1, whereupon they encountered difficulties in attaching it to the new motor. A third spacewalk proved necessary, and

Gennadi Strekalov on the boom of Mir's crane.

on 22 May they ran cables to drive the panel and feed its output into the complex's power system. The panel that they moved was the one that projected out to the left of Kristall. This had to be retracted before Kristall could to be swung onto the righthand docking port (its assigned place in the final configuration). The panel had been designed to be detachable to enable it to be redeployed and it could be moved because it was accessible to the crane, which was why the first crane had been mounted on the left side of the base block – it had all been carefully planned years earlier. Once the newly installed panel was extended, and its output verified, the two men returned to supervise the retraction of Kristall's other panel. Although this jammed with 25 per cent of its length still exposed, it was left in this state because it would not interfere with the Shuttle. (This panel was not accessible to the crane, and so could not be transferred until a crane was installed on the righthand side of the base block – the attachment had been affixed, but the boom had yet to be installed).

SPEKTR

Progress-M 27 departed early on 23 May to release the front port for Spektr, which had been launched on 20 May. Preparatory to reconfiguring the modules to accept it, Kristall was powered down and two new batteries installed to ensure that (denied its own solar panels) it would have sufficient power for its Ljappa arm to swing it onto the axial port – which was done on 27 May. The next day, Dezhurov and Strekalov donned spacesuits, sealed and depressurised the multiple docking adapter and moved a Konus drogue from the lower port to the righthand port – onto which Kristall was swung two days later. On 1 June, following its slow propellant-efficient rendezvous, Spektr docked on the axis without incident. The next day, Dezhurov and Strekalov moved the drogue assembly from the Kristall port back to the lower port and Spektr was swung down. Spektr had been designed to be positioned opposite Kvant 2, and had similar solar panels, but late on in its manufacture the far end of the module had been redesigned and instead of a cluster of instruments it had an extra pair of solar panels to overcome the diminished output of the base block's ageing panels. After two days of checking Spektr's systems, on 5 June the new panels were commanded to deploy. Although one unfolded immediately, a clamp failed to release the other, and its partially unfolded segments remained near the tip of the conical support. The panel's motor was rocked on 8 June to attempt to release the clamp, but it remained stuck. Dezhurov and Strekalov were requested to make a spacewalk to release it, but Strekalov refused because no one knew how the panel would react once it was free – it might disable one of the spacewalkers. While the cosmonauts unloaded cargo, the engineers worked out how to safely release the clamp. This deployment was a high priority, because a fault had prevented one of Kvant 2's panels tracking the Sun and one of the Kristall panels was unavailable. With 126 square metres of panel and an output of 16 kW, Spektr had often been referred to as "the power module". In fact, it had initially been called Optizon, because its mission had been remote-sensing of the Earth's environment. Despite having been stripped down to accommodate extra solar panels, it had several instruments: the Balkan-1 lidar measured the altitudes of cloud structures; the Faza and Feniks spectrometers studied the Earth's atmosphere and surface, respectively; the Astra-2 spectrometer measured the constituents of the gaseous environment at orbital altitude; and the Taurus and Grif X-ray and gamma-ray detectors measured emissions resulting from the complex's passage through the ionosphere. It also had a suite of spectrometers (Volkov, Svet, Ryabina-4P and KR-05) and radiometers (Yusa and Neva-5). In addition, the Mir InfraRed Atmospheric Spectrometer (MIRAS), developed as a joint venture by France and Belgium with Russia's Space Research Institute, was to scan absorption lines in the atmosphere at sunrise and sunset to identify constituents, in order to map their distribution and to monitor secular variations. Its data was to be downlinked daily, to contribute to an ongoing project using other satellites and aircraft to study the interaction between solar illumination and the atmosphere. It had been intended to install it on the Mir 2 base block, but when it became evident that this would not fly, the instrument was rebuilt to be carried as cargo by Spektr, transferred internally to Kvant 2, assembled in the airlock, and installed outside Spektr during a spacewalk. One advanced feature of Spektr was the Pelican remote manipulator. This 2-metre-long twin-segment arm had a gripper on its end. It was set between the solar panels, and was to be used to extract small packages from a scientific airlock and affix them to a number of external anchors, each of which had a socket for a power, control and telemetry umbilical. It was to enable exposure cassettes to be deployed and retrieved without requiring a cosmonaut to go outside (eliminating trivial spacewalks was important because each Orlan-DMA spacesuit was certified for just ten excursions). A standardised experiment package had been developed. Once this was in place, four pallets could be unfolded individually, as required, to expose samples.

Spektr's arrival was welcome news to Thagard, as any further delay in delivering the 880 kilograms of American apparatus it carried would have made impractical the programme planned for the second part of his tour. The new equipment included an ergometer, a centrifuge, a thermoelectric freezer, and several laptops. As soon as he was certain that the freezer worked, he transferred into it the samples that had been kept in the much smaller European Space Agency cabinet. On 6 June he passed the 84-day record set 21 years earlier by the crew of the final Skylab mission. The final preparation for the arrival of Atlantis was made on 10 June. As there were modules in place on both the lower and righthand ports, it was not necessary to depressurise the multiple docking adapter to transfer the Konus to enable Kristall to undock and swing itself onto the axis in order to ensure that its androgynous port would be clear of any projections that might inhibit the Shuttle's freedom of movement.

A CASE OF OVERCROWDING

After several false starts, Atlantis lifted off on 27 June for mission STS-71 (which was coincidentally the 100th American space flight to carry astronauts). The Orbiter Docking System (ODS) was mounted near the front of the payload bay. Set in a

With Kristall on the front port, Mir is ready to receive STS-71.

twin-triangular truss, it was basically two interconnected tubes, one leading from the mid-deck hatch to the tunnel to the Spacelab module mounted at the rear of the bay, and the other running up, through an airlock, to the androgynous docking system. Developed to enable Buran to dock with Mir, this docking system had been bought from Energiya. It had been hoped to take receipt of it in July 1994, but permission to import the pyrotechnic system was delayed, and it did not arrive until September. Atlantis had been chosen for the simple reason that when the project was agreed at the end of 1993 it was undergoing refurbishment, so the modifications necessary to accommodate the ODS were made immediately. When the docking system arrived, it was added. Although the Spacelab was outfitted as a life-sciences laboratory, some scientists who were not directly involved complained that this cooperation with the Russians was being undertaken at the expense of the science programme in general, because it had commandeered life-sciences and microgravity Spacelab flights, but this failed to acknowledge the potential for both areas of research likely to result from maintaining a succession of astronauts on Mir. In particular, because the Shuttle was limited to a fortnight in space, NASA would now be able to utilise its sophisticated apparatus to characterise the fully adapted state of the Mir residents, who had spent over 100 days in space.

Flown by Robert Gibson and Charles Precourt, Atlantis rendezvoused with Mir on 29 June. It approached from below, as had Discovery, but it did not pass by and ascend to dock by drawing back along Mir's velocity vector (the V-bar approach). Instead it ascended the radius vector (R-bar) that came straight up from the centre of the Earth. This enabled Atlantis to exploit the gravity gradient of one vehicle orbiting lower than the other to brake its very slow climb in the final phase without firing its upward-directed thrusters and spraying Mir with efflux. In fact, Atlantis had to fire its *downward*-directed thrusters intermittently to maintain its closure rate. However, the great advantage of this approach was that it was fail-safe – if the Shuttle were to be disabled in the final approach, gravity would draw it *away* from Mir, ruling out a collision.[2] Although Atlantis was delivering Anatoli Solovyov and Nikolai Budarin as Mir's next resident crew, Bonnie Dunbar, having backed-up Thagard, was sufficiently fluent in Russian to handle VHF communications with Mir. With Kristall on its front port, Mir was turned to aim the androgynous port towards the ascending Shuttle. At 300 metres Gibson paused, closed to 100 metres, paused again, closed to 10 metres and waited for Mir to fly in range of the Russian communications network so that he could attempt the docking – as this had to be accomplished before Mir lost contact with Kaliningrad, he had about 15 minutes. A camera aimed up the centreline of the ODS gave him a view of a standoff target in the centre of the Kristall port. The video downlinks were again shown live in split-screen format. When Atlantis paused just a few metres out, Mir was put into free-drift to ensure that its attitude-control system would not compete with the Shuttle to control the joined mass. Atlantis closed at a rate of 3 centimetres per second, offset by no more than 2 centimetres from the axis of the approach cone, and was oriented with an angular error in all three axes of less than 0.5 degree, so it was as nearly perfect as anyone could have hoped. Precisely on time, while above Lake Baikal, the triple-petals of the androgynous system meshed and their capture latches engaged. There was a momentary shudder as the 100-tonne vehicles jostled one another and springs within the mechanism damped the residual relative motions. After about fifteen minutes in this soft-docked state, Atlantis fired its thrusters to drive the guide-ring against Kristall to ensure proper alignment before the twelve latches in the collars engaged to achieve a rigid connection and establish a hermetically sealed tunnel between the two vehicles.

An hour later, Gibson opened the top hatch of the ODS and greeted Dezhurov in the narrow tunnel. In the Kaliningrad control room, Russian Space Agency chief Yuri Koptev and NASA administrator Daniel Goldin celebrated the historic moment. The two crews then congregated in the base block for a 'photo opportunity' that showed that Mir had not been meant to accommodate 10 people. The two crews reconvened in the Spacelab module the following morning and, after a brief ceremony, got down to business. Dunbar and Ellen Baker were to perform medical tests on Mir's retiring crew. They received a thorough medical examination using the

---

[2] Unfortunately, by some inexplicable oversight, the mission patch depicts Atlantis making a V-bar approach!

Vladimir Dezhurov (rear) welcomes Robert Gibson into Mir's Kristall module.

life-science equipment that included a treadmill, a bicycle, and a lower-body negative pressure chamber. In the meantime, their successors transferred cargo, which included a heavy-duty cutter for the clamp that had snagged Spektr's solar panel. As a result of the reaction in its power-generating fuel cells, Atlantis had a great deal of water to donate to Mir (this 'waste product' was usually vented to space). It was pumped into small containers, taken onboard Mir, and emptied into the Rodnik tank. Although Mir recycled 60 per cent of its water, only that recovered by the air conditioning condenser was potable, so a top-up of half a tonne of water was very welcome. Finally, Solovyov and Budarin placed their Kazbek couch liners and Sokol suits into Soyuz-TM 21, and Dezhurov, Strekalov and Thagard transferred theirs to Atlantis, to formally mark the handover. The results transferred to Atlantis included frozen quail chicks, the European Space Agency's exposure cassettes, Thagard's biomedical samples, semiconductors, film and biological experiments. The Russians exploited the Shuttle's payload capacity to return 'expired' elements of the Salyut 5B computer and items which were normally discarded with empty cargo ferries, to enable engineers to reassess the service life of each item. This aspect of the joint mission relied on quartermastering skills to track all the items transferred in each direction, and ensure that items being returned were properly loaded consistent with the Shuttle's centre of mass (a laser bar-code reader was under development to assist future operations). Late in the preparations for the mission, it had been decided that rather than keep the combined 'stack' orientated as it was at the time of docking (with Atlantis below Mir), the Shuttle would reorient the complex at the start of each day to maximise Mir's power generation. However,

Norman Thagard (left) and Gennadi Strekalov, with an IMAX movie camera.

because the differential gravity field would tend to restore the most stable position, maintaining this solar-inertial attitude would involve frequent adjustments. The stack was to be returned to the gravity gradient each evening, and be left there overnight. When it was found that maintaining the solar-inertial attitude was consuming much more propellant than predicted, it was discovered that this was because the Shuttle computer's mass model was unable to deal accurately with such a large 'attached' mass and overcompensated, oscillating each side of the optimal alignment. The long-term solution was an improved mass model, but on this occasion the tolerance was relaxed because the minor deviations did not degrade Mir's power. The inadequacy of the computer's mass model had never been suspected. With this, and a variety of other engineering data, NASA was racing up the learning curve of how to operate a Shuttle in conjunction with a massive orbital structure. Taking the technical risk out of operations planned for the International Space Station was the primary objective of the Shuttle–Mir programme. The hatches were closed on 1 July to enable Atlantis to perform manoeuvres to test the integrity of the ODS (it did not lose its hermetic seal) and to observe the dynamics of the solar panels (they wobbled, alarmingly) and then the tunnel was reopened.

Late on 3 July, as the spacefarers bade their farewells and congregated in their respective vehicles, Atlantis pumped up the complex's atmosphere to a pressure of 15.4 psi as a parting gift, thus obviating the need to send up air with the next cargo ferry. The final closing of the hatches must have been a sad moment for Dunbar. As Thagard's backup, she had hoped to serve with Solovyov and Budarin on Mir, and be retrieved by Atlantis upon its return later in the year, but this plan had had to be cancelled when it became clear that leaving an astronaut onboard would conflict with the 'extended mission' planned by the European Space Agency (the limiting factor being the return capacity of the Soyuz lifeboat). However, as NASA had booked a continuous slot thereafter, the presence of one of its astronauts would limit other international visitors to handover periods. The demand for access was so great that the problem was how to fit everyone in! How fortunate it was that the Mir complex had not been abandoned in the turmoil after the collapse of the Soviet Union. A few minutes before Atlantis was to depart on 4 July, Solovyov and Budarin undocked Soyuz-TM 21 from the rear port, withdrew 100 metres, and flew around to one side of the complex in order to film the Shuttle's departure. Once the docking latches had released, the spring-loaded mechanism pushed the vehicles apart. If the latches had failed to disengage, explosive bolts would have been fired; and if the bolts failed to completely separate the two components, Atlantis had tools to enable spacewalkers to release the 96 bolts that held the docking system on the ODS. Of course, fouling Kristall's port would almost certainly have brought the Shuttle–Mir programme to a premature conclusion. Not all went as planned, however. Because Mir's Salyut 5B computer was not configured to control such an asymmetric stack, it had played a passive rôle while Atlantis was docked. The attitude-control system was reactivated as soon as the Shuttle left, but it became confused by the translation imparted by the springs of the ODS, attempted to recover, gave up, and shutdown. Although this left Mir in 'free drift', Soyuz-TM 21 was readily able to redock.

A case of overcrowding

Atlantis about to undock from Mir.

Having withdrawn 500 metres, Atlantis made its own photographic fly-around of the complex before departing. It returned to the Kennedy Space Center on 7 July. For re-entry, Dezhurov, Strekalov and Thagard wore NASA pressure suits. A row of reclining frames was erected on the mid-deck for them, so that they would not have to endure sitting upright at the start of their readaptation to gravity – the time that they would be most at risk of orthostatic intolerance, and most likely to black out. Contrary to instructions, Thagard climbed out of his couch and walked away to the recovery van. The delay in launching Spektr, and the resultant delay in sending up Atlantis, had meant that instead of 90 days, he had spent 115 days in space. But despite these scheduling problems, he had managed to carry out his entire research programme, and NASA now had detailed medical data on how the body would adapt during a three-month tour onboard the International Space Station. In his debriefing, he made several points: Mir was roomy, and very habitable, but it definitely had the look and feel of a locker room that had been lived in for a decade; he had got on well with Dezhurov and Strekalov, but he suffered cultural isolation because often several days would pass without hearing English on the radio; he had yearned for his family; being a 'lab rat' in such a rigorous biomedical study was no joke; the requirement to record food intake was a disincentive to eating; and he had dreaded the prospect of being told that he would have to extend his mission to six months. After four Shuttle flights and a tour on Mir, Thagard retired from NASA and returned to academia.

While Atlantis and Mir had been docked, the payload that it was to deliver on its next visit (a Russian docking module) arrived at the Kennedy Space Center. Even NASA's fiercest critics had to admit that this hastily-arranged programme was off to an excellent start. Although the main objectives were for America and Russia to learn to work together in space and to reduce the technical risk in building and operating a joint facility, the opportunity for astronauts to serve tours on Mir prompted NASA to review the science programme planned for the International Space Station, and it identified several dozen experiments that could be undertaken on Mir, at little or no extra cost to the individual project budgets, and these were brought forward. Clearly, although the first docking was primarily an engineering test flight, later Shuttle–Mir flights would facilitate the build-up of NASA science on Mir.

A SHORT MAINTENANCE TOUR

During their brief tour, Solovyov and Budarin were to devote most of their time to engineering work. On 14 July, they went out to inspect the solar panel on Kvant 2 that had ceased to track the Sun, and discovered that it was obstructed by a piece of apparatus that had been improperly stowed on an earlier spacewalk. They then used the crane to swing down to the end of Spektr, to sever the clamp that had fouled one of its panels. Although the panel unfolded, the fore-and-aft segments at its tip stood perpendicular, degrading its output by 20 per cent. It was decided to leave the panel in this state as they had no access to the affected segments. The task on 19 July was to affix the MIRAS instrument to the end of Spektr, but Solovyov's cooling system malfunctioned almost immediately and he had to plug an umbilical into the airlock's facilities. They had to cancel the deployment because it would take two men to haul the 225-kilogram, 2.6-metre-long cylindrical package. Ironically, this operation had been advanced a month to facilitate an earlier start on the atmospheric observations. Budarin retrieved cassettes (including the Trek cosmic-ray detector that had been out since 1991) and set up new ones. To round out this frustrating excursion, the hatch did not form a hermetic seal, and it took almost an hour to fix it and repressurise the airlock. Two days later, Budarin used the crane to swing Solovyov and the MIRAS from the airlock to the unpressurised conical compartment at the far end of Spektr. They affixed the instrument to its scan platform by three clamps, and plugged in its power and communications umbilicals. When it failed to respond to commands from the ground they reseated the connectors, but this had no effect. Somewhat dejected, on their way back they inspected Kristall's partially extended remaining solar panel. Once inside, they checked the internal connectors and discovered that one was not properly seated, so they reset this and MIRAS promptly reported in to the anxious ground controllers. Progress-M 28 arrived at the front port on 22 July with a cargo that included a gyrodyne and some 350 kilograms of apparatus for the forthcoming European Space Agency visit. Once this was unloaded, Solovyov and Budarin set to work on maintenance tasks, and replaced a gyrodyne in Kvant 2. Only in August did they find time to resume materials processing, using the Gallar furnace.

EUROPE'S LONG MISSION

Progress-M 28 left on 4 September. Soyuz-TM 22 arrived the next day, with Yuri Gidzenko, Sergei Avdeyev and Thomas Reiter, a German researcher flying for the European Space Agency. Matching the earlier slippage, this launch had been delayed two weeks to allow Solovyov and Budarin time to finish their maintenance chores – they left in Soyuz-TM 21 on 11 September but Reiter remained on Mir for what he expected to be a 135-day tour ending with the next handover, which was scheduled for mid-January. This was Gidzenko's first mission, but Avdeyev had served a tour on Mir in 1992. The new residents were to make spacewalks to mount apparatus on Spektr. They had hoped to receive the Priroda module in November, but no sooner had they settled in than its launch was postponed by four months. Then they were told on 17 October that their tour would have to be extended by six weeks. Financial restrictions precluded paying overtime to the workers who were building the rocket that was to deliver their successors, and without a stock of rockets such delays in production were becoming a major source of concern. The European Space Agency welcomed the extension of Reiter's mission, and announced that it wished to book a third visit, in 1998, but this would have to be squeezed in after the final Shuttle–Mir mission and another French visit prior to mothballing Mir so that attention could be redirected to the International Space Station.

A total of 400 kilograms of apparatus had been ferried up in Progress-M 28 and Spektr for Reiter's programme, he had carried 10 kilograms with him as luggage,

Sergei Avdeyev (left), Yuri Gidzenko and Thomas Reiter.

and another 85 kilograms was to follow on the next ferry craft. Although his programme included 18 biomedical, 10 technology, and 8 materials-processing experiments, the highlight was to be a spacewalk. At a more mundane level, Reiter, like Merbold and Thagard, was to collect samples of saliva, urine and blood, and freeze them for later analysis, and his biomedical observations were to be correlated with the data from a radiation dosimeter. One key aspect of his programme concerned loss of bone mass, and he tested methods of minimising it by simulating the effect of walking in gravity by periodically stimulating a heel and by placing an ankle under compression, and he fortnightly tested his heels using the Ultrasonic Bone Densitometer (UBD) to assess any benefit. (One heel was the experiment, and the other was to act as the control.) Most of his biomedical experiments reinforced previous studies of the cardiovascular system, visual acuity, psycho-motoric functions, the dynamics of posture (using the ANBRE skin-tight limb-motion monitoring suit and the ELITE four-camera system), and the respiratory system (using the RMS-2).

Progress-M 29 docked at the rear on 10 October, and delivered the samples for Reiter's furnace experiments and the ESEF exposure cassettes that he was to deploy on his spacewalk. The Orlan-DMA suits were unpacked on 18 October, adjusted to fit their new users, tested, and then replenished. On 20 October, Avdeyev and Reiter made a five-hour excursion. Avdeyev used the crane to swing

Sergei Avdeyev and Thomas Reiter spacewalk with the assistance of a crane.

Reiter down to Spektr to attach the ESEF cassettes to fixtures near the MIRAS scan platform. Three of the four cassettes were passive traps to accumulate ambient particulate debris, and the fourth was an active system to record the speed, mass and trajectory of each impact. Umbilicals were connected so that the clam-shell covers of the traps could be opened by remote control from within Mir. Each cassette was to be opened during a specific period. They were to be closed while vehicles were manoeuvring nearby, in order to prevent contamination (the Astra mass spectrometer had long ago shown that this was a significant source of local pollution). On their way back, Avdeyev and Reiter replaced the cassette in the Komza experiment. The next day, it was announced that because his mission was to be substantially extended, Reiter would make a second excursion early in the new year.

On 1 November, the Vozdukh in Kvant 1 failed. Its rôle was to prevent a build-up of carbon dioxide. Until it could be repaired, Mir would have to rely on canisters of lithium hydroxide (as the base block had prior to the delivery of the regenerative unit) but these had to be replaced when they became saturated, and there were only sufficient for 30 days. Over the next few days, the Vozdukh problem was traced to a loss of pressure in the primary coolant loop running between Kvant 1 and the base block. Prior to removing the carbon dioxide, the Vozdukh first dehumidified the air, which released heat that was removed by the coolant loop. Denied active cooling, the Vozdukh became ineffective. Once the cover had been taken off the hydraulic pump in Kvant 1, a 2-litre free-floating blob of glycol was exposed. This leak was traced to a pipe feeding the main pump. The leak was stemmed by a quick-setting putty-like material. After a pressurisation test, the system was reactivated. Tools and materials were added to the manifest for the next cargo ferry, to make a permanent repair. To increase the margin of safety provided by the backup system, it was decided to add lithium hydroxide canisters to the next Shuttle manifest.

THE DOCKING MODULE

Launched on 12 November as STS-74, Atlantis was flown by Kenneth Cameron and James Halsell. As there were no Mir residents to be retrieved, there was no need for the life-sciences Spacelab, and the payload bay held the 4.2-tonne Docking Module (DM) supplied by Energiya, which had an androgynous port at each end. Two days later, Chris Hadfield used the remote manipulator to unstow the 4.6-metre-long 2.2-metre-diameter module and place it directly above the ODS. As he did so, the actual transfer was compared with a computer model created by the Space Vision System based on using obliquely angled video cameras to monitor reference markings.[3] The arm was then placed into "limp" mode, and Atlantis fired its downward thrusters to nudge the extended guide-ring of the ODS up against the DM to soft-dock. Once the residual motions had been damped, the mechanism was retracted to achieve a rigid

---

[3] When refined, this system was to produce an animated bore-sighted viewpoint to assist in mating modules with the International Space Station, on which direct viewing will often be impractical.

connection and establish a hermetically sealed tunnel, then the arm was withdrawn. If there had been a problem, Jerry Ross and William McArthur would have made a spacewalk to overcome it. Atlantis rendezvoused with Mir on 15 November. It flew the same R-bar approach as before, and, also as before, the Shuttle was responsible for all the manoeuvres. This time, however, Kristall was mounted on the radial port. With the DM on top of the ODS (giving 4 metres of additional clearance) the solar panels projecting from Kvant 2, Spektr and the base block were not a problem – it had been the lack of clearance between the panels and the Shuttle's cabin that had prompted Kristall's transfer to the axial port for the first docking. The complex was oriented so that Kristall's port faced Atlantis. Cameron did not have a direct view of the docking system on top of the DM, so the remote manipulator arm was fixed in position above and to the side of it to provide a side view during the final few metres of the approach. (No further 'blind dockings' would be needed after the DM had been installed.) To provide a little extra clearance for the DM, the base block's panels had been turned face-on to the Kristall module. Although Atlantis had never manoeuvred in such a confined space before, Cameron made it seem easy. When the hatches were opened, Cameron and Gidzenko shook hands through the tunnel, and then everybody gathered in the base block. With the national flags of Russia, America, Canada (for Hadfield) and Germany (for Reiter) on display, it was evident that Mir had indeed become a major international resource. In fact, Yuri Koptev noted that because Mir had been underfunded by 180 billion roubles in the current financial year, continued operations would not have been possible without the 350 billion roubles raised from the fee-paying missions.

A view from STS-74 as the Docking Module is manoeuvred to mate with the axial port of the Kristall module.

For the first time, the Shuttle really was serving as a 'space truck' delivering a module and cargo. The cargo to Mir comprised 250 kilograms of food, 450 kilograms of water, 20 lithium hydroxide canisters (to provide an extra margin of safety in case the coolant loop failed before a permanent repair could be made, without encroaching on the manifest of the next cargo ferry), some 300 kilograms of items for NASA's research, and various items (including a guitar) for the residents. The 375 kilograms of cargo retrieved included processed materials, computer disks of experiment data, frozen biomedical samples, and expired and broken hardware for examination by the engineers who had built it. In all, 275 items were transferred onto Mir and 195 items retrieved. Atlantis undertook manoeuvres to re-evaluate its ability to control such an offset centre of gravity. This time it was not necessary to orient Mir in solar-inertial attitude, because the Sun angle was better. The hatches were closed 17 November, but Atlantis did not undock until the following day, leaving the DM on Kristall. The Shuttle withdrew to perform a photographic fly-around before departing, and landed two days later. Docking with Mir had started to seem routine! The fact that the DM had been attached so effortlessly boosted NASA's confidence that it would be able to install modules 'in the blind' on the International Space Station. This tricky flight, said Shuttle manager Tommy Holloway, had "far exceeded expectation". Dan Goldin said that Mir was "proving to be an ideal test site for vital engineering research", and that the missions were "already paying back benefits", by providing "proximity and docking operations" and by "simulating an early construction flight".

Atlantis left behind more than the DM. Two bulky boxes containing solar panels were mounted on it. One was a retractable panel deleted from Priroda (the launch of which had again slipped, this time from the first to the second quarter of 1996). The other had 42 square metres of the most powerful type of American transducer in a Russian frame. This Cooperative Solar Array had an output of 6 kW. These panels were to be erected on spacewalks to overcome Mir's long-term power shortage, but would not be accessible until a crane was mounted on that side of the complex.

BACK TO THE ROUTINE

Gidzenko and Avdeyev resumed engineering work. The gyrodynes were deactivated for several days to allow preventive maintenance. In November, Reiter carried out a number of materials experiments by making semiconductors, alloys, and glasses in the European Space Agency's Titus furnace. On 8 December, the multiple docking adapter was depressurised to move the Konus drogue from the right to the left side in preparation for Priroda. Progress-M 29 left on 19 December, and Progress-M 30 arrived the next day. In addition to the usual consumables, it contained 62 kilograms of cargo for Reiter's extended tour, and a selection of Christmas presents. After a snort of brandy to see in the New Year, Reiter set up another Titus experiment, this time to study the thermal properties of undercooled melts. Gidzenko and Avdeyev repaired the coolant pipe using a recently delivered bypass and bottle of glycol. On 8 February 1996 Reiter made his second spacewalk. Gidzenko pushed the YMK out of the airlock and strapped it to the framework immediately outside, so that it would no longer clutter up the airlock, then he and Reiter retrieved two of the ESEF cassettes from Spektr. Gidzenko tried to retrieve Kristall's redundant Kurs antenna, but its bolts were too tight and he had to leave it. (It had been intended to send it back so that the engineers could examine it for signs of exposure.) By 20 February, the base block had been in orbit for ten years. Soyuz-TM 23 was launched the next day, and then, in daily succession, Progress-M 30 left and Soyuz-TM 23 docked to deliver Yuri Onufrienko and Yuri Usachev. Whilst this was Onufrienko's first flight, Usachev had served a tour on Mir in 1994. Most of the handover was devoted to the formidable task of bringing the newcomers up to date on where things were stored. (As an illustration of this problem, Reiter had been unable to find his centrifuge for two months.)

Sergei Avdeyev (right) uses the Tchibis negative-pressure leggings to increase his cardiovascular capacity in preparation for returning to full gravity.

Gidzenko, Avdeyev and Reiter left in Soyuz-TM 22 on 29 February. The 'long' visit that the European Space Agency had booked had turned into a standard six-month tour. Unlike Thagard, Reiter had welcomed the news of his extension. He left most of his biomedical samples and the ESEF cassettes to be returned by the next Shuttle. Analysis of the active ESEF experiment revealed that the complex passed through a stream of debris twice a day, at which time it suffered 5,000 microscopic impacts in an interval of only one minute. This was useful data for the designers of the micrometeoroid blanket that was to protect the International Space Station. By the end of 1995 the Russian Space Agency had decided to postpone its contribution to that complex, in order to continue to operate Mir through the turn of the century, and NASA had agreed to extend the Shuttle–Mir programme to assist. At this news, Japan said it would like to have a series of life-sciences and radiation experiments conducted on Mir, Germany confirmed it would like to send up an experiment, and China (which was planning its own human spaceflight programme using technology developed for the Soyuz spacecraft) expressed an interest in an acclimatisation visit. There was clearly nothing like a successful track record to encourage interest.

NASA MOVES IN

During their five-month tour on Mir starting in February 1996, Yuri Onufrienko and Yuri Usachev were to perform five spacewalks and to commission Priroda, the launch of which was now due in April. On 15 March, on their first excursion, they mounted a crane on the previously fitted anchor in order to provide coverage on the righthand side of the base block, as a preliminary to retrieving a solar panel delivered on the DM and mounting this on Kvant 1 opposite the one transferred from Kristall. On 24 March Kevin Chilton and Richard Searfoss mated Atlantis with the DM. For the STS-76 mission, the payload bay contained a Spacehab module which had been built for microgravity experiments and converted into a cargo-carrier. It carried 740 kilograms of apparatus for NASA's programme (including the television cameras and lighting kit for the Glovebox) and 980 kilograms of items for Mir (including food, a transformer and a gyrodyne). In addition, it had a rack of microgravity experiments and a freezer in which to return the rest of Reiter's biomedical samples. This time, a total of 500 kilograms of scientific results and assorted apparatus was retrieved from Mir. Linda Godwin and Rich Clifford went out on 27 March. Since this was the first time that astronauts had spacewalked from a Shuttle that could not chase after them if a tether snapped, their backpacks were augmented by SAFER manoeuvring units. And as it was the first time that Shuttle spacewalkers had worked in the immediate vicinity of the Mir complex, they were prohibited from straying beyond the far end of the DM so as not to disturb the instrumentation on its surface. After dismantling the DM's now-redundant camera, they connected clamps to handrails to accommodate the Mir Environmental Effects Package (MEEP) cassettes. They then tested a new portable foot-restraint and a tether system intended for use when assembling the International Space Station. Shannon Lucid, a veteran of four previous Shuttle missions, remained on Mir when the DM hatches were sealed on 28 March. Although Atlantis was due back in early August to collect her, she had her own Kazbek couch liner and Sokol suit just in case she had to return to Earth in the Soyuz 'lifeboat'. She began her tour by documenting the NASA equipment on the complex – an onerous task with which her hosts could sympathise. Following Thagard's example, Lucid set up her sleeping bag in the Spektr module, where she did most of her work.

Shannon Lucid in the Spektr module.

PRIRODA

Priroda was finally launched on 23 April. Unlike its predecessors, it pursued a rapid rendezvous in order to minimise the period of independent flight, because it had only batteries for power. Because previous modules had had to make several approaches to Mir before finally managing to dock, Priroda had extra-large propellant tanks, and in case it encountered difficulties in the final phase it had been fitted with the TORU remote-control system. Although one of its power buses tripped off-line two days into the rendezvous, robbing it of half its power, it docked without incident on 26 April and still had sufficient power to swing onto the lefthand port the following day, thereby completing the Mir complex.

Priroda had a comprehensive suite of remote-sensing instruments for monitoring industrial pollutants, energy and mass-exchange processes at sea level, the height of waves, and the mean temperature and vertical structure of cloud formations. It was also capable of yielding wind direction and speed, mapping thermal variations across the ocean, multispectral studies, and serving as a relay station for Project Centaur by uploading information from automated geophysical stations at remote sites. Most of its built-in apparatus had been developed collaboratively by the members of the Intercosmos organisation. The Ikar had three sets of microwave radiometers (Ikar-N was aimed at the nadir, the Ikar-D scanning radiometer sampled obliquely and Ikar-P offered a panoramic view) and monitored 11 wavelengths. Istok-1 was a 64-channel multispectral infrared radiometer to study the oceans. Greben was an altimeter to measure mean sea level along the ground track. The Ozon-M spectrometer measured ozone and aerosols in the upper atmosphere. The medium-resolution (MSU-KS) and high-resolution (MSU-E) optical scanners provided views of clouds. The Travers synthetic-aperture radar (which was deployed after the module was in place) could penetrate cloud to provide a medium-resolution ground-imaging capability. The Moz multichannel spectrometer studied the oceans by reflected insolation. In addition, Germany supplied the MOMS-2P optical multispectral imager, which could be used in various modes, including high-resolution mapping of surface relief, and carried its own GPS-based navigation to accurately position each image. France supplied a lidar which could detect tropospheric aerosols, and thereby complement the stratospheric capability of Ozon-M. Since this had originally been developed for Salyut 7, it was named Alissa – l'Atmosphere, Lidar Sur Salyut. Between them, Spektr and Priroda had a formidable remote-sensing capability, particularly where simultaneous studies by complementary sensors were synergistic. Over land, the microwave radiometers measured soil moisture and the energy balance of the surface, and the imaging radar saw terrain through cloud. In the oceans, the narrow sensors would detect plankton, ocean currents and processes operating near the ocean–atmosphere interface, and the lidar would see the vertical structure of the atmosphere and of tropospheric aerosols. Knowledge of any of these parameters in a narrow time-frame on a global scale was valuable, and the ability to correlate these parameters was considerably more so. The modules had been designed as a pair to provide broadly based and correlated data to achieve an understanding of global hydrological cycles. In both cases, however, some instruments had been deleted in order to accommodate NASA apparatus. In the case of Priroda, this cargo comprised 284 kilograms for biomedical studies and some 400 kilograms for a variety of microgravity experiments, including the Mir Electric-Field Characterisation Experiment (to determine the ambient emissions in the 400 MHz to 18 GHz radio range, likely to interfere with other apparatus), the High-Temperature Liquid-Phase Experiment (HTLPE), the isothermal furnace supplied by Canada for the Queen's University Experiment in Liquid Diffusion (QUELD), and samples for the Optizon Liquid-Phase Sintering Experiment (OLIPSE) that was to be run in the Optizon furnace in the Kristall module. Before Priroda's remote-sensing apparatus could be commissioned, its cargo had to be unloaded. But the first task was to strip out the 168 chemical storage batteries that had sustained it during its rendezvous. (In the process, it was discovered that a faulty relay was the reason for one of its power buses tripping off-line during the rendezvous.) Each battery was unbolted and sealed in an individual plastic bag to contain any gaseous emissions. Progress-M 31 docked on 7 May, was unloaded in record time, and the batteries were stowed in its orbital module for disposal. As they inspected their new toys, the cosmonauts noted that it felt as if Christmas had come early that year. On 12 May Priroda was declared fit for scientific work.

Lucid's first task was to set up the SAMS in Priroda to assess its microgravity environment. The centrepiece of NASA's apparatus was a Glovebox for procedures requiring physical isolation. Television cameras and lighting apparatus that had been delivered by Atlantis were installed in the Glovebox to document experiments. The Microgravity Isolation Mount (MIM) supplied by Canada used magnetic levitation to isolate experiments from ambient vibrations in the 0.01–100 Hz range. Lucid set it up inside the Glovebox so that residual accelerations transmitted onto the surface of a liquid could be videotaped while the SAMS recorded the ambient accelerations to evaluate the efficacy of the MIM. It had once been feared that highly sensitive microgravity experiments would be practicable only on a *free-flying* space platform, but devices like the MIM offered the prospect of being able to do so in the 'noisy' environment of an inhabited complex. Lucid's other research tasks involved the use of boundary-layer processes to determine the diffusion coefficients of semiconductors, binary-metals and glasses; the process of melting; embryo studies of quail egg development; monitoring ambient radiation, air and water quality; and a variety of biomedical tests. Her favourite task, however, was photography for geological, ecological and environmental studies, as it gave her an opportunity to use the high-fidelity porthole in Kvant 1.

Shannon Lucid (left), Yuri Usachev and Yuri Onufrienko alongside the glovebox in the Priroda module.

Unlike a Shuttle mission, in which every activity is carefully assigned a slot on a timeline, Lucid was free to set her own pace. She worked through a four-day task list that was updated daily by Houston, and she checked off items as and when she managed to achieve them. The only significant problem she encountered early in her tour was when a card in the MIPS failed, denying her this downlink capability until a replacement could be sent up on the next cargo ferry. The MIPS was proving to be very useful. Not only did this enable data on computer disks to be dumped on an ongoing basis (instead of being saved for return at the end of the mission), but it also enabled Lucid to keep in touch with her colleagues using e-mail, which ameliorated the sense of cultural

Shannon Lucid exercising in the Mir base block.

isolation Thagard had suffered. In fact, the support system had been improved in Kaliningrad and Houston, and Lucid was more effectively integrated into a distributed scientific team. On the other hand, she came to value the autonomy of having to operate on her own during the periods that the complex was out of radio communication. Lucid later urged the International Space Station's mission planners to remember that while it was standard to design experiments to require very little crew involvement, because time on a Shuttle was at a premium, an astronaut on a long tour needs to feel more involved in the work in order to sustain interest, and she recommended two-way video conferencing with the designers of the experiment as a standard procedure.

While Lucid was setting up the NASA experiments, Onufrienko and Usachev had an extremely busy time spacewalking. On 21 May they used the newly installed crane to retrieve the Cooperative Solar Array from the DM and install it on Kvant 1, on the motor that had been in place for some time. This outing ended with a piece of theatre: the inflation of a 1.2-metre-long replica of a soft-drink can that was filmed against black space for use in a Pepsi advertisement. Four days later, they returned and manually deployed the solar panel using a hand crank. NASA wished to monitor the output and rate of degradation of this new type of transducer to evaluate computer predictions. When the redeployable panel already on Kvant 1 degraded to the degree that it was ineffective, it was to

Yuri Onufrienko and Yuri Usachev inflated an oversized Pepsi can for a commercial.

be jettisoned and replaced by the other panel on the DM. On 30 May, they put the MOMS-2P on a scan platform at the end of Priroda. On 6 June, they replaced the Komza cassettes on Spektr, deployed the SKK-11 cassette on Kvant 2, and affixed two micrometeoroid packages (PIE and MSRE) for NASA to the anchor of the now-vacant Trek experiment. On 13 June they set out to deploy the Travers radar on Priroda. When the large framework dish had attempted to deploy on a boom projecting from the side of the module (as similar antennas had on the automated Almaz satellites) it had become snagged. After freeing the antenna, they made their way to Kvant 1 and dismantled the Rapana girder, moved it out of the way by strapping it to the base of the Sofora, and erected the Strombus girder in its place (but in this case inclined aft at 11 degrees). Onufrienko and Usachev were probably not too surprised on 21 June to hear that they would have to extend their tour by 40 days because financial problems had delayed the assembly of the rocket that was to deliver their successors. The Russian Space Agency also announced that in order to save on rocket costs, tours of duty would be standardised at six months, which was the maximum permitted by the Soyuz-TM's service life.

On 12 July, Lucid was told that she, too, would have to extend her stay on Mir. Post-flight inspection of the solid-rocket boosters from STS-75 in June had revealed that the field-joints had suffered serious hot-gas penetration, and the boosters for the forthcoming Shuttle–Mir mission were to be replaced. When the resident handover had been postponed, Lucid had been disappointed that she would not be present for the visit by the French researcher Claudie André-Deshays, but the delay of her own replacement to late September meant that she would be present after all. With things going well, on 15 July she exceeded Thagard's 115-day record for an American in space. The postponement of Atlantis had ramifications. Without the consumables it was to deliver, Mir would not be able to sustain the expanded crew during the period of the handover, so another ferry had to be sent. Its launch on 24 July was aborted less than a minute from ignition, when a propellant sensor failed. But Progress-M 32 finally lifted off on 31 July. Progress-M 31 undocked on 1 August, and its successor arrived the following day. It had to be unloaded immediately, as (on the revised schedule) the handover crew would require the same port. When Gennadi Manakov received his pre-flight medical on 9 August he showed a heart irregularity, and so he and Pavel Vinogradov were stood down and their backups assigned the forthcoming flight. In mid-August, Lucid ran an extensive series of combustion experiments for NASA. She burned eight types of fuel and 80 candles with various characteristics in the Glovebox, and filmed them for later analysis by specialists in combustion. The study of microgravity combustion was basic science, but if it could yield insight applicable on Earth it would have significant commercial consequences. Lucid's other results included 40 QUELD samples, 70 OLIPSE samples and a large number of computer disks, video tapes and rolls of film. On 18 August Progress-M 32 left, and the following day Soyuz-TM 24 arrived with Valeri Korzun, Alexander Kaleri and Claudie André-Deshays. As the expanded Mir complex offered an unprecedented degree of privacy, André-Deshays set up home in Priroda. Lucid had finished her own research and was using her extended stay to set up the BioTechnology System to be used by her successor in protein crystal and cell culture experiments. She planted wheat in the Svet cultivator in the Kristall module for the Greenhouse experiment, collected data for the Anticipatory Postural Activity experiment to investigate posture while undertaking various activities, wore the Belt-Pack Amplifier System which contained sensors to measure muscle stimulation, and used the Metabolic-Gas Analyser System to analyse her expiration in exercise using an ergometer.

Shannon Lucid views wheat in the Svet cultivator.

Meanwhile, André-Deshays had priority on the power supply. As a rheumatologist specialising in neurological medicine for aerospace, André-Deshays was ideally qualified for the cardiovascular and neurosensory tests that formed the basis of the programme. She conducted experiments designed to investigate the rapid changes to the cardiovascular system that occur in the early phase of adaptation, in particular how the body senses blood pressure and regulates its flow. This involved donning the Physiolab harness which carried the various sensors. The neurosensory research involved strapping the subject into the Cognilab instrumented chair, which measured the body's response to muscular stimulation in different conditions. She also had technology experiments: as a follow-up to the Alice experiment by Tognini, she had Alice-2, which used a furnace that accurately controlled the temperature of a fluid and a CCD camera that recorded the fine-scale phenomena at the 'critical point' phase transition; and Dynalab (a variation on the Resonance theme), which measured the propagation of vibrations through the complex.

Claudie André-Deshays.

On 2 September André-Deshays departed in Soyuz-TM 23 with Usachev and Onufrienko. She later opined that two weeks was not enough time to settle down to life on Mir and fulfil an experiment programme. Within a month, France requested two extended visits to Mir in 1999. Progress-M 32, which had been station-keeping a few kilometres behind Mir, moved back in and docked at the rear on 3 September; it was the first time that a cargo ferry had moved from one end of the complex to the other. Although its dry cargo had been unloaded, its propellant had not, so it had to return to replenish the base block's tanks.

Lucid spent her brief time with the new residents preparing to return to Earth, but on 7 September she claimed Kondakova's 169-day woman's record, and on 17 September, with STS-79 already on its way, she also took Reiter's 179-day visitors' record. The docking on 19 September was complicated by the fact that the recently deployed Cooperative Solar Array projected within 3 metres of Atlantis's nose, but Bill Readdy and Terry Wilcutt readily manoeuvred into the gap. The time that Lucid and John Blaha transferred their Kazbek couch liners and Sokol suits marked the official exchange. This was the first time that *astronauts* had handed over to one another in space. Just as the residents did during their handovers, Lucid briefed her successor on where everything was and how best to work on the complex. Atlantis had a double-length Spacehab module in its bay loaded with long-overdue supplies. Tom Akers, the 'loadmaster', evaluated a new method of handling the cargo. Instead of transferring individual items from the lockers as previously, items had been pre-loaded into locker-sized canvas bags that were easily handled, and the transfer was tracked with the bar-code reader. Of the 2,250 kilograms transferred onto Mir, the

John Blaha hugs Shannon Lucid. Valeri Korzun and Alexander Kaleri use models to plan Atlantis's arrival. STS-79 commander Bill Readdy transfers a gyrodyne. Tom Akers, in Spacehab, logs canvas bags containing cargo.

largest item was a gyrodyne. A total of 1,000 kilograms was retrieved. In addition to Lucid's results (in 20 canvas bags), an expired Orlan suit was being returned so that its degradation could be assessed, together with a gyrodyne for refurbishment and later return. Stowing the return cargo turned out to be trickier than expected, because it had to be weighed and located consistent with centre-of-mass requirements – there was nothing like doing something for real to learn how to do it. Atlantis left on 24 September, and after conducting experiments using apparatus in the Spacehab, returned to the Kennedy Space Center. Immediately upon landing, Lucid was imaged with a nuclear magnetic resonance scanner to record the density of her skeleton and key musculature. To her surprise, she rapidly readapted to gravity. The Shuttle–Mir project manager, Frank Culbertson, said that she had "set the standard" for NASA's work on Mir. She was subsequently welcomed into the select group of astronauts to be awarded the Congressional Space Medal of Honor.

On Mir, Blaha opined that there was no better way for NASA to prepare for the

John Blaha in Priroda.

future than to have a cadre of astronauts learn to live and work in space by spending tours on Mir, as everything from the exercise regime to the logistics system was new to the agency. His programme built on Lucid's, and added several new experiments. One of the packages delivered by Atlantis was a 'powered transfer' with mammalian cartilage cells for the tissue-growth experiment in the BioTechnology System that he started even before the Shuttle departed, and was to run throughout his tour. Other new experiments included the Diffusion-controlled Crystallisation Apparatus for Microgravity using a semi-permeable membrane to grow protein over a long period, with their growth being filmed, and the Binary-Colloid Alloy Test that was set up in the Glovebox to study the crystallisation of alloys of colloids – in the first month, he processed rapid-growth samples of different relative concentrations, each taking a day, prior to a three-month test. He had a number of technological tests to perform to provide information for outfitting the modules of the International Space Station. One was an instrumented foot-restraint that monitored how much his body moved whilst he worked 'in place'. Another was a push-off pad that measured the force he imparted in moving around in the complex. The Passive Accelerometer System measured low-intensity continuous effects due to air drag and differential gravity, in order to further characterise the microgravity environment. This augmented data from a set of accelerometers and strain gauges that he installed for the Mir Structural Dynamics Experiment to measure the transient stresses during manoeuvres and docking operations, and thermal effects due to flying into and out of the Earth's shadow. Once set up, this apparatus was to gather data whenever NASA microgravity experiments were running. Blaha was very impressed with Mir. "This is an incredible space station," he reported as he began his second month.

The resupply ship had been scheduled for mid-October, but its rocket was late. When Progress-M 33 arrived on 22 November, it delivered heavy-duty power cables (for a forthcoming spacewalk) and two Japanese experiments, one of which required sampling the bacteria and mould that grew in the nooks and crannies of Mir (with a view to eliminating such colonies from the International Space Station) and the other involved exposing human DNA, silkworms, soil bacteria and yeast of cosmic rays to

Mir with all four radial modules in place, and the Docking Module on the end of Kristall.

determine the effects on cell structure and to reveal any resulting genetic damage. In fact, Blaha was sampling the air and water, and incubating smears from exposed surfaces for bacteria and fungi. Although the base block had been inhabited for a decade, it was in remarkably good shape in this respect – the rich green mould that had broken out on Salyut 4 was just an unpleasant memory.

TESTING TIMES

The Russian Space Agency's budget was in steep decline (by this point, in real terms, it was worth 20 per cent of its 1989 value), and most of it went to operating Mir. Yuri Koptev warned that unless the slashing of his budget ceased, his agency would not be able to make up the shortfall to accommodate the firm bookings by fee-paying guests after the contract with NASA expired, and Mir might have to be vacated earlier than planned. (It had been hoped to continue to occupy Mir until 1999 and then have the final crew prepare it for de-orbiting early the following year.)

On 2 December 1996, Korzun and Kaleri conducted a spacewalk to complete the installation of the Cooperative Solar Array. To date, the power it delivered had only been enough to enable NASA to monitor its efficiency. A 22-metre cable was to be strung across to the base block, to the socket for the dorsal panel (which had degraded to the degree that it could now barely deliver 1 kW, even under ideal illumination, and so it had been decided to disconnect it in order to reuse its socket). The job was finished a week later, and after the power grid had been reconfigured the new panel was brought on line. While outside, the cosmonauts had also retrieved the Rapana girder from its stowage point at the base of the Sofora, and mounted it as an extension of Strombus in order that the sensors at its far end would be as far as possible from the body of the complex, and had moved the Kurs transponder on Kristall to the DM in order to restore the system's angular coverage. By this point, the surface of Kvant 1 was covered with a variety of installations, and spacewalkers had to take care not to disturb anything. On their first excursion Korzun and Kaleri accidentally yanked out the plug for the ham-radio antenna, so on the second they had traced the cable and reseated its connector. If it had not been for the late launch of Progress-M 33, these activities would have occurred much earlier; consequently, a side effect of the delays in the rocket factory was that Mir was denied much needed power for several months. On 6 December, the wheat that Lucid had planted in the Svet in August completed its cycle and yielded grain, so Blaha 'harvested' it. This was the first time that a *staple* had matured. Although the wheat and most of the grain was frozen, to be returned to Earth, a few of the seeds were immediately planted to start a second crop that would be uprooted and taken back to Earth at the end of Blaha's tour so that they could be subjected to a comprehensive biochemical analysis. During a press conference at the end of the year, Blaha was asked whether he was eager for the Shuttle to retrieve him on schedule; he replied laconically, "if it gets here, it gets here". Clearly, like Lucid, he was enjoying his tour. STS-81 lifted off on its first attempt. Contrary to the early fears of the critics, no launch attempt had been scrubbed during the final moments of the countdown.

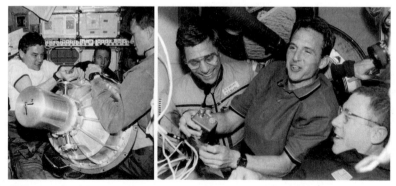

Valeri Korzun and Brent Jett work on a gyrodyne in Spacehab. John Blaha, Jerry Linenger (centre) and STS-81 commander Mike Baker.

Mike Baker and Brent Jett tested upgraded software as they docked Atlantis on 15 January 1997, and the stresses were measured by the Mir Structural Dynamics Experiment sensors that had been installed by Blaha. In addition to the now-routine cargo, the Spacehab module carried three storage batteries and a gyrodyne. Marsha Ivins supervised the transfer activities by Jeff Wisoff and John Grunsfeld. Having handed over to Jerry Linenger, Blaha left on Atlantis on 20 January. On landing in Florida two days later, he was "absolutely stunned" at the strength of gravity and, in contrast to his predecessors, he obliged the doctors and let himself be carried off the Shuttle. As a physician, Linenger's programme was primarily biomedical, but he had several fluid physics and materials-processing assignments. In addition, he was to become the first astronaut to make a spacewalk using an Orlan suit, by deploying a new experiment and retrieving two that had been set up the previous year.

Soyuz-TM 25 docked at the front port on 12 February to deliver Vasili Tsibliev, Alexander Lazutkin and Reinhold Ewald, who was to accompany the handover. As Ewald had been the backup for the European Space Agency's mission in 1992, his biomedical programme built upon that work – although this was a German-sponsored visit. It turned out to be rather more stressful than intended. When there were more than three people on board, the oxygen output of the Elektron was supplemented by 'burning' Vika canisters. On 24 February, Lazutkin routinely inserted a canister into the unit in Kvant 1 and rejoined his colleagues at the dining table in the base block. A moment later, dense smoke emerged from the tunnel. The canister had split, releasing oxygen into the electronics and starting a fire. Although the cosmonauts were on the scene within seconds with portable extinguishers, the combustion was sustained for 10 minutes. The danger was that the jet of flame would melt a hole in the pressure hull, depressurising the complex, with the ferry at the rear inaccessible. By the time the fire burned itself out, the entire unit had been reduced to soot-blackened scrap. For an hour or so, the men wore full-face masks and portable oxygen bottles while the air conditioning system extracted the worst of the smoke. Although eye irritation was reported, Linenger decided that there was no lung damage by smoke inhalation. Nevertheless, for the next several days they wore filter masks to avoid breathing in particulates. The Russian press report described this as a "small fire", but it was the worst fire to date. On an earlier station, it would likely have resulted in evacuation, but Mir could not lightly be abandoned, and the mess was tidied up. The split canister was extracted from the ruined Vika and sealed in a plastic bag for return on the next Shuttle for examination. (Such canisters had been used for many years on commercial aircraft without incident, so the manufacturer was eager to find out why this one had failed so

Alexander Kaleri, Jerry Linenger and Valeri Korzun (front) with Vasili Tsibliev, Reinhold Ewald and Alexander Lazutkin (rear).

spectacularly.) When Korzun, Kaleri and Ewald departed in Soyuz-TM 24 on 2 March, they returned samples of air and water to be tested for contaminants.

An experiment was conducted when Progress-M 33 made its return on 4 March. Usually, such a ship would rendezvous using its Kurs system and then pause at the 200-metre point in order to line up for a straight-in approach. The Kurs system was supplied by a Ukrainian company, which, because the system served a key function, had raised its price. The cash-starved Russians hoped that the advent of the TORU would enable the Kurs to be deleted, thereby simultaneously saving mass, increasing payload and cutting costs. To date, however, the TORU had been used only to steer errant ferries during the final approach. Tsibliev, sitting at the TORU controls in the base block, was to try to steer Progress-M 33 in from a range of several kilometres. Without the Kurs radar to give range and range-rate information, he would have only the image from the television camera mounted on the ferry's nose. He would have to judge range by the angular size of Mir on the grid overlaying the screen, and estimate range-rate by how rapidly this image changed. He was to command braking burns at set points on the approach. Although this may appear straightforward, performing a rendezvous in this way would be no mean feat. In the event, the video link failed and he aborted the test. However, the ferry continued in, and flew uncomfortably close by. Since it had insufficient propellant to set up for another rendezvous, Kaliningrad commanded the ship to withdraw and de-orbit itself. It was decided to have Tsibliev make the test with the next ferry, once its docked mission was over.

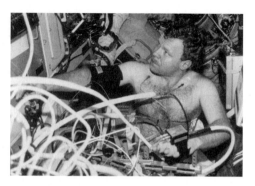

Reinhold Ewald at work.

Vasili Tsibliev.

The next day, the Elektron unit in Kvant 2 had to be deactivated after a bubble of air blocked the flow of water in its electrolysis canal. When the older unit in Kvant 1 was activated, it failed to extract the liberated hydrogen (for venting) as efficiently as it should, and had to be switched off. This was a serious problem because the only means of producing oxygen was the Vika in Kvant 2. Although there were sufficient canisters for two months, if this were to fail there was bottled oxygen only for a few days as an emergency reserve. After the fire there was understandable reluctance to rely on the Vika but there was no alternative, and a stock of fire extinguishers were stationed nearby as a precaution. As the canisters would not last until the next visit by Atlantis, it was evident that unless the next cargo ferry arrived on time in early April and brought additional canisters and parts to repair the Elektron, the complex would have to be evacuated. One lesson learned from previous stations was that an empty station could be disabled by a fault that could easily have been fixed if a crew had been present, so to vacate Mir would be to invite trouble. As if to underline this risk, when a sensor failed on 19 March the attitude-control system set up a runaway three-axis roll in attempting to correct what it took to be an unexpected motion. The computerised gyrodynes had to be shut down, the rotation cancelled manually using thrusters, and the complex stabilised in the gravity gradient while the problem was investigated. The automated system was finally reactivated using a backup sensor. A spare sensor was installed in place of the failed one, which then became the backup. As Mir was currently in direct communication with Kaliningrad for only 10 minutes on favourable passes, it is doubtful that control could have been reasserted from the ground.

On 2 April a problem in the thermal regulation system temporarily disabled the Vozdukh in Kvant 1, which required active cooling to function properly. Although the coolant loops had been designed to be leak-proof, they had been designed for a five-year life. However, the base block and Kvant 1 had been in space for more than twice that time, and where they ran alongside power cables they had succumbed to electrochemical corrosion. For some time the cosmonauts had been using a sealant to plug leaks in Kvant 1, but the system had finally lost pressure. With the Vozdukh out of service, they resorted to using lithium hydroxide canisters to scrub the air of carbon dioxide. A stock of canisters was added to the list for the next cargo ship. When the complex became hot and sweaty, the exercise regime was eased to preclude a build up of carbon dioxide.

Progress-M 34 slipped into the rear port on 8 April without incident. In addition to the usual components for the atmospheric and environmental systems, it brought parts to repair the least damaged of the two Elektrons, a stock of 60 Vika canisters, a tank with 50 kilograms of oxygen (to be consumed prior to resuming reliance on the Vika) and 15 lithium hydroxide canisters – enough to see the complex through to the next Shuttle visit, which could deliver a heavier payload. Although Linenger had only a month to go before being succeeded by Michael Foale, Lazutkin and Linenger were advised that they would have to extend their tour by six weeks, since the rocket that was to launch their successors was late. The Elektron in Kvant 2 was soon repaired, and reactivated. The old one in Kvant 1 was removed for return to Earth by the next Shuttle, which would deliver a replacement. Life on board remained unpredictable. On 13 April the toilet ceased to 'process' urine for feeding to the Elektron. Lazutkin spent much of the next week plugging coolant leaks. The fact that some sections of pipe were difficult to access made it a time-consuming task. On 21 April he reported an allergic reaction to the glycol. Nevertheless, the Vozdukh was able to be restarted. It was decided to have a later Shuttle deliver a second Vozdukh, to be installed in the base block as a backup. Tsibliev and Linenger went out on 29 April to retrieve the PIE and MSRE packages from the DM and to deploy the Advanced Materials Exposure Experiment. Despite the fire, and having been drummed into helping with the clean up and subsequent maintenance tasks, Linenger had made progress with his programme – indeed, by processing 50 QUELD samples he was working through the samples assigned to his successor.

Jerry Linenger wears his Sokol suit.

CRISIS

Atlantis arrived on schedule on 17 May, as mission STS-84. Its Spacehab contained 500 kilograms of apparatus for NASA's science programme and 1,200 kilograms of Russian material. The Russian cargo included a stock of lithium hydroxide canisters to build up a reserve in case the Vozdukh had to be deactivated for a prolonged period, a refurbished gyrodyne, the parts to repair the toilet, and a new 120-kilogram Elektron to replace the one in Kvant 1. The donation of 300 litres of water from the Shuttle's fuel cells was welcome, because the potable water on Mir was deemed to be contaminated with glycol extracted by the air conditioner. One particularly eager visitor was Yelena Kondakova, who was making her second trip to Mir, but her first flight on a Shuttle. When Atlantis left on 22 May, in addition to retrieving Linenger, it took 400 kilograms of accumulated NASA research results, the old Elektron and the remains of the Vika unit that had caught fire, which were both being returned for examination.

Michael Foale.

In addition to his busy programme of protein crystallisation, Earth observations, life sciences, materials processing and engineering experiments, Foale was to assist in maintenance tasks, help to unload the cargo ferry expected in early June, look after Mir while the residents made two spacewalks in July, and tag along while Soyuz-TM 25 was flown to the rear port in preparation for the August handover. Having immersed himself in Russian culture, he readily made himself the third member of the crew, as opposed to being the foreign visitor. The repair of the environmental systems got off to an excellent start, but the final coolant leak in Kvant 1 was not sealed until 4 June, after which the module's thermal regulation system was pressurised and returned to service. By the end of the following week, the residual glycol spills had been cleaned up. Prior to making a start on installing the new Elektron in Kvant 1, Tsibliev had to perform the postponed TORU test. Loaded with accumulated trash, Progress-M 34 undocked on 24 June. As Tsibliev steered it back in the next day, he lost control, and the 7-tonne spacecraft smashed into the Spektr module, badly mangling a solar panel and puncturing a conformal radiator. The twisting of the panel's mounting, and the force on the hull transmitted through the struts supporting the radiator, combined to puncture the module and the air began to vent to space. At the time of the impact, Foale was on his way to aim a laser rangefinder through the small porthole at the rear of Kvant 1 in order to provide Tsibliev with closing-rate data. Lazutkin was looking for the ferry through one of the base block's portholes.[4] Lazutkin suddenly ordered Foale to retreat to the Soyuz. As Foale passed through the multiple docking adapter *en route* to the Soyuz, there was "a loud bang". Lazutkin, who was still in the base block with his feet anchored to the floor, felt the shockwave propagate through the structure. The bang gave way to the hiss of escaping air. However, as the inner ear is sensitive to a drop in pressure it was apparent that the leak was very slow. Lazutkin had caught sight of the errant ferry a few seconds before it struck, and knew that it had hit the Spektr module, which had to be the source of the leak. It was therefore imperative that its hatch be closed as soon as possible. But this would not be simple to do. In addition to the air tubes that snaked through the hatches, cables had been strung between the modules to establish an integrated power grid linking the various solar panels to the storage batteries located about the complex. As there was no time to de-install the cables, they would have to be cut. A cutter was stored in the docking adapter for such an emergency. Cutting the cables was in itself a risky task, because severing a cable might cause a short circuit, which meant that the eventual recovery would involve methodically verifying the state of the complex's systems. It took several minutes to clear the aperture and fit the flat-plate hatch cover. By that time, the pressure was about 15 per cent below normal, which was not serious, and further monitoring showed this to be stable.

With the station secure, it was time to tackle the immediate consequences. The quartet of solar panels on Spektr had contributed almost half of Mir's power, but, with the cables cut, this power was no longer being fed into the power distribution system. The complex had survived the decompression, but it now faced a power crisis. As it was vital that the batteries should not be drained, the apparatus in the other radial modules was immediately switched off. Nevertheless, a crisis was in the making because the complex needed power to be able to continue making power. The power output from a solar panel is related to the angle of insolation by a

---

[4] It was later determined that the mass of the trash loaded into the orbital module of the cargo ship had been miscalculated by as much as a tonne, with the result that the braking burns that Tsibliev commanded via the TORU had not been effective, and consequently it arrived early and closed much more rapidly than expected.

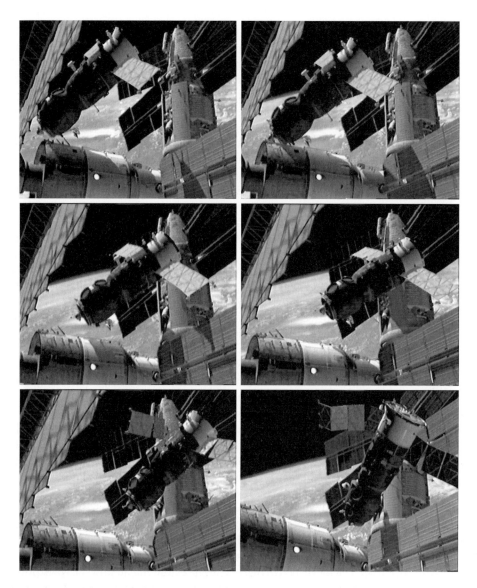

A sequence of graphics illustrating how Progress-M 34 collided with the Spektr module. Courtesy of Dave Woolard.

sinusoidal function; it produces its peak output only when it is face on. As Mir flew around its orbit, it generally kept the same orientation with respect to the Earth below, so the panels had to be turned to track the Sun, and it took power to turn the motors. Each panel had its own motor. If the batteries were overly depleted, there would be insufficient power to turn the panels, which in turn would deny power to recharge the batteries. It was a runaway process; a spiral to oblivion. The vital thing was to switch off everything in order to conserve the batteries, and then – with half the generating capacity lost – endeavour to remain on the safe side of the cusp in the power curve. The situation was complicated by the fact that the orientation of the complex was being controlled by the gyrodynes. As these consumed power and were sensitive to fluctuations, they had to be switched off, which left the complex in free-drift. It was at this point that Mir flew into communications range with Kaliningrad. The first the flight controllers knew that there was a problem was when the telemetry stream failed to materialise; the transmitter had been turned off along with everything else. In any case, there was nothing the ground could do – the crew were effectively on their own.

Soyuz-TM 25 had been powered up immediately, just in case it proved to be necessary to make a hasty escape, and now it was the only part of the complex not crippled by the power loss. Its thrusters were used to stabilise the complex and then to reorientate it so that the majority of its remaining solar panels faced the Sun. It took several hours of continuous adjustment to fully charge the batteries; only then could Mir's control systems be reactivated. For the next few days, as they inspected the station, the three men worked by torchlight, and established a sleep roster so that they would not be taken by surprise by a sudden deterioration. And there were still the slow-to-develop problems that could, in the end, force abandonment. With power low, the cooling system was ineffective. The soaring temperature was not just uncomfortable for the crew; the Vozdukh carbon dioxide scrubber overheated and had to be turned off, which forced reliance on the limited supply of lithium hydroxide canisters. The Elektron suffered, so waste water could not be electrolysed to make oxygen, which forced reliance on the sole-remaining Vika, and this meant having masks and fire extinguishers to hand. The toilet could be used, but the urine reprocessing system could not be used, because the tank for the Elektron was full. And so it went on. Nevertheless, the situation was improving with each day. It was by no means clear, however, that they would not be forced out. One thing was sure: if Mir was vacated, it was unlikely to be reoccupied. The spacecraft and its crew were in a symbiotic relationship; just as the crew relied on it for their survival, it relied on their presence. But basic survival was not the objective, Mir was a laboratory, and if it was to have a long-term future, then, at the very least, the undamaged solar panels on Spektr would have to be brought back on line.

It had been intended that after Progress-M 34 had redocked, verifying the TORU test, it was to have been discarded to clear the port for its successor – which was already on the pad. This launch, set for 27 June, was postponed to give time to work out a way to restore the power, and to fabricate the necessary material and tools. It was concluded that although it might be feasible to run external cabling from Spektr's panels to sockets outside the base block, it would be simpler to try to reconnect them internally. With Spektr exposed to vacuum, however, the hatch would have to remain closed. How could new cables be run through a sealed hatch? An ingenious scheme was conceived. One of the Konus drogues in the docking adapter would be modified to serve as an air-tight electrical junction. There were two of these drogues. One was kept permanently on the axial port for dockings. The other, which was detachable, had been moved around the radial ports as necessary to facilitate the movements of the add-on modules, which swung themselves around on their short Ljappa arms. This Konus had last been used to accept the Priroda module. It comprised the hollow drogue, which was essentially a conical guide plate, and the bulbous end cap that contained the clamp for the mechanism at the tip of the probe. The two parts were connected by a ring of bolts. The plan was to remove the end cap and the clamp, and to bolt on a new unit with a set of electrical sockets on each side. If this could be mounted on Spektr's hatch, then the situation should be retrievable. It would require an internal spacewalk, however, to install this hatch and reroute the necessary cables in the stricken Spektr module. Trials in the hydrotank by Anatoli Solovyov and Pavel Vinogradov (who were to be Mir's next residents) demonstrated that the procedure was manageable, so the apparatus was loaded into Progress-M 35, which was launched on 5 July. As this closed in to dock two days later, the view from its camera clearly revealed the extent to which one of Spektr's solar panels had been twisted by the collision, but the other three appeared undamaged. Kaliningrad had hoped to be able to effect repairs within a week, but when Tsibliev and Lazutkin reviewed the procedure they requested additional time to prepare, and so the internal spacewalk was pushed back a week.

Spektr's damaged solar panel.

The fact that the Konus was not already on Spektr's hatch meant the preparatory work of swapping the junction plate for the probe mechanism could be done in a shirt-sleeved environment, which would make the job of working on the two dozen bolts more manageable. A spacesuited rehearsal without depressurising the docking adapter was scheduled for 15 July, and the real thing was set for 18 July. With power restored to the maximum available, the next crew would be able to concentrate on external activities designed to locate and plug Spektr's leak, so that the module could be repressurised and, with a little luck, much of its apparatus salvaged. The extent to which this proved feasible would determine NASA's future on Mir. The module housed half of the agency's science apparatus. It was questionable whether there would be sufficient scientific yield to justify continuing the Shuttle–Mir programme if Spektr had to be written off. If the errant ferry had hit and depressurised Kristall, with its androgynous docking system, the programme would, of necessity, have been curtailed. Foale would then have been obliged to return with Tsibliev and Lazutkin in the Soyuz, because Atlantis would *not* be returning in September. A great deal, therefore, was riding on the outcome of this makeshift repair.

Although the preparations progressed well, the mounting stress evidently took its toll on the commander. On 12 July, Tsibliev reported feeling a heart arrhythmia, and when this was confirmed the following day by an EKG he was ordered to take a combination of heart medication and tranquillisers for 10 days. On 15 July Vladimir Solovyov, the lead flight director, decided to postpone the repair exercise, and asked NASA if Foale could support Lazutkin in this task. Foale had backed up Linenger in the preparations for Linenger's spacewalk, and therefore was familiar with the procedures for operating the Orlan suit. On 16 July, while rehearsing the process of disconnecting the cabling in the docking adapter, Lazutkin inadvertently unplugged the cable of the recently installed sensor and the attitude-control system, denied this input, put Mir into free drift, with the result that the power output from the solar panels promptly fell and the complex was rapidly overwhelmed by another power crisis. Once again, Tsibliev used the thrusters of the Soyuz ferry to reorient the complex to enable the solar panels to feed power to recharge the batteries – a process that took two days. Although the cosmonauts argued to go ahead with the repair on 25 July, in order to provide sufficient power to enable Léopold Eyharts to accompany the coming three-week handover, they were ordered on 21 July to leave the task to their successors. As a consequence of this decision, the French visit was postponed to the *next* handover, in early 1998. This slippage would propagate on through the visiting programme. Mir was booked up with commercial activities to the end of the century, but it was far from clear that the complex would last long enough to sustain the demand for its services.

RECOVERY

Progress-M 35 undocked on 6 August and settled into a parallel orbit. The next day, Soyuz-TM 26 arrived with Anatoli Solovyov and Pavel Vinogradov. After a brief handover designed not to tax the environmental systems, Tsibliev and Lazutkin left in Soyuz-TM 25 on 14 August. Solovyov, Vinogradov and Foale undocked the next day to swap ends. Instead of pulling back and waiting while the complex rotated, this time Mir was held stable and the Soyuz flew around to line up to approach the front port, pausing *en route* to enable Foale to shoot a video of the Spektr module for damage assessement. Progress-M 35 returned on 18 August. Moments before it was to dock, Mir's attitude-control computer shut down, leaving the complex in free drift. Solovyov, ready at the TORU, promptly took control and steered the ferry in.

The first task on the list was to reconnect the undamaged solar panels on Spektr. On 22 August Foale retreated to the descent module of the Soyuz, and Solovyov and Vinogradov donned their Orlan suits. With all six of its hatches closed, the docking adapter was cramped. They got off to a poor start. Firstly, a leaky hatch seal slowed the depressurisation of the compartment, and when this was finally achieved it was discovered that one of Vinogradov's gloves was leaking and the compartment had to be repressurised to attend to this. It had been intended to venture into Spektr while in contact with Kaliningrad, but these delays meant that when they were finally able to remove the flat-plate covering, they were out of range. Vinogradov entered the module first, moving feet first. As the powerless module was dark, he took recently delivered lamps. There were white crystals floating in the module and the apparatus was coated with a thin layer of frost. Vinogradov plugged the various cables into the sockets on the Spektr side of the modified Konus in order to be able to draw on the power produced by the solar panels. Solovyov then entered the module as well, and they sought some indication of the puncture, but because the interior was crammed with apparatus it was not possible to inspect the pressure shell thoroughly. On their way out, they retrieved several items for Foale. The power grid was reconfigured the next day to use the additional power. Unfortunately, the commands sent through the modifed Konus to steer the panels were ineffective, and they could not be made to track the Sun for maximum output.

On 6 September, Foale accompanied Solovyov on a spacewalk from the airlock on Kvant 2. Solovyov mounted one of the cranes and Foale manoeuvred him to the crumpled radiator on Spektr. Solovyov used a cutter to open the thermal insulation and inspect the struts which attached the radiator to the hull, but although the struts were bent there was no sign of a puncture. They then inspected the damaged solar panel, which was holed, distorted, and canted significantly off axis. Solovyov shot video to enable the engineers to evaluate whether the leak was in the stressed motor housing. They had brought handrails to be affixed to assist a later attempt to seal the puncture, but on seeing no sign of the leak they stored these nearby. Using a telescoping pole with a hook at its tip, Solovyov rotated the undamaged arrays to a better orientation for generating power. Back at the airlock, Foale retrieved one of the experiments that had been set up by Linenger. Afterwards, Vladimir Solovyov told reporters, "We had the crew look at *seven* of the most seriously damaged areas of the Spektr module, but nothing suspicious was found that could be named as causing the breach."

Valeri Ryumin, the Russian Shuttle–Mir director, announced on 12 September that he intended to fly on the final Shuttle to inspect the complex's condition, and that preparations would begin in 1998 to de-orbit Mir in early 1999, which was bad news for those hoping to visit the complex after NASA left.

When Mir's principal computer crashed on 14 September, a data processor was cannibalised for parts to restore it, and (as before, after a period in free-drift) it took several days to fully recover from the power shortage. It was decided to send up a refurbished computer on the next Shuttle as an interim backup, and a new one on the next cargo ferry – the launch of which would be delayed until the computer became available. On 18 September Foale, citing the historic call by John F Kennedy in 1961 for the nation to prove itself by doing things that were hard rather than things that were easy, implored NASA not to curtail the Shuttle–Mir programme. Mir's computer crashed again on 22 September. Rejecting remarks by Congressional critics that Mir was now too dangerous, on 25 September Dan Goldin gave the go-ahead for the next handover. A few hours later, Atlantis lifted off for mission STS-86. Two days later, Jim Wetherbee, who had made the rendezvous with Mir in February 1995, got the chance to close the final 10 metres for a docking. It had been intended that Wendy Lawrence should succeed Foale, but the fact that she was too short to wear the Orlan suit had prompted the decision on 30 July to send her backup, David Wolf, as he would be able to serve as a reserve spacewalker.[5] However, Lawrence still flew to supervise the cargo transfers. Also visiting were cosmonaut Vladimir Titov (making his second Shuttle flight), Frenchman Jean-Loup Chrétien (who had ridden a Soyuz to Mir in 1998) and Scott Parazynski, who had been rejected for Shuttle–Mir by the Russians for being too tall for the Soyuz (which he would have had to use in an emergency). The installation of the refurbished computer took several days, during which time the complex was stabilised by the Shuttle. On 1 October Parazynski and Titov made a spacewalk to retrieve the MEEP experiment, which had been placed on the DM during the STS-76 visit. Since the damage to Spektr was on the opposite side of the complex, they were unable to search for the puncture, but they affixed a 55-kilogram 'cap' to the DM for retrieval by the residents in the event that it was determined that the leak was in the mount of the damaged solar panel – in which case the panel and its motor would be jettisoned and the cap inserted to seal the cavity. (The cap had been delivered by Shuttle because it was too bulky to pass through the hatch of a cargo ferry.) After undocking on 3 October Atlantis manoeuvred into position to watch as air was pumped into Spektr, and Titov reported particles "streaming" out from the somewhere near the motor mount, but the specific source was not identified. Nevertheless, this was a welcome observation because it suggested that the leak theory was correct.

Progress-M 36 lifted off on 5 October. When Progress-M 35 was commanded to undock the next day, one of its latches failed to disengage. Once the rear transfer tunnel had been repressurised, the cosmonauts entered the ship to retract that latch manually, and the ship left on 7 October. Its successor (whose rendezvous had been extended) slipped into the vacant port on 8 October to deliver another computer as a backup. The Shuttle had delivered several new storage batteries, and over the next few days these were installed in the base block and Kvant 2 as replacements for old ones that were failing to hold their charge. Meanwhile, Wolf made a start on his research programme. On 20 October, however, he retreated to the descent module of Soyuz-TM 26 to enable the multiple docking adapter to be depressurised. Yet again this process was foiled, because an equalisation valve was inadvertently left open. Once Solovyov and Vinogradov gained access to Spektr, they tackled the issue of the command link to the motors of the still-functioning solar panels which, the engineers had decided, did not work because the circuit running through the Konus was connected to the box of electro-

Pavel Vinogradov (left), David Wolf and Anatoli Solovyov.

---

[5] If this exchange had not been made, Wolf would have succeeded Lawrence, and been backed up by Andy Thomas.

nics that controlled the motors – which had evidently suffered as a result of the frigid temperatures in the airless module. Instead, the cables from the motors were disconnected from the electronics box and run to the Konus (which had several spare sockets). It was tricky work, because the connectors were not readily accessible, and they found that their tool was too short to reach one behind some apparatus, so although they managed to rewire two of the panels, the third (which was one of the two at the end of the module) had to be left immobile. Over the next few days, they strung cables from the other side of the Konus to the now redundant box of electronics in Kristall to enable Spektr's motors to be steered remotely.

The egress on 3 November was delayed by over an hour due to a malfunction in the telemetry from one of the Orlan suits. Once out, Solovyov and Vinogradov went to Kvant 1. Wolf, who could this time remain in the base block, retracted the solar panel that had been relocated there from Kristall. The spacewalkers then dismounted the panel from the motor, and swung it on a crane to a storage location. On returning to the airlock, they discovered that the outer hatch would not form a hermetic seal. The 'C'-clamp that had been utilised to hold the hatch shut since it was damaged in 1990 was in place, but the seal had deteriorated. As had their predecessors, they used the inner compartment as an emergency airlock. Three days later (with the telemetry system repaired) they went out again, retrieved the spare solar panel that had been delivered on the side of the DM, mounted it on the vacant motor, and plugged it in. It failed to deploy, so the spacewalkers cranked it out manually. The replacement of the old panel by a brand new one would help to overcome the power shortage. On their way back, they paused to cut out a section of the unused dorsal panel for return to Earth to enable its state of deterioration to be studied. Back in the airlock, they installed ten clamps around the hatch, but it still had a slow leak.

While Wolf continued to work on his programme, Solovyov and Vinogradov set about upgrading the environmental systems by installing the second Vozdukh in the base block, as a backup to the one in Kvant 1 that had been operating for a decade. On 21 November, the computer crashed when several channels that provided status information on the gyrodynes mysteriously failed. This time, the complex had fully charged batteries and plenty of power generation. The refurbished computer that had been installed during STS-86's visit was replaced by the new computer delivered by Progress-M 36 in the hope that this would restore the complex to full health. Nevertheless, Viktor Blagov announced that the spacewalk that had been scheduled on 5 December to enable Solovyov and Wolf to retrieve NASA experiments would be postponed to early January 1998, and made only after Solovyov

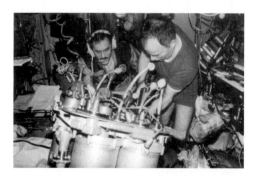

Pavel Vinogradov (left) and Anatoli Solovyov work on a Vozdukh.

and Vinogradov had repaired the airlock hatch. As an indication of the improved power situation, on 1 December the Optizon furnace in Kristall was used for the first time since the collision. To allow Wolf more time to complete his experiments, the launch of STS-89, on which he was to return to Earth, was slipped a week. On 17 December Progress-M 36 undocked, withdrew several kilometres, and released the Inspektor, a small-scale prototype of a manoeuvrable vehicle being developed by Germany for use in conjunction with the International Space Station. The 72-kilogram package incorporated a star tracker, a television camera and a set of thrusters. Remotely controlled from Mir, it was first to make a fly-around of the ferry and then return to Mir for another fly-around prior to departing. However, its star tracker failed and, unable to orient itself, it had to be abandoned. The next day, the base block's air conditioning system developed a freon leak, and the condensate recovery system had to be shut down lest the potable water be contaminated, and as a result the humidity shot up. Progress-M 37's arrival on 20 December was particularly welcome, because in addition to a replacement seal for the airlock hatch, it brought Christmas and New Year presents. To everyone's surprise, the new computer crashed on 1 January

Pavel Vinogradov (left), David Wolf and Anatoli Solovyov celebrate Christmas.

1998, and the next few days were spent recovering from the after-effects. On 8 January Solovyov and Vinogradov suited up in Kvant 2's middle compartment, then depressurised it and entered the airlock to inspect its outer hatch prior to going out to retrieve the Advanced Materials Exposure Experiment that had been deployed by Linenger. On their return, they closed and clamped the hatch and went straight into the inner compartment. Afterwards, they repressurised the airlock to monitor the seal, which vented over the next few days. On Earth, however, Viktor Blagov said that because the inner compartment functioned as an airlock, the task of replacing the seal would be tackled by the next crew. Wolf went out with Solovyov on 14 January and used a spectroreflectometer to characterise the physical condition of the complex's exterior. The instrument measured reflectance across the spectrum from ultraviolet to infrared. It displayed its results on a screen for instant feedback, and stored the full data for subsequent download to Earth. He progressively worked his way down the length of Kvant 2, placing the instrument on the module's radiator to take a reading. (This was part of a risk-mitigation exercise aimed at predicting the lifetime of the radiators intended for the International Space Station.) With this done, Wolf wound up his programme. Having brought the Canadian Protein Crystallisation Experiment with him and put it on the Microgravity Isolation Mount, he had grown 36 different proteins during his tour.

As Atlantis had been returned to Rockwell in California in early November for a refit, STS-89 was flown by Endeavour, which docked on 24 January to deliver Andy Thomas, who was to fly NASA's final tour on Mir. When a faulty sensor indicated that the Shuttle had a propellant leak the next day, attitude control was temporarily assigned to Mir while a software patch was uploaded. Endeavour left on 29 January, having retrieved Wolf and his researches. Progress-M 37 left on 30 January and took up a parallel orbit, and Soyuz-TM 27 docked the next day with Talget Musabayev, Nikolai Budarin and Léopold Eyharts, the Frenchman whose handover visit had been postponed. While the two resident crews undertook maintenance tasks, the two new researchers started work on their respective programmes.

Léopold Eyharts.

As the VDU on the end of the Sofora had almost exhausted its propellant gas, the computer was reprogrammed on 9 February to use the thrusters on Priroda for roll-control as an interim measure until Musabayev and Budarin could replace the VDU. Eyharts left with Vinogradov and Solovyov in Soyuz-TM 26 on 19 February, Soyuz-TM 27 was relocated to the front port the next day, and Progress-M 37 redocked at the rear on 23 February. As Thomas planted seeds in the Astroculture apparatus and grew tissue cultures in the BioTechnology System, his colleagues prepared for a spacewalk on 3 March. Having suited up in Kvant 2's middle compartment and advanced into the airlock, they were stymied upon finding that one of the ten 'C'-clamps had been so tightened by their predecessors (in an effort to seal the hatch) that Budarin broke a wrench attempting to release it, and so the excursion had to be postponed until a stronger wrench could be delivered on the next cargo ship. While they waited, they repaired the leak in the air conditioning system using parts delivered by STS-89, to enable the condenser to restore the humidity to a more comfortable level. Progress-M 37 left on 15 March. When Progress-M 38 arrived on 17 March Kaliningrad noted an anomaly in its Kurs system, so Musabayev docked it using the TORU. The central section of this ferry had been modified to carry a replacement VDU. On 1 April, after the stiff 'C'-clamp was released, Musabayev and Budarin affixed anchors to provide access to Spektr's damaged solar panel. As the damage had been noticed to be 'creeping', the frame was to be rigidised by a 1.5-metre-long brace which had a clamp on each end. However, setting up the work site took so long that they did not have time to install the brace – which was done on 6 April. On emerging again on 11 April, they made their way to the top of the Sofora girder, disconnected the umbilical from the VDU, released the VDU and pushed it away from the complex. On 17 April they disassembled the Strombus and Rapana girders and stowed them at the base of the Sofora. The mechanism in the central section of the ferry that was to ease out and tilt the VDU failed, so the spacewalkers did this manually. On 22 April, they hinged the Sofora and angled it down so that its far end faced the VDU, connected the VDU to it, and then angled it back up again. Once the umbilical had been connected, Thomas activated the heater in the unit and the computer was instructed to cease utilising Priroda's thrusters for roll-control. With over 30 hours of spacewalks to monitor, this had been an exciting time for Thomas.

The replacement of the VDU notwithstanding, on 28 April it was confirmed that future Progress ferries would progressively lower the complex's orbit, leading to its de-orbiting late in 1999. However, the *Duma* implored President Yeltsin to retain it. On 8 May Viktor Blagov said that Mir ought to be retained until the assembly of the International Space Station was well underway. Meanwhile, Thomas worked on his programme and Musabayev and Budarin undertook maintenance. Progress-M 38 left on 15 May, and Progress-M 39 took its place on 17 May – with an eagerly awaited guitar for Musabayev as part of its cargo. The computer crashed again at the end of the month, but the recovery was eased by the fact that at that time Mir was orbiting in continuous sunlight. As the computer was being replaced, Thomas packed up his experiments.

Andy Thomas.

STS-91 was dispatched on 2 June for the final Shuttle visit, and Charles Precourt docked Discovery on 4 June. After making his inspection, Valeri Ryumin told his colleagues in Kaliningrad that they had no appreciation of the situation on Mir. It was "very difficult" to work because "everything is hanging on the walls". Since a crew of two or three was fully occupied in keeping the complex operating, it would require a six-person crew to exploit its capabilities. "We have to

Valeri Ryumin and Nikolai Budarin.

think about how to prevent the International Space Station from becoming so clogged," he emphasised. After the final load of stores was transferred to Mir, unwanted items were retrieved and stowed in the Spacehab module. On 8 June Musabayev presented Precourt with a symbolic key to be delivered to the International Space Station once the assembly of that complex began. After making a slow fly-around to film Mir, Discovery halted 80 metres in front to enable Thomas to watch for a tracer gas that was pumped into Spektr, but he saw nothing that would indicate the site of the leak.

During 979 days spent on Mir, astronauts had experienced a mixture of routine operations and crisis. "The Shuttle–Mir programme has been very useful in giving our astronauts good training in crisis management," wryly observed Representative James Sensenbrenner, the senior Republican on the House's space committee.

A final view of Mir by STS-91.

9

The final chapter

MIR AFTER NASA

"With two, there is plenty of room," observed Talget Musabayev after Discovery's departure concluded NASA's two-year occupancy of Mir. After setting up a round of materials experiments in the Optizon and Gallar furnaces, he and Nikolai Budarin set about compiling an inventory of the apparatus on board the complex. NASA had hoped that Mir would be decommissioned after the Shuttle–Mir programme, but the Russians wished to host commercial missions. After a meeting with the government on 2 July 1998, Yuri Koptev, the head of the Russian Space Agency, confirmed that as there was no funding to continue to operate Mir; the complex would be de-orbited in June 1999. Meanwhile, as Russia had failed to pay for the electricity supplied to its cosmodrome, the Kazakhs denied it power, and as a result the launch of the next crew was slipped ten days, to 13 August. When the routine preparatory test of the Kurs on the rear of the complex on 10 August indicated an anomaly, the undocking of Progress-M 39 was advanced to 12 August in order to exercise the system during the ship's withdrawal. Although this indicated that the system worked, a spare box of electronics was loaded onto Soyuz-TM 28, which lifted off on time. Meanwhile, Progress-M 39 manoeuvred into a storage orbit. In addition to the next resident crew of Gennadi Padalka and Sergei Avdeyev, the ferry's arrival on 15 August delivered Yuri Baturin who, prior to training to become a cosmonaut, was a space physicist who had served as President Yeltsin's national security adviser and Defence Council Secretary. After conducting experiments during the handover, Baturin departed with Musabayev and Budarin in Soyuz-TM 27 on 25 August. On his return, Baturin opined that Mir should be retained for

Yuri Baturin (left), Gennadi Padalka and Sergei Avdeyev.

two years beyond the planned mid-1999 de-orbiting. Soyuz-TM 28 was moved to the front on 27 August, and Progress-M 39 slipped back into the rear port on 1 September, after which it was filled with the trash accumulated during the handover. The first task for Padalka and Avdeyev was to make an internal spacewalk in the Spektr module to reseat the power connectors, some of which were providing only intermittent flow. After several days devoted to resizing and refurbishing the Orlan suits, the operation was conducted late on 15 September and took only 30 minutes.

Meanwhile, funding issues had delayed the fabrication of the 'base block' for the International Space Station to such a degree that on 30 September the dispatch of its commissioning crew was posponed to early 2000, and this prompted further calls to continue to operate Mir, but on 2 October the Russian Space Agency reaffirmed that it would not fund Mir beyond its planned de-orbit date. Progress-M 39 undocked on 26 October and Progress-M 40 took its place later that day. It had twice the usual amount of propellant, to begin the process of lowering Mir's orbit. Nevertheless, the ferry's first action was to *raise* the orbit slightly in order to counter the increased air drag resulting from the 'inflation' of the atmosphere by a recent increase in the level of solar activity. In early November, NASA rejected a Russian suggestion that the first module of the International Space Station be docked on the front of Mir in order to enable the commissioning crew to live on Mir until the new base block was ready for launch. Late on 10 November, Padalka and Avdeyev made a spacewalk to deploy experiments that had been delivered by Progress-M 40. In addition to setting up a Japanese experiment to test a thin-film solar transducer being considered for use on communications satellites, they deployed the Comets package supplied by France to collect micrometeoroids during the passage of the Earth through the Leonid meteor stream in mid-November, which was expected to be particularly rich this time. At the peak of the stream, the cosmonauts spent the night in their Soyuz, in case the complex was holed, which it was not. The first module of the International Space Station was launched on 20 November. Although the Russian Space Agency was eager to de-orbit Mir, Yuri Semenov, the head of the Energiya Corporation, which *owned* Mir, said that he was seeking non-governmental funding to continue to operate the complex. In space, Padalka and Avdeyev worked through November and into December using apparatus delivered by previous researchers, to add value to those experiments. On 1 December, Energiya leaked that it was talking to an investor about funding Mir beyond its nominal de-orbit date. Reinhold Ewald had installed a 'server' to enable Mir to access the internet, but the software did not become available until mid-1998. On 12 December, a 'slow scan' camera donated by radio hams was installed to enable snapshots to be downloaded and posted on the internet. Progress-M 40 raised Mir's orbit again on 24 December to overcome the still-higher-than-usual air drag. Over the next week, the crew celebrated Christmas on Mir and marked the start of what was expected to be its final year.

In order to accommodate the commercial visitors within the timescale available, it had been decided in July 1998 to send up both the French and Slovak researchers in February on what was nominally to be the final mission to Mir. Viktor Afanaseyev would deliver Jean-Pierre Haigneré and Ivan Bella. Although Bella would return with Padalka, Haigneré would serve a full tour along with Afanaseyev and Avdeyev, who would have to serve a double tour. The plan was for them to get Mir ready for de-orbiting and depart a week before the docked tanker made the final manoeuvre. The rationale for their remaining until the 'last minute' was to ensure that the complex was not left unattended for any length of time in the run up to its demise, during which it would have to remain stabilised.

On 3 January 1999 Semenov confirmed that a private investor (whose name he refused to reveal) was interested in funding further operations. On 22 January Prime Minister Yevgeni Primakov signed a decree stating that *if* the company secured the necessary funding – estimated at $250 million per annum – the order to de-orbit Mir would be rescinded and the company would be free to operate Mir on a commercial basis for a further three years.

With a Znamya affixed to its nose, Progress-M 40 undocked on 2 February and withdrew 400 metres. The segmented mirror began to unfurl, but then snagged on the Kurs antenna, which had inadvertently not been retracted. An attempt to reverse the deployment in order to retract the antenna and start again simply ripped the material of the mirror, so the package was jettisoned to enable the ship to depart and de-orbit itself. Soyuz-TM 28 was relocated from the front to the rear ports on 8 February.

The prospect of private investment resulted in an open-ended plan in which the crew would leave on 1 June as planned if the funds did not materialise, otherwise they would stay and hand the complex over to Sergei Zalyotin and Alexander Kaleri in August. However, when Semenov reported on 11 February that the prospective investor had withdrawn, Afanasayev was told that he was to decommission Mir and vacate it on 23 August, to enable it to be de-orbited five days later. A cash crisis held up the manufacture of his spacecraft but Soyuz-TM 29 lifted off on 20 February. Although nominally a commercial flight, Bella's fee was written off by the Russians as part of a Soviet-era debt repayment. Bella returned with Padalka in Soyuz-TM 28 early on 28 February, and Haigneré remained with Afanasayev and Avdeyev.

Jean-Pierre Haigneré (left), Viktor Afanasayev and Ivan Bella.

Haigneré had visited Mir in 1993, so was well prepared for a full tour. He found the complex to be rather warmer than before, but more comfortable. His programme used the existing apparatus to build upon earlier experiments. He began by planting more wheat in the Svet and started a smelting experiment in the Titus furnace. As previously, however, the essence of his research was biomedical. Progress-M 41 arrived on 4 April. On 16 April Afanasayev and Haigneré spacewalked to retrieve the Comets micrometeoroid collector, which they replaced with another French experiment that was to expose organic material to the space environment. They were to have tested a dispenser that was to be used to seal the leak in Spektr, if ever that was found, but the valve of the dispenser failed to emit the sealant. On 1 June it was announced that the de-orbiting of the complex had been postponed six months in order to allow Energiya time to seek an alternative investor. On 20 June Avdeyev exceeded the record of 748 days spent in space, in his case accumulated over three missions. The single-mission endurance record was still held by Valeri Poliakov.

Progress-M 41 left on 17 July and Progress-M 42 arrived the next day. Its cargo included a package for a spacewalk, a special analogue computer to look after Mir while it was vacant, and two batteries to guarantee that this would have power. The last major event of the mission was to be the deployment of the prototype of an antenna to be utilised on communications satellites. On 23 July Afanasayev and Avdeyev took the tightly packed cylindrical package out, attached it mid-way up the Sofora girder and plugged in its umbilical. When Haigneré issued the command for it to unfurl, nothing happened. After 40 minutes of prodding by the spacewalkers, with no result, it was left *in situ*. On their way back, they retrieved the recently deployed French cassette. The engineers reasoned that as the antenna package had not been hermetically sealed, its deployment mechanism had become locked by water vapour that froze when it was taken outside, and that a few days in harsh sunlight should be sufficient to de-gas it. Accordingly, when the spacewalkers returned on 28 July, the antenna responded and immediately unfurled into a 5.2-by-6.4-metre ellipsoidal dish just over 1 metre deep, after which it was jettisoned. On their way back, they retrieved more experiments. As if to emphasise Mir's vulnerability, on 20 July Kaliningrad sent a false command that shut down the main computer. Opportunity was taken to install the analogue computer, as this could not be done while the digital computer was running because it had to be wired directly to the gyrodynes. During this hiatus, the orientation of the complex was controlled by Progress-M 42. After the accumulated trash was loaded into Progress-M 42 on 10 August, the crew began preparations for their own departure. As a bonus, they made *two* observations of the solar eclipse of 11 August: on one orbit they saw the Moon's shadow crossing the English Channel heading for the Continent,[1] and the next time around they caught up with it crossing the Indian subcontinent. During their final week, the cosmonauts mothballed each of the modules in turn and sealed its hatch, starting with the DM on 23 August, then Priroda on 25 August and Kvant 2 and Kristall on 26 August. After packing their

Viktor Afanasayev and Sergei Avdeyev attempt to deploy the antenna affixed to the Sofora girder.

---

[1] The author was on a ship in the English Channel and was briefly immersed in the Moon's shadow – he did not notice Mir passing overhead, however.

spoils the next day, Afanasayev offered his final thoughts to the television camera in the base block and joined his colleagues in Soyuz-TM 29, which undocked from the front port early on 28 August. A few hours later, the capsule landed on the Kazakh steppe northeast of Arkalyk and came to rest on its side, with its occupants hanging from their couches. On emerging, Haigneré was greeted by Claudie André-Deshays, his cosmonaut wife, and after suffering no difficulties in weightlessness, he promptly vomited.

MIRCORP

Having been inhabited continuously since 8 September 1989 – a few days short of a decade – Mir flew on in its automated flight regime. It was in a circular orbit at 355 kilometres, but this would rapidly decrease. Progress-M 42 was to refine the rate of decay for de-orbiting.

On 7 September Kaliningrad shut down Mir's gyrodynes and its main computer, leaving the complex under the control of the special computer. A week later, Koptev said that since Energiya's efforts to secure private investment had come to nothing, the complex *would* be de-orbited early in the New Year. Nevertheless, Energiya did not give up, and on 10 January 2000 it said that it had made a deal with the venture capital company Gold & Appel, which, for $20 million, would buy the *rights* to use Mir for a variety of commercial ventures ranging from research to tourism. "Mir is too valuable a resource to be thrown away," explained Walt Anderson, the chairman of Gold & Appel. After drawing together funding from a number of venture capital sources, Anderson formed MirCorp (with Energiya owning a 60 per cent share) and appointed Jeffrey Manber as its president. A week later, Koptev reported that Mir would be reoccupied in April by a crew who would refurbish it to support another phase of activity. Dan Goldin was "shocked and disappointed". He was concerned that the factories turning out rockets and spacecraft were not geared up to support *both* the International Space Station and renewed Mir operations – it was true that the assembly of the ISS was running late, but the pace was quickening and the base block was set for launch in July, with the commissioning crew going up in October, so this was no time to divert resources. Nevertheless, Progress-M1 1, the first of the new type of ferry built to support the ISS, lifted off on 1 February. Progress-M 42 undocked from Mir's rear port the next day, and the new ship took its place the day after that. During a series of manoeuvres over the next week, the ferry boosted Mir back up to its normal operating altitude. At a press conference held in London on 17 February MirCorp (which was based in Amsterdam) announced that it had signed a leasing agreement for "a new era of space commercialisation", starting by offering wealthy tourists a week on Mir in return for a fee of $20 million.

The plan was to launch Soyuz-TM 30 in early April with Sergei Zalyotin, Alexander Kaleri and the movie actor Vladimir Steklov, who was to shoot footage for use in a film – in fact, he had entered training before the MirCorp deal was signed. However, on 16 March it was announced that Steklov would not fly, because the necessary funding was not in place. Mir's computer was reactivated on 20 March

Alexander Kaleri (left) and Sergei Zalyotin.

and the environmental systems restarted to prepare the base block for inhabitation. Soyuz-TM 30 was launched on 4 April. As Goldin had feared, this was the first of the upgraded type developed for the ISS. After the Kurs had brought the ferry within 10 metres of the front port two days later, Zalyotin took over to be sure of docking. As they methodically tested the base block's systems, air was pumped in by Progress-M1 1 to make up for that lost while Mir was vacant – all space stations slowly leak air, but in this case the rate was anomalously high, and on 17 April was traced to the pressure equalisation valve for Spektr, which was promptly capped. Once the other modules had been checked out, Zalyotin reported, "Mir is in good shape, and we see no reason why its useful life should not be extended." To mark the new phase of activity, a MirCorp banner was placed prominently in the field of view of the base block's television camera. With that, the cosmonauts made a start on a miscellany of maintenance chores. Progress-M1 1 took away the trash on 26 April, and Progress-M1 2 arrived early on 28 April. The next day, MirCorp said that it was about to sign up its first customers, but did not identify them. On 3 May the cosmonauts resized and serviced the Orlan suits. A week later, they tested them in the airlock. Their first task on emerging on 12 May was to test the dispenser of sealant, whose valve had been replaced. There was no longer any expectation of reclaiming the Spektr module, but the engineers wished to test the specially designed tool, which worked by sealing a puncture on a test rig. Next, they made their way to Kvant 1 to inspect the

Sergei Zalyotin.

Cooperative Solar Array, which, in March, had been impaired when its motor ceased to turn the panel to track the Sun. When Kaleri reported that a long section of insulation on a cable had been burned, it became evident that there had been a massive short circuit. As a bonus for the Japanese company that had sent up a thin-film solar transducer for long-term evaluation, they retrieved this package on their way back. Meanwhile, MirCorp said that it was working on a number of projects. One plan was to make Mir the first orbital 'portal' for the internet. It was also negotiating with an Italian company, ItaliMir, to fly Carlo Viberti, an engineer, to undertake scientific experiments. In the past, it had typically taken 18 months to train a visiting researcher, so the training cycle represented a severe problem for any plan to open Mir to non-specialists. Zalyotin and Kaleri's mission was open-ended, in that depending on how the fund-raising went, they would either depart in June or remain to handover to the next crew in August or September. However, on 8 June it was announced that the launch of Salizhan Sharipov and Pavel Vinogradov had been postponed to the end of the year, and that they would hand over to their successors in April 2001. On 13 June, having loaded Progress-M1 2 with trash, Zalyotin and Kaleri sealed it, returned Mir to its automated regime and left early on 16 June. Mir, Zalyotin opined, was "good for another few years".

Three days later, MirCorp reported that its first space tourist would be Dennis Tito, a 59-year-old American who, after working at the Jet Propulsion Laboratory in the 1960s on trajectories for NASA's planetary probes, had founded Wilshire Associates, which was a company of investment management consultants. He was to accompany the handover of Mir. By mid-July, however, the schedule had slipped, with the launch of the next crew in "early 2001" and Tito's visit in "mid 2001". On 18 July, MirCorp said that it hoped to reinstate ongoing habitation, with a tourist defraying the cost of the spacecraft for each handover. One innovative scheme to raise public interest was a deal struck with Mark Burnett, the producer of 'Survivor TV', to create a show 'Destination Mir' in which ten contestants would compete for a flight to Mir. Progress-M1 2 left on 15 October, and Progress-M 43 took its place on 20 October to look after the vacant complex pending a decision on its future. Although MirCorp announced that it was intending to offer public shares in early 2001, Semenov was now sceptical, saying that Mir had "run its course" and the company had "missed the train". Mir's fate was sealed on 1 December, when the company missed a payment. On 12 December MirCorp said it was withdrawing from the plan. Energiya reported that it would de-orbit the complex early in the New Year.

A FIERY END

There was a scare on 26 December when contact was temporarily lost due to a brief power drain. Viktor Blagov reported that if a problem developed that threatened the planned ground-controlled de-orbit procedure, a crew would be sent to supervise the set up for the manoeuvre. Gennadi Padalka and Nikolai Budarin were placed on alert for a short-notice launch. In early January Mir's main computer and gyrodynes were powered up and its rear Kurs tested. After undocking on 25 January, Progress-M 43 flew in formation, ready to redock in the event that its successor failed to dock, but Progress-M1 5 took the vacated port without incident on 27 January.[2] To enable it to de-orbit the complex, this ship had been fitted with additional propellant tanks. On 20 February 2001 the base block marked its 15th anniversary in space. It was now down to 275 kilometres and falling by almost 1 kilometre per day, and this rate would increase as the air drag increased. Despite a suggestion by the *Duma* that Mir be raised into a storage orbit in order to preserve the option of reoccupying it, Yuri Koptev confirmed that it would be de-orbited in March. The timing was controlled

---

[2] There is a gap in the sequence of the Progress-M1 series because some of them went to the International Space Station.

by the rate of decay, which by mid-March had increased to 2 kilometres per day. One way or another the complex was coming down. Left alone, it would re-enter on 27 March, but as this might rain debris down onto a populated area the plan was to de-orbit it into the ocean early on 23 March. As Mir passed over the eastern part of the Russian tracking network, Progress-M1 5 fired its thrusters for some 20 minutes to form a 188-kilometre perigee over the Pacific. On the next pass, they were fired again to lower the perigee to 158 kilometres. On the next pass, the radars measured the trajectory to enable the final manoeuvre to be calculated, and on its next pass the ferry fired its main engine to lower the perigee sufficient to ensure that the complex would make contact with the atmosphere soon after passing beyond Japan, at which time the solar panels and other appendages were ripped off. Over the mid-Pacific the radial modules came off. As its demise would be visible from Fiji, a CNN crew was dispatched. They reported a series of rumbling sonic booms and saw parallel trails with brilliant heads as the various parts of the complex burned up. At 140 tonnes, it was the heaviest object ever deliberately de-orbited. Whatever debris survived fell into the ocean some 3,000 kilometres east of New Zealand. Nothing was recovered. Mir, as the Russian communiqué put it, had "ceased to exist".

Mir burns up over Fiji.

Kaliningrad at the time of Mir's demise.

SPACE TOURIST

In Kaliningrad, the flight controllers congratulated themselves on a job well done and turned their attention to the International Space Station. Zvezda, the base block, had been launched in July 2000. The commissioning crew, Bill Shepherd, Yuri Gidzenko and Sergei Krikalev followed in Soyuz-TM 31 on 31 October and left on Discovery as mission STS-102 in March, having handed over to Yuri Usachev, James Voss and Susan Helms. As the ferry was to be replaced in April, Energiya, as the majority shareholder in MirCorp, decided, much to NASA's irritation, to honour the contract with Tito and send him on this ferry-exchange mission. So it was that a month after Mir's demise, the International Space Station received its first

Dennis Tito.

space tourist. He was launched with Talget Musabayev and Yuri Baturin in Soyuz-TM 32. On docking on 30 April he was all smiles as he floated in through the hatch. Before it consented to his flight, NASA had insisted that he sign a declaration that he would be personally liable for any damage that he caused, and that he would not sue in case of personal injury. Furthermore, unless supervised by an ISS crew member, he was to remain in the Russian part of the complex. In practice, he spent most of his time admiring the Earth slowly passing below, and made himself useful by doing mundane chores for his hosts. NASA-TV did not air his press conferences. The three visitors departed in Soyuz-TM 31 on 5 May, and landed just after dawn on the Kazakh steppe the next day.

Table 9.1 Mir docking operations

Spacecraft	Docking		Port	Undocking		Days
Soyuz-T 15	14 Mar 1986	1638	front	5 May 1986	1612	51.98
Progress 25	21 Mar 1986	1416	rear	20 Apr 1986	2324	30.38
Progress 26	27 Apr 1986	0126	rear	22 Jun 1986	2225	56.87
Soyuz-TM 1	23 May 1986	1412	front	29 May 1986	1323	5.96
Soyuz-T 15	26 Jun 1986	2346	front	16 Jul 1986	1307	19.55
Progress 27	18 Jan 1987	1027	rear	23 Feb 1987	1429	36.17
Soyuz-TM 2	8 Feb 1987	0228	front	30 Jul 1987	0034	171.92
Progress 28	5 Mar 1987	1543	rear	27 Mar 1987	0807	21.68
Kvant 1	9 Apr 1987	0436	rear	permanently docked		—
Progress 29	23 Apr 1987	2105	rear	11 May 1987	0711	17.42
Progress 30	21 May 1987	0953	rear	19 Jul 1987	0420	58.77
Soyuz-TM 3	24 Jul 1987	0731	rear	31 Jul 1987	0328	6.83
Soyuz-TM 3	31 Jul 1987	0348	front	29 Dec 1987	0855	151.21
Progress 31	6 Aug 1987	0228	rear	22 Sep 1987	0358	47.06
Progress 32	26 Sep 1987	0408	rear	10 Nov 1987	0609	45.08
Progress 32	10 Nov 1987	0847	rear	17 Nov 1987	2225	7.57
Progress 33	23 Nov 1987	0439	rear	19 Dec 1987	1116	26.27
Soyuz-TM 4	23 Dec 1987	1551	rear	31 Dec 1987	1210	7.85
Soyuz-TM 4	31 Dec 1987	1229	front	17 Jun 1988	1018	168.90
Progress 34	23 Jan 1988	0309	rear	2 Mar 1988	0640	38.14
Progress 35	26 Mar 1988	0122	rear	5 May 1988	0536	40.17
Progress 36	15 May 1988	0613	rear	5 Jun 1988	1512	21.37
Soyuz-TM 5	9 Jun 1988	1957	rear	18 Jun 1988	1411	8.76

Table 9.1 continued

Spacecraft	Docking		Port	Undocking		Days
Soyuz-TM 5	18 Jun 1988	1427	front	6 Sep 1988	0255	79.52
Progress 37	21 Jul 1988	0234	rear	12 Aug 1988	0832	22.25
Soyuz-TM 6	31 Aug 1988	0941	rear	8 Sep 1988	0505	7.81
Soyuz-TM 6	8 Sep 1988	0525	front	21 Dec 1988	0633	104.04
Progress 38	12 Sep 1988	0522	rear	23 Nov 1988	1513	72.41
Soyuz-TM 7	28 Nov 1988	2016	rear	22 Dec 1988	0945	23.56
Soyuz-TM 7	22 Dec 1988	0959	front	27 Apr 1989	0328	125.73
Progress 39	27 Dec 1988	0840	rear	7 Feb 1989	0946	42.04
Progress 40	12 Feb 1989	1330	rear	3 Mar 1989	0446	18.64
Progress 41	18 Mar 1989	2245	rear	21 Apr 1989	0546	33.29
Progress-M 1	25 Aug 1989	0919	front	1 Dec 1989	1202	98.11
Soyuz-TM 8	8 Sep 1989	0225	rear	11 Dec 1989	1123	94.37
Kvant 2	6 Dec 1989	1521	front	8 Dec 1989	–	2
Kvant 2	8 Dec 1989	–	upper	permanently docked		–
Soyuz-TM 8	11 Dec 1989	1143	front	19 Feb 1990	–	60
Progress-M 2	22 Dec 1989	0841	rear	9 Feb 1990	0533	48.87
Soyuz-TM 9	13 Feb 1990	0938	rear	21 Feb 1990	0656	7.88
Soyuz-TM 9	21 Feb 1990	0715	front	28 May 1990	1548	96.36
Progress-M 3	3 Mar 1990	0305	rear	27 Apr 1990	–	55
Progress 42	8 May 1990	0245	rear	27 May 1990	1109	19.35
Soyuz-TM 9	28 May 1990	1612	rear	4 Jul 1990	0208	36.41
Kristall	10 Jun 1990	1447	front	11 Jun 1990	–	1
Kristall	11 Jun 1990	–	lower	27 May 1995	0328	1810
Soyuz-TM 9	4 Jul 1990	0234	front	9 Aug 1990	0809	36.23
Soyuz-TM 10	3 Aug 1990	1546	rear	10 Dec 1990	0548	128.58
Progress-M 4	17 Aug 1990	0926	front	17 Sep 1990	1643	31.30
Progress-M 5	29 Sep 1990	1627	front	28 Nov 1990	0915	59.70
Soyuz-TM 11	4 Dec 1990	1257	front	26 Mar 1991	–	111
Progress-M 6	16 Jan 1991	1935	rear	16 Mar 1991	1547	58.84
Soyuz-TM 11	26 Mar 1991	–	rear	26 May 1991	1013	61
Progress-M 7	28 Mar 1991	1603	front	7 May 1991	0300	39.46
Soyuz-TM 12	20 May 1991	1830	front	28 May 1991	1410	7.82
Soyuz-TM 12	28 May 1991	1452	rear	10 Oct 1991	0352	134.54
Progress-M 8	1 Jun 1991	1345	front	16 Aug 1991	0217	75.52
Progress-M 9	23 Aug 1991	0454	front	30 Sep 1991	0454	38.00
Soyuz-TM 13	4 Oct 1991	1038	front	15 Oct 1991	0410	10.73
Soyuz-TM 13	15 Oct 1991	0554	rear	14 Mar 1992	1443	151.36
Progress-M 10	21 Oct 1991	0641	front	20 Jan 1992	1014	91.15
Progress-M 11	27 Jan 1992	1231	front	13 Mar 1992	1144	45.97
Soyuz-TM 13	14 Mar 1992	1510	front	25 Mar 1992	–	11
Soyuz-TM 14	19 Mar 1992	1532	rear	9 Aug 1992	–	143
Progress-M 12	22 Apr 1992	0320	front	28 Jun 1992	–	67
Progress-M 13	4 Jul 1992	2055	front	24 Jul 1992	1249	19.66
Soyuz-TM 15	29 Jul 1992	1151	front	1 Feb 1993	0600	186.76
Progress-M 14	18 Aug 1992	0421	rear	21 Oct 1992	1946	64.64

Table 9.1 *continued*

Spacecraft	Docking		Port	Undocking		Days
Progress-M 15	29 Oct 1992	2206	rear	4 Feb 1993	0345	97.23
Soyuz-TM 16	26 Jan 1993	1032	Kristall	22 Jul 1993	–	177
Progress-M 16	23 Feb 1993	2318	rear	26 Mar 1993	0950	30.44
Progress-M 16	26 Mar 1993	1007	rear	27 Mar 1993	0721	0.88
Progress-M 17	2 Apr 1993	0912	rear	11 Aug 1993	1936	131.43
Progress-M 18	24 May 1993	1225	front	3 Jul 1993	1958	40.31
Soyuz-TM 17	3 Jul 1993	2024	front	14 Jan 1994	0737	194.47
Progress-M 19	13 Aug 1993	0400	rear	12 Oct 1993	2059	60.71
Progress-M 20	15 Oct 1993	0225	rear	21 Nov 1993	0538	37.13
Soyuz-TM 18	10 Jan 1994	1450	rear	24 Jan 1994	0612	13.64
Soyuz-TM 18	24 Jan 1994	0801	front	9 Jul 1994	–	166
Progress-M 21	30 Jan 1994	0656	rear	23 Mar 1994	0420	51.89
Progress-M 22	24 Mar 1994	0940	rear	23 May 1994	0457	59.80
Progress-M 23	24 May 1994	1000	rear	2 Jul 1994	1247	39.12
Soyuz-TM 19	3 Jul 1994	1755	rear	2 Nov 1994	1330	121.82
Progress-M 24	2 Sep 1994	1745	front	4 Oct 1994	2156	32.17
Soyuz-TM 20	6 Oct 1994	0328	front	11 Jan 1995	1157	97.35
Soyuz-TM 19	2 Nov 1994	1405	rear	4 Nov 1994	1025	1.85
Progress-M 25	13 Nov 1994	1204	rear	16 Feb 1995	1605	95.17
Soyuz-TM 20	11 Jan 1995	1223	front	22 Mar 1995	~0400	69.65
Progress-M 26	17 Feb 1995	2122	rear	15 Mar 1995	0527	25.34
Soyuz-TM 21	16 Mar 1995	1046	rear	4 Jul 1995	1454	110.17
Progress-M 27	12 Apr 1995	0101	front	23 May 1995	0340	41.11
Kristall	27 May 1995	~0500	front	30 May 1995	–	3
Kristall	30 May 1995	–	right	10 Jun 1995	–	11
Spektr	1 Jun 1995	0458	front	2 Jun 1995	~2100	1.67
Spektr	2 Jun 1995	~2300	lower	permanently docked		–
Kristall	10 Jun 1995	–	front	17 Jul 1995	~0630	37
STS-71	29 Jun 1995	1700	Kristall	4 Jul 1995	1510	4.92
Soyuz-TM 21	4 Jul 1995	1538	rear	11 Sep 1995	0727	68.66
Kristall	17 Jul 1995	~0800	right	permanently docked		–
Progress-M 28	22 Jul 1995	0840	front	4 Sep 1995	0910	44.02
Soyuz-TM 22	5 Sep 1995	1428	front	29 Feb 1996	1020	176.82
Progress-M 29	10 Oct 1995	2332	rear	19 Dec 1995	1215	69.53
STS-74	15 Nov 1995	0928	Kristall	18 Nov 1995	1116	3.07
Progress-M 30	20 Dec 1995	1910	rear	22 Feb 1996	1026	63.63
Soyuz-TM 23	23 Feb 1996	1723	rear	2 Sep 1996	0820	190.63
STS-76	24 Mar 1996	0534	DM	29 Mar 1996	0408	4.94
Priroda	26 Apr 1996	1543	front	27 Apr 1996	–	1
Priroda	27 Apr 1996	–	left	permanently docked		–
Progress-M 31	7 May 1996	1254	front	1 Aug 1996	2047	86.33
Progress-M 32	3 Aug 1996	0203	front	18 Aug 1996	1334	15.48
Soyuz-TM 24	19 Aug 1996	1852	front	7 Feb 1997	–	173
Progress-M 32	3 Sep 1996	1335	rear	20 Nov 1996	2244	78
STS-79	19 Sep 1996	0615	DM	24 Sep 1996	0533	4.97

288   **The final chapter**

Table 9.1 *continued*

Spacecraft	Docking		Port	Undocking		Days
Progress-M 33	22 Nov 1996	0358	rear	6 Feb 1997	1514	75
STS-81	15 Jan 1997	0655	DM	20 Jan 1997	0515	4.93
Soyuz-TM 24	7 Feb 1997	–	rear	2 Mar 1997	–	23
Soyuz-TM 25	12 Feb 1997	1851	front	14 Aug 1997	–	2
Progress-M 34	8 Apr 1997	2131	rear	24 Jun 1997	1422	–
STS-84	17 May 1997	0534	DM	22 May 1997	0505	4.98
Progress-M 35	7 Jul 1997	1000	rear	6 Aug 1997	1546	–
Soyuz-TM 26	7 Aug 1997	2102	rear	15 Aug 1997	1730	–
Soyuz-TM 26	15 Aug 1997	1814	front	19 Feb 1998	0853	–
Progress-M 35	18 Aug 1997	1653	rear	7 Oct 1997	1504	–
STS-86	27 Sep 1997	2258	DM	3 Oct 1997	2028	–
Progress-M 36	8 Oct 1997	2008	rear	17 Dec 1997	0902	–
Progress-M 37	22 Dec 1997	1322	rear	30 Jan 1998	1553	–
STS-89	24 Jan 1998	2314	DM	29 Jan 1998	1955	–
Soyuz-TM 27	31 Jan 1998	2054	rear	20 Feb 1998	1147	–
Soyuz-TM 27	20 Feb 1998	1232	front	25 Aug 1998	0605	–
Progress-M 37	23 Feb 1998	1242	rear	15 Mar 1998	2216	–
Progress-M 38	17 Mar 1998	0331	rear	15 May 1998	2241	–
Progress-M 39	17 May 1998	0351	rear	12 Aug 1998	1328	–
STS-91	4 Jun 1998	2058	DM	8 Jun 1998	2100	–
Soyuz-TM 28	15 Aug 1998	1457	rear	7 Aug 1998	0944	–
Soyuz-TM 28	7 Aug 1998	1007	front	8 Feb 1999	1423	–
Progress-M 39	1 Sep 1998	0835	rear	26 Oct 1998	0303	–
Progress-M 40	26 Oct 1998	0844	rear	2 Feb 1999	1300	–
Soyuz-TM 28	8 Feb 1999	1439	rear	28 Feb 1999	0152	–
Soyuz-TM 29	22 Feb 1999	0836	front	28 Aug 1999	0117	–
Progress-M 41	4 Apr 1999	1647	rear	17 Jul 1999	1520	–
Progress-M 42	18 Jul 1999	2153	rear	2 Feb 2000	0612	–
Progress-M1 1	3 Feb 2000	1102	rear	26 Apr 2000	2033	–
Soyuz-TM 30	6 Apr 2000	1032	front	16 Jun 2000	0125	–
Progress-M1 2	28 Apr 2000	0129	rear	15 Aug 2000	–	–
Progress-M 43	20 Oct 2000	–	rear	25 Jan 2001	0819	–
Progress-M1 5	27 Jan 2001	0833	rear	< deorbited 0807, 23 Mar 2001 >		

Dates are Moscow Time.

Table 9.2 Mir crewing

Cosmonaut		Arrival	Departure	Days
Leonid Kizim	CDR	Soyuz-T 15	Soyuz-T 15	125.00
Vladimir Solovyov	FE	Soyuz-T 15	Soyuz-T 15	125.00
Yuri Romanenko	CDR	Soyuz-TM 2	Soyuz-TM 3	326.48
Alexander Laveikin	FE	Soyuz-TM 2	Soyuz-TM 2	174.14
Alexander Viktorenko	CDR	Soyuz-TM 3	Soyuz-TM 2	7.96
Alexander Alexandrov	FE	Soyuz-TM 3	Soyuz-TM 3	160.26
Mohammed Faris	CR	Soyuz-TM 3	Soyuz-TM 2	7.96
Vladimir Titov	CDR	Soyuz-TM 4	Soyuz-TM 6	365.94
Musa Manarov	FE	Soyuz-TM 4	Soyuz-TM 6	365.94
Anatoli Levchenko	CR	Soyuz-TM 4	Soyuz-TM 3	7.91
Anatoli Solovyov	CDR	Soyuz-TM 5	Soyuz-TM 4	9.84
Viktor Savinykh	FE	Soyuz-TM 5	Soyuz-TM 4	9.84
Alexander Alexandrov (B)	CR	Soyuz-TM 5	Soyuz-TM 4	9.84
Vladimir Lyakhov	CDR	Soyuz-TM 6	Soyuz-TM 5	8.85
Valeri Poliakov	CR	Soyuz-TM 6	Soyuz-TM 7	240.90
Abdul Mohmand	CR	Soyuz-TM 6	Soyuz-TM 5	8.85
Alexander Volkov	CDR	Soyuz-TM 7	Soyuz-TM 7	151.46
Sergei Krikalev	FE	Soyuz-TM 7	Soyuz-TM 7	151.46
Jean-Loup Chrétien	CR	Soyuz-TM 7	Soyuz-TM 6	24.75
Alexander Viktorenko	CDR	Soyuz-TM 8	Soyuz-TM 8	166.30
Alexander Serebrov	FE	Soyuz-TM 8	Soyuz-TM 8	166.30
Anatoli Solovyov	CDR	Soyuz-TM 9	Soyuz-TM 9	179.06
Alexander Balandin	FE	Soyuz-TM 9	Soyuz-TM 9	179.06
Gennadi Manakov	CDR	Soyuz-TM 10	Soyuz-TM 10	130.86
Gennadi Strekalov	FE	Soyuz-TM 10	Soyuz-TM 10	130.86
Viktor Afanasayev	CDR	Soyuz-TM 11	Soyuz-TM 11	175.08
Musa Manarov	FE	Soyuz-TM 11	Soyuz-TM 11	175.08
Toehiro Akiyama	CR	Soyuz-TM 11	Soyuz-TM 10	7.92
Anatoli Artsebarski	CDR	Soyuz-TM 12	Soyuz-TM 12	144.64
Sergei Krikalev	FE	Soyuz-TM 12	Soyuz-TM 13	311.84
Helen Sharman	CR	Soyuz-TM 12	Soyuz-TM 11	7.88
Alexander Volkov	CDR	Soyuz-TM 13	Soyuz-TM 13	175.12
Takhtar Aubakirov	CR	Soyuz-TM 13	Soyuz-TM 12	7.92
Franz Viehböck	CR	Soyuz-TM 13	Soyuz-TM 12	7.92
Alexander Viktorenko	CDR	Soyuz-TM 14	Soyuz-TM 14	145.60
Alexander Kaleri	FE	Soyuz-TM 14	Soyuz-TM 14	145.60
Klaus-Dietrich Flade	CR	Soyuz-TM 14	Soyuz-TM 13	7.92
Anatoli Solovyov	CDR	Soyuz-TM 15	Soyuz-TM 15	188.90
Sergei Avdeyev	FE	Soyuz-TM 15	Soyuz-TM 15	188.90
Michel Tognini	CR	Soyuz-TM 15	Soyuz-TM 14	13.90
Gennadi Manakov	CDR	Soyuz-TM 16	Soyuz-TM 16	179.03
Alexander Poleshchuk	FE	Soyuz-TM 16	Soyuz-TM 16	179.03
Vasili Tsibliev	CDR	Soyuz-TM 17	Soyuz-TM 17	196.74
Alexander Serebrov	FE	Soyuz-TM 17	Soyuz-TM 17	196.74
Jean-Paul Haigneré	CR	Soyuz-TM 17	Soyuz-TM 16	20.67

290 The final chapter

Table 9.2 *continued*

Cosmonaut		Arrival	Departure	Days
Viktor Afanasayev	CDR	Soyuz-TM 18	Soyuz-TM 18	182.02
Yuri Usachev	FE	Soyuz-TM 18	Soyuz-TM 18	182.02
Valeri Poliakov	CR	Soyuz-TM 18	Soyuz-TM 20	437.75
Yuri Malenchenko	CDR	Soyuz-TM 19	Soyuz-TM 19	125.96
Talget Musabayev	FE	Soyuz-TM 19	Soyuz-TM 19	125.96
Alexander Viktorenko	CDR	Soyuz-TM 20	Soyuz-TM 20	169.22
Yelena Kondakova	FE	Soyuz-TM 20	Soyuz-TM 20	169.22
Ulf Merbold	CR	Soyuz-TM 20	Soyuz-TM 19	31.52
Vladimir Dezhurov	CDR	Soyuz-TM 21	STS-71	115.36
Gennadi Strekalov	FE	Soyuz-TM 21	STS-71	115.36
Norman Thagard	CR	Soyuz-TM 21	STS-71	115.36
Anatoli Solovyov	CDR	STS-71	Soyuz-TM 21	75.48
Nikolai Budarin	FE	STS-71	Soyuz-TM 21	75.48
Yuri Gidzenko	CDR	Soyuz-TM 22	Soyuz-TM 22	179.03
Sergei Avdeyev	FE	Soyuz-TM 22	Soyuz-TM 22	179.03
Thomas Reiter	CR	Soyuz-TM 22	Soyuz-TM 22	179.03
Yuri Onufrienko	CDR	Soyuz-TM 23	Soyuz-TM 23	192.84
Yuri Usachev	FE	Soyuz-TM 23	Soyuz-TM 23	192.84
Shannon Lucid	CR	STS-76	STS-79	188.20
Valeri Korzun	CDR	Soyuz-TM 24	Soyuz-TM 24	196.73
Alexander Kaleri	FE	Soyuz-TM 24	Soyuz-TM 24	196.73
Claudie André-Deshays	CR	Soyuz-TM 24	Soyuz-TM 23	15.76
John Blaha	CR	STS-79	STS-81	128.23
Jerry Linenger	CR	STS-81	STS-84	132.17
Vasili Tsibliev	CDR	Soyuz-TM 25	Soyuz-TM 25	184.92
Alexander Lazutkin	FE	Soyuz-TM 25	Soyuz-TM 25	184.92
Reinhold Ewald	CR	Soyuz-TM 25	Soyuz-TM 24	19.69
Michael Foale	CR	STS-84	STS-86	144.58
Anatoli Solovyov	CDR	Soyuz-TM 26	Soyuz-TM 26	197.73
Pavel Vinogradov	FE	Soyuz-TM 26	Soyuz-TM 26	197.73
David Wolf	CR	STS-86	STS-89	127.84
Andrew Thomas	CR	STS-89	STS-91	140.64
Talgat Musabayev	CDR	Soyuz-TM 27	Soyuz-TM 27	207.53
Nikolai Budarin	FE	Soyuz-TM 27	Soyuz-TM 27	207.53
Léopold Eyharts	CR	Soyuz-TM 27	Soyuz-TM 26	20.69
Gennadi Padalka	CDR	Soyuz-TM 28	Soyuz-TM 28	198.69
Sergei Avdeyev	FE	Soyuz-TM 28	Soyuz-TM 29	379.62
Yuri Baturin	CR	Soyuz-TM 28	Soyuz-TM 27	11.82
Viktor Afanasayev	CDR	Soyuz-TM 29	Soyuz-TM 29	188.84
Jean-Pierre Haigneré	FE	Soyuz-TM 29	Soyuz-TM 29	188.84
Ivan Bella	CR	Soyuz-TM 29	Soyuz-TM 28	7.92
Sergei Zalyotin	CDR	Soyuz-TM 30	Soyuz-TM 30	72.82
Alexander Kaleri	FE	Soyuz-TM 30	Soyuz-TM 30	72.82

Dates are Moscow Time. Only genuine Mir crew members are listed; the members of Shuttle crews who visit for a few days are not included.

Table 9.3 Mir spacewalks

Date	Hours	Activity
11 Apr 1987	3.6	Romanenko and Laveikin opened a radial docking port, inspected the rear docking system to discover why Kvant 1 could not achieve hard-dock, found a bag of rubbish within its drogue, tugged this out, then watched the probe retract to complete the docking.
12 Jun 1987	1.9	Romanenko and Laveikin opened a radial docking port, began to erect a solar panel on the motor built into the roof of the base block.
16 Jun 1987	3.25	Romanenko and Laveikin opened a radial docking port, completed erecting the solar panel, plugged in its cables, then affixed sample cassettes.
26 Feb 1988	4.4	Titov and Manarov opened a radial docking port cover and retracted one of the solar panel segments to replace it with a new one which included a section which produced telemetry to monitor degradation in performance.
30 June 1988	5.2	Titov and Manarov opened a radial docking port, opened the thermal protection blanket at the rear of Kvant 1 to gain access to a malfunctioning X-ray telescope, but the tool supplied snapped when they tried to use it to release the detector.
20 Oct 1988	4.2	Titov and Manarov opened a radial docking port, used a new tool to release and replace the X-ray detector, to repair the telescope, then affixed an anchor near the docking adaptor in preparation for a French apparatus; they wore the new Orlan-DMA suits for the first time.
9 Dec 1988	5.9	Volkov and Chrétien opened a radial docking port, set up the Echantillon cassette, attached the ERA apparatus to the anchor, deployed it, tested its vibration modes, then finally jettisoned it.
8 Jan 1990	2.9	Viktorenko and Serebrov opened a radial docking port, installed two star trackers on Kvant 1 in order to help to orientate the to-be-expanded complex.
11 Jan 1990	2.9	Viktorenko and Serebrov opened a radial docking port, retrieved the Echantillon cassette, dismantled the anchor used by the ERA apparatus, set up exposure cassettes and then swapped the drogue from the upper to the lower port in the multiple docking adaptor.
26 Jan 1990	3.1	Viktorenko and Serebrov exited Kvant 1's airlock, affixed an anchor just outside, dismantled the Kurs antenna at the other end of the module, set up exposure cassettes, and then erected the scan platform.
1 Feb 1990	5.0	Viktorenko and Serebrov exited the airlock and Serebrov tested the YMK autonomous manoeuvring unit.
5 Feb 1990	3.75	Viktorenko and Serebrov exited the airlock and Viktorenko reran the YMK test.

Table 9.3 continued

Date	Hours	Activity
26 July 1990	3.5	Solovyov and Balandin inspected the airlock hatch to discover why it would not shut, then went to remove the ladder off their ferry, attached it to Kristall for storage, then returned and were able to close the airlock hatch only by extreme eff-ort.
30 Oct 1990	3.75	Manakov and Strekalov attempted to repair the airlock hatch, but realised that the hinge needed to be replaced.
7 Jan 1991	5.3	Afanasayev and Manarov replaced the hinge to repair the airlock hatch, affixed an anchor to the base block, retrieved sample cassettes and removed a camera from the scan platform.
23 Jan 1991	5.6	Afanasayev and Manarov exited the airlock, installed a crane on the new anchor, then retrieved sample cassettes.
26 Jan 1991	6.4	Afanasayev and Manarov exited the airlock, installed mounts on each side of Kvant 1 (preparatory to transferring Kristall's solar panels), fitted laser-reflectors (for a rangefinder that was to be used by Buran during its rendezvous), and then installed the Sprut-5 spectrometer.
26 Apr 1991	3.6	Afanasayev and Manarov exited the airlock, inspected a faulty Kurs antenna on Kvant 1, retrieved sample cassettes, replaced the camera in the scan platform, then tested a mechanical joint.
25 Jun 1991	5.0	Artsebarski and Krikalev exited the airlock, replaced the broken Kurs antenna, and set up a mechanical joint.
28 Jun 1991	3.5	Artsebarski and Krikalev exited the airlock, set up a charged-particle spectrometer and the Trek cosmic ray detector, then retrieved the mechanical joint.
15 Jul 1991	5.8	Artsebarski and Krikalev exited the airlock and mounted a work platform in Kvant 1.
19 Jul 1991	5.5	Artsebarski and Krikalev exited the airlock, affixed the Sofora package to the mount, then started to erect the girder.
23 Jul 1991	5.6	Artsebarski and Krikalev exited the airlock and built more of the Sofora girder.
27 Jul 1991	6.8	Artsebarski and Krikalev exited the airlock, finished erecting the Sofora girder and placed a Hammer & Sickle on the end of it.
20 Feb 1991	4.2	Volkov had to remain near the airlock when the cooling unit in his suit failed, so Krikalev dismantled obsolete fixtures from the Sofora girder site, then retrieved an experimental solar cell.
8 Jul 1992	2.1	Viktorenko and Kaleri exited the airlock and cut through thermal insulation on Kvant 2 to install two new gyrodynes.
3 Sep 1992	3.9	Solovyov and Avdeyev exited the airlock, tilted and then locked the Sofora girder down across the cargo ferry at the rear of the complex, then cranked the thruster pack up out of the ferry.
7 Sep 1992	5.2	Solovyov and Avdeyev exited the airlock, ran an umbilical along the Sofora girder, affixed metal straps to the thruster pack, and retrieved the Hammer & Sickle.

Table 9.3 *continued*

Date	Hours	Activity
11 Sep 1992	5.8	Solovyov and Avdeyev exited the airlock, attached the thruster pack to the Sofora girder, then swung it back up.
15 Sep 1992	3.6	Solovyov and Avdeyev exited the airlock, put a Kurs antenna on Kristall's axial androgynous port, and then retrieved sample cassettes.
19 Apr 1993	5.5	Manakov and Poleshchuk exited the airlock, attached a motor to the (previously installed) framework on the side of Kvant 1, plugged it into the power supply, then found that one of the handles of the crane was no longer there!
18 June 1993	4.6	Manakov and Poleshchuk exited the airlock, fitted a replacement handle, then installed the second motor and plugged it in.
16 Sep 1993	4.3	Tsibliev and Serebrov exited the airlock, affixed a platform on Kvant 1 and then mounted the Rapana container on it.
20 Sep 1993	3.25	Tsibliev and Serebrov exited the airlock, erected the Rapana girder and set up sample cassettes.
28 Sep 1993	1.9	Tsibliev and Serebrov exited the airlock, set up some sample cassettes and retrieved others, and then began to video the state of the complex's surface.
22 Oct 1993	0.6	Tsibliev and Serebrov exited the airlock, deployed a meteoroid package, and continued recording the complex's state.
29 Oct 1993	4.2	Tsibliev and Serebrov exited the airlock, completed recording the complex's state, checked the base of the Sofora girder, and retrieved sample cassettes.
9 Sep 1994	5.1	Malenchenko and Musabayev exited the airlock, inspected the front port for damage by the impact of Progress-M 24, then Kristall for damage following the collision with Soyuz-TM 17, affixed an anchor on the base block (for a second crane), then set up several sample cassettes.
13 Sep 1994	6.0	Malenchenko and Musabayev exited the airlock, inspected the fittings of Kristall's solar panels, swung to Kvant 1 to inspect the motors fitted to receive them, inspected the Sofora girder, then retrieved experiments previously mounted on the Rapana girder.
12 May 1995	6.25	Dezhurov and Strekalov exited the airlock and retracted the left-side solar panel on Kristall.
17 May 1995	6.8	Dezhurov and Strekalov exited the airlock, and released, transferred and attached the panel to the motor on the left-side of Kvant 1.
22 May 1995	5.25	Dezhurov and Strekalov exited the airlock, finished stringing cabling from the solar panel to the base block, then partially retracted the remaining panel.
28 May 1995	0.4	Dezhurov and Strekalov depressurised the docking adaptor, then moved the radial drogue from the lower to the right port (for Kristall).

Table 9.3 *continued*

Date	Hours	Activity
2 Jun 1995	0.5	Dezhurov and Strekalov depressurised the docking adaptor, then moved the radial drogue from the right to the lower port (for Spektr; it was moved back a few days after Spektr docked, but with both modules on, doing so did not require depressurising the docking adaptor).
14 Jul 1995	5.6	Solovyov and Budarin exited the airlock, inspected a faulty solar panel on Kvant 2, released a clamp to deploy a stuck panel on Spektr, then inspected the right-side radial port for any sign of damage.
19 Jul 1995	3.2	Solovyov had to remain near the airlock when the cooler in his suit failed (they had intended to deploy the MIRAS package), so Budarin retrieved sample cassettes (including Trek), and set up new ones.
21 Jul 1995	5.6	Solovyov and Budarin exited the airlock and affixed the MIRAS package to a scan platform on the end of Spektr.
20 Oct 1995	5.25	Avdeyev and Reiter installed the ESEF cassettes on Spektr (near MIRAS) and exchanged the cassette in the Komza experiment.
8 Dec 1995	0.6	Gidzenko and Avdeyev depressurised the docking adaptor, then moved the radial drogue from the right to the left port (for Priroda).
8 Feb 1996	3.1	Gidzenko and Reiter retrieved two of the ESEF cassettes, put in a new one, then tried (and failed) to retrieve the mounting of a redundant antenna for examination.
15 Mar 1996	5.8	Onufrienko and Usachev installed a second crane (on the right side of the base block).
27 Mar 1996	6.1	Godwin and Clifford left Atlantis, retrieved a camera from the DM and deployed the four-cassette MEEP.
21 May 1996	5.3	Onufrienko and Usachev retrieved the Cooperative Solar Array from the DM, mounted it on the motor on the righthand side of Kvant 1 and filmed a Pepsi commercial.
25 May 1996	5.7	Onufrienko and Usachev extended the Cooperative Solar Array.
30 May 1996	4.3	Onufrienko and Usachev attached the MOMS multispectral imager outside Priroda.
6 Jun 1996	3.6	Onufrienko and Usachev swapped the cassettes in the Komza package, and deployed a materials cassette and two micrometeoroid traps.
13 Jun 1996	3.5	Onufrienko and Usachev dismantled the Rapana truss and left it alongside the Sofora girder, then erected the Strombus truss in its place. On the way back, they completed the deployment of the stuck Travers radar dish on Priroda (its automated mechanism having failed).
2 Dec 1996	5.9	Korzun and Kaleri started to string cables to the base block for the Cooperative Solar Array into the power system, then retrieved the Rapana and remounted it on top of the Strombus so that the sensors at its end would be as far from the station as possible.

Table 9.3 *continued*

Date	Hours	Activity
9 Dec 1996	6.6	Korzun and Kaleri completed laying the power cable to the base block and then transferred the Kristall Kurs antenna to the DM.
29 Apr 1997	4.8	Tsibliev and Linenger retrieved the PIE and MSRE, and then deployed the AMEE.
22 Aug 1997	3.3	Solovyov and Vinogradov depressurised the multiple docking adapter to affix the modified Konus and entered Spektr to wire it up.
6 Sep 1997	6.0	Solovyov and Foale inspected Spektr's exterior and oriented ts solar panels.
1 Oct 1997	5.0	Parazynski and Titov left Atlantis to retrieve the MEEP from the DM.
20 Oct 1997	6.6	Solovyov and Vinogradov depressurised the multiple docking adapter to reconfigure the cables on the modified Konus.
3 Nov 1997	6.1	Solovyov and Vinogradov retracted and deleted the lefthand solar panel on Kvant 1 (originally off Kristall).
6 Nov 1997	6.2	Solovyov and Vinogradov retrieved the spare solar array from the DM and installed it on Kvant 1's lefthand side.
8 Jan 1998	4.1	Solovyov and Vinogradov retrieved the AMEE.
14 Jan 1998	3.8	Solovyov and Wolf to use a spectroreflectometer to assess the condition of the radiators on Kvant 2.
3 Mar 1998	1.2	Musabayev and Budarin depressurised the airlock but could not open the outer hatch and had to cancel the planned excursion.
1 Apr 1998	6.7	Musabayev and Budarin affixed anchors to provide access to Spektr's damaged solar panel.
6 Apr 1998	4.4	Musabayev and Budarin fitted a brace to Spektr's damaged solar panel.
11 Apr 1998	6.4	Musabayev and Budarin dismounted the VDU from the Sofora girder.
17 Apr 1998	6.6	Musabayev and Budarin made preparations to install the new VDU.
22 Apr 1998	6.4	Musabayev and Budarin mounted the new VDU on the Sofora girder.
15 Sep 1998	0.5	Padalka and Avdeyev depressurised the multiple docking adapter to reseat electrical connectors.
10 Nov 1998	5.9	Padalka and Avdeyev deployed the Comets micrometeroids sampler (for the Leonids).
16 Apr 1999	6.3	Afanasayev and Haigneré retrieved the Comets package.
23 Jul 1999	6.1	Afanasayev and Avdeyev were frustrated in their attempt to deploy an antenna on the Sofora girder.
28 Jul 1999	5.4	Afanasayev and Avdeyev deployed the antenna on the Sofora girder.
12 May 2000	4.8	Zalyotin and Kaleri tested a tool that was to dispense sealant, inspected the Cooperative Solar Array and retrieved experiments.

10

In retrospect

THE INCREDIBLE SOYUZ

The key point to acknowledge in assessing Soviet human spaceflight is that since its lunar programme faltered in the late 1960s it has doggedly pursued the development of a space station capable of continuous habitation. To properly assess this effort, it is first necessary to review the evolution of the Soyuz spacecraft. This was designed in the early 1960s as a general purpose vehicle to be launched by the *Semyorka* rocket that had been used for Sputnik and Vostok. In a sense, Soyuz was Korolev's first 'spaceship', because the Vostok had simply drifted in the orbit in which it was inserted by its rocket until it fired its retrorocket to initiate a simple ballistic descent. The Soyuz, in contrast, was to be capable of *manoeuvring*, both to rendezvous with another vehicle and to refine its trajectory during its re-entry of the atmosphere. It was to have a modular design, so that it would be capable of undertaking a *variety* of missions. In addition to the descent capsule and the service module (which contained the engines), an 'orbital module' ameliorated the camped state of the descent module by providing living space for the crew. Specialised forms of each module could be developed to satisfy different mission requirements. For example, one version of the descent module had a strengthened heat shield to enable it to survive atmospheric re-entry at the higher speed of a direct return from the Moon. The orbital module could carry different docking systems and be fitted with a variety of apparatus depending on the specific mission. Similarly, the power of the engine could be tailored to given mission requirements. This meant that the *lightest* version of each module could be employed for each specific vehicle. The Earth-orbital version could be launched by the *Semyorka*, but to dispatch the lunar version on a loop around the Moon required Chelomei's Proton (topped by one of Korolev's own stages, borrowed from the vast rocket that he had designed for the lunar landing mission). Because the circumlunar version (flown unmanned in the guise of Zond) would not enter lunar orbit, it had a stripped-down service module that required only to make mid-course corrections to refine the return trajectory, and since its trajectory was computed from Earth-based radar tracking it did not need to be able to navigate independently in deep space. The orbital module was deleted for the simple reason that it made the vehicle too heavy for the Proton. Although Korolev's objective had been to fly a cosmonaut around the Moon before the Americans, this particular race was narrowly lost in December 1968 (when, in fact, Apollo 8 performed the even more demanding mission of going into lunar orbit) and, with its propaganda value lost, this part of the programme was curtailed. The spacecraft for the full lunar mission would have incorporated onboard navigational systems for use in lunar orbit, and its service module would have had a more powerful engine to boost it out of lunar orbit to head for home. This form was tested (without a crew) several times in the early 1970s – as indeed was the lander, also based on Soyuz technology – but all the manoeuvring trials were performed in Earth orbit. To test other aspects of the lunar mission, the 'standard' Soyuz flew a number of missions (without a crew) in the late 1960s to prove first-orbit rendezvous (to simulate the returning lander rendezvousing with its mothership), and later cosmonauts spacewalked from one vehicle to another (as was to be done in lunar orbit to transfer to and from the lander).

Soyuz rendezvous in Earth orbit relied on ground radar to track the orbits and on the control centre to compute the necessary manoeuvres – onboard radar was used only during the final few kilometres of the approach. In contrast, American Gemini astronauts took pride in using onboard inertial and computer systems to derive their own manoeuvres. Apollo's system (supplemented by sextant sightings) was capable of navigating to the Moon and back on its own. In this respect, therefore, the Soyuz was relatively primitive. On the other hand, it was standard practice to provide each form with *only* the facilities for its mission; the Earth-orbital variant did not require this independence. Similarly, since the lunar plan called for an external crew transfer, the docking system was required only to engage the vehicles for a short period, and a simple mechanism was bolted to the front of the orbital module. With the decision to build a space station, a docking system with a collar to establish a hermetic seal was developed so that the crew could transfer via an internal tunnel. After the deaths of the Soyuz 11 crew due to a pressure leak, both the descent and service modules were modified. As the mission of such a 'ferry' was to perform a single rendezvous and subsequent de-orbit, the propellant tanks that had been fitted to facilitate *extensive* orbital manoeuvring were deleted, and the solar panels were replaced by rechargeable batteries. To protect the crew from the kind of accident that had killed the Soyuz 11 crew, cosmonauts were to wear lightweight pressure suits. However, the crew had to be reduced to two in order to accommodate the new life support systems. This form had a two-day endurance. Another form had the simplified engine and the improved safety features, but retained the solar panels in order to fly independent missions. In addition to the Apollo–Soyuz link-up (which had an androgynous docking system), this version flew with a telescope (on Soyuz 13) and an Earth-resources camera (on Soyuz 22) mounted on the orbital module. The Soyuz-T was introduced in the late 1970s. It had solar panels so that a vehicle that experienced difficulty rendezvousing would not be obliged to break off and return to Earth. This had the Argon computer of Salyut 4 and the integrated propulsion system of Salyut 6, and the fuel saved by jettisoning the orbital module prior to the de-orbit manoeuvre enabled it to carry a heavier payload, which enabled the third seat to be restored. Soyuz-T was intended for Salyut 7, but the astonishing longevity of Salyut 6 led to its early introduction. The Soyuz-TM developed for Mir had improved communications, a more advanced rendezvous system, and even greater payload, but some of this was lost when it was decided to revert to the old practice of disposing of the orbital module only after the de-orbit burn, in order not to deprive the crew of its facilities in the event that they encountered difficulties with this manoeuvre and had to remain in space another day. The orbital module itself was modified several years later to incorporate a 'blister' porthole for better forward vision while manoeuvring in close to Mir, with its many projections.

The modularity of the Soyuz was exploited to create the Progress logistics ferry. The descent module's avionics were relocated to the service module, the descent module was replaced by a simple casing housing tankage for 'wet' cargo, and storage racks were installed in the orbital module to carry 'dry' cargo.

Although Sergei Korolev designed his spaceship to perform a variety of rôles, it is doubtful that even he would have expected it still to be in front-line service in the new millennium. Despite Soyuz killing two crews and nearly stranding another in orbit, Chelomei's three-seat spacecraft, which was more capable but needed the Proton to launch it, was never adopted; Korolev's is the *only* rocket to have been trusted to carry cosmonauts. The *Semyorka* was designed to fire a heavy nuclear warhead on a ballistic arc. Its diameter and lifting power dictated the dimensions and mass of the Soyuz spacecraft, which defined the crew capacity, which defined the requirements of the environmental system of the space stations, whose dimensions and mass were dictated by the Proton rocket. Thus, it is evident, the structure and operation of the Mir complex derive from technical compromises made long ago. The new Energiya rocket offered the prospect of assembling a large orbital station for a crew of dozens but, on its cancellation, there was no rôle for the 13-tonne Zenit-launched six-person spacecraft that was to have serviced that facility, so this, too, was cancelled, leaving only the venerable Soyuz to support the new International Space Station. Although it is natural to consider the Soyuz to be 'old technology', especially in comparison to the American Space Shuttle which, more than any other spacecraft, looks the part, the Soyuz is an excellent example of the evolutionary law of *the survival of the fittest*.

AN EVOLUTIONARY PROGRAMME

The Soviet space station programme began as a reconnaissance platform, but because there was an urgency to introduce it, and because Chelomei's spacecraft was at an early stage of development, the project was split so that the Korolev Bureau could rebuild a *hybrid* variant using proven systems. The first station, therefore, appeared to be a straightforward extension of Soyuz technology. Although the process of step-by-step *evolutionary* development predominated, it is evident with hindsight that there was a bold strategic plan at work to develop a continuously inhabited station and, had it not been for an unfortunate series of accidents early on, this goal *might* have been achieved many years before it finally was.[1] The stages in this process are now clear.

The designers of the first Salyut stations, which had only a single docking port, had to install everything required for the mission into the vehicle prior to its launch. Whereas Chelomei had envisaged a large resupply ferry, a Soyuz could not transport much extra equipment or consumables. As a result, to maximise the scientific programme, the crew's facilities were basic and their tenancy was limited by the consumables. In fact, the life of a station was limited not only by the food and water available, but also by its supply of propellant, which was consumed both in altering its orientation for experiments and optimising the output of its solar panels, and also in countering orbital decay. Its habitability was further limited by the ability of the environmental system to regulate the internal temperature and maintain a breathable atmosphere. If, therefore, the crew ran out of air, food or water, or the station ran out of propellant, or the thermal control system broke down, or toxins accumulated, it would become uninhabitable. However, the ability of the first Salyut stations to support prolonged habitation never became the issue. The stripped-down ferry introduced following the Soyuz 11 accident had batteries instead of solar panels. These so limited the time available for the rendezvous that if a problem developed the ferry had to abort and return to Earth – simply reaching the station represented a significant milestone, and all too often this was not achieved. The early Salyuts were intended to sustain two periods of occupancy. Salyut 1 would have been revisited if the Soyuz 11 accident had not occurred. Upon the failure to revisit Salyut 3, there was insufficient time to mount another mission within the station's operational life. The first three years of the programme can be characterised as a series of lost opportunities. It was not until Salyut 4 that a station was reoccupied, and by the time it was finally vacated its environmental systems, which had sustained a total of three months of inhabitation, were so worn out that the humidity had risen to such an extent that a rich mould had coated the walls. Nevertheless, Salyut 4 *more* than made up for all the problems that had afflicted its predecessors.

The first dockings between two Soyuz spacecraft, undertaken in the late 1960s, in preparation for the lunar programme, were made automatically, but dockings with Salyuts were undertaken manually once the automated system had brought the ship to the 200-metre point. However, an automated system for the final approach and docking was under development for a variant of the Soyuz configured to ferry cargo, and when this was tested by Soyuz 15, its repeated failure frustrated the attempt to reoccupy Salyut 3. The significance of this trial was that even if the resupply ferry *had* been available, it could *not* have serviced Salyut 3 as a crew would need to be on the station to unload the cargo, and the station had only one port. It was, however, a clear indication of what was to come. It was also typical of the Soviet evolutionary approach to technology, of refining proven hardware in order to build up the

---

[1] The Soyuz 11 deaths were unrelated to the station, after all; the next station was lost in a rocket malfunction, and the first military one was crippled by a system not on Korolev's configuration.

required operational capability. The addition of the second docking port on Salyut 6, and the introduction of the Progress cargo ferry, transformed the programme – a dozen ships replenished its consumables and brought components to overhaul its environmental system, and thereby enabled it to sustain *five* crews. Salyut 6 had been designed to support occupation for half of an 18- to 24-month period (it was hoped to achieve a 90-day, a 120-day and a 175-day mission), but in the event it lasted twice as long.

Table 10.1 Space station occupancy

Station	Main Expeditions (days)	Total
Salyut 1	23	23
Salyut 3	16	16
Salyut 4	30, 63	93
Salyut 5	49, 18	67
Salyut 6	96, 140, 175, 185, 13, 75	684
Salyut 7	211, 149, 237, 168, 50	815
	Total =	1698

A Progress ferry had a capacity of just over 2 tonnes, split more or less evenly between 'wet' and 'dry' cargo. Experience with Salyut 6 revealed that only about 15 per cent of this mass directly supported science activities; the rest was to replenish consumables for the station and its crew. With a total daily mass requirement of 16 kilograms, resupply flights were made at two-monthly intervals. Often the dry cargo was unloaded promptly, over a period of several days, but at other times (because the craft was retained until the docking port was needed) it was done incrementally, which had the advantage that the ferry's orbital module could serve as a warehouse, and thereby relieve the cramped situation in the station. Prior to being jettisoned, the ferry's orbital module would be filled with discarded food packets, dirty clothes and expired components from the environmental system, to be destroyed when the craft de-orbited itself and burned up in the atmosphere. Despite the early difficulties with automated dockings, the Progress ships came and went like clockwork.

When the endurance record exceeded the in-orbit life of the Soyuz, the second docking port of the second-generation Salyuts enabled a ferry to be replaced. Rather than send up a new empty Soyuz to dock automatically in the manner of a Progress, it was decided to assign crews, and send up guest researchers to work on the already inhabited station for periods of a week or so prior to taking away the old ferry – one Salyut 6 crew on a six-month flight hosted four visiting crews.

Key to the second-generation Salyut's success was the unification of the orbital manoeuvring and attitude-control systems and the installation of plumbing to enable the propellants to be replenished. Unfortunately, the refuelling operation could only be undertaken at the rear port, and this imposed the operational requirement that a Soyuz delivered to the rear had to be relocated to the front once the old one had left in order to keep the rear clear for cargo ships. This was unfortunate, but, despite the potential for failure, every fly-around was completed without incident. Salyut 6, therefore, was a triumphant success in every respect.

Salyut 7 was structurally similar to its predecessor, but more of its systems had been made accessible so that these could be serviced. The new Soyuz-T had a longer service life, so for Salyut 7 the timetable for visiting missions could be relaxed. The goals were two-fold: (1) to extend the endurance limit, and (2) to have a succession of crews hand the station over from one to the next, as a step towards continuous occupancy. The record was extended (although only by two months), but a series of technical issues precluded the first orbital handover until just before the station was decommissioned. Salyut 7 lasted six months longer than its illustrious predecessor, and was inhabited for a slightly higher percentage of its time, so it was by no means a failure, but it was dogged by a series of problems. In addition to requiring a bypass to be installed on its main engine, at one point it had to be salvaged from deep freeze following a power failure while it was unoccupied. In overcoming such problems, the programme advanced far beyond the state of the art inherited from Salyut 6, which, by and large, had suffered no serious failures. Salyut 7's major contribution to the programme was demonstrating that spacewalkers could perform intricate engineering tasks. Whereas it had once been believed that a station that was leaking propellant would have to be abandoned, it now seemed that no repair was too difficult.

In addition, exploiting the capabilities of the Soyuz-T, every effort was made to overcome rendezvous mishaps that previously would have led to immediate aborts. Even when the Igla transponder was torn off Soyuz-T 8, the radar tracking network and the Kaliningrad flight controllers enabled it to fly in from its transfer orbit to the 200-metre point. Commanders were trained to take over if the automated docking system failed, but it was expected that the Igla would provide range and range-rate. Despite being denied these cues, Vladimir Titov made a manual approach and would probably have docked if he had not entered the Earth's shadow, which prompted an abort in order to preclude a collision in the darkness. The disruption to Salyut 7's programme was severe, but the lessons learned from this fully manual approach provided the basis for Vladimir Dzhanibekov's successful docking with the station when it was derelict. The early 1980s, therefore, provided ample opportunities for overcoming adversity, and so when Mir was launched in 1986 confidence was running at an all-time high.

CONSTRUCTING THE MIR COMPLEX

Mir got off to an extremely impressive start. Its commissioning crew checked it out, then flew to Salyut 7 to finish their predecessors' programme before returning once more with as much of the mothballed station's apparatus as their Soyuz ferry could carry. Given the prior use of Cosmos 1443 to resupply Salyut 7, it is surprising that such a vehicle was not sent up to Mir early on, rather than the second Progress ferry (the first had been required to replenish Mir's propellant after its climb to operating altitude – a service that the big module was not equipped to provide). Since the base block was outfitted as a habitat, a full load of standard science apparatus would have facilitated a viable programme as the cosmonauts awaited the first of the specialised laboratories. In the event, when they exhausted the scope of the salvaged equipment, Kizim and Solovyov were obliged to return to Earth, which pre-empted the hope of continuously inhabiting Mir.

The process of assembling Mir (evidently delayed by almost a year) began with Kvant 1, which was docked at the rear of the base block. It *had* to be placed there, because to have put it on the front axis would have blocked access to the radial ports on the multiple docking adapter, and inhibited further build-up. However, mounting it at the rear blocked the base block's main engines and, as Kvant 1 did not have any engines of its own, its addition made the complex reliant on docked cargo ferries for orbital manoeuvring. Kvant 1's presence, therefore, seemed rather awkward. It was, in fact, something of an interloper. It had been intended for the front of Salyut 7, and, if *that* had been as reliable as its predecessor, it undoubtedly would have been launched after the departure of the Cosmos 1686 laboratory. Kvant 1 had a set of gyrodynes with which to orient the complex with sufficient pointing accuracy to run the X-ray telescopes that it incorporated. Adapting it for the rear of Mir had involved adding plumbing to transfer fluids from a Progress docked at its rear into the base block – as otherwise its presence would have restricted Mir to fluid replenishment only via the front port (a facility that had not been available on Salyut 7, which was why there was no such plumbing in the module's initial configuration). Sending Kvant 1 to Mir was, therefore, a case of exploiting 'off the shelf' hardware. The Korolev Bureau, which had designed Kvant 1, had originally planned to build more such modules, but it was ordered instead to develop a suite of modules based on Chelomei's TKS, without the descent capsule. There were docking ports for four such modules on the front of the complex. Designed to enhance the crew facilities, Kvant 2 added environmental and waste management systems, a shower and a second toilet, as well as an airlock. As a new module was to dock on the axial port and then use its Ljappa arm to swing itself onto the requisite radial port, at such times the resident ferry had to be at the rear. Because this prevented resupply, it was clear that the new module could not be left on the axis for long. Kvant 2 was to sit on the upper port. Once it was in place, the complex would remain 'L'-shaped until the next module arrived and converted it into a 'T'-shape. While asymmetric, the complex's orientation would be very difficult to control, and it would be impractical to adjust its orbit to overcome atmospheric drag. In order to keep this awkward period as brief as possible, this phase of the assembly could not start until the first *two* modules were ready. Kvant 2 and its partner were a matched pair with the same general configuration. The final pair of modules, while of a different design, formed a second matched pair. At this point (around mid-1988), it appears that there was a change in the plan, and it was decided to bring forward the 'technology' module in the expectation that its furnaces would produce sufficient semiconductor material to compensate the cost of further operations, and this meant that Kvant 2 could not be sent up until the first of the *second* pair became available. A succession of crews flew in the expectation of commissioning these new modules, but it was December 1989 before Kvant 2 was added, and upgrading the computer to manage the asymmetric configuration was more difficult than expected. Calibrating the computer's mass model consumed so much fuel that a resupply tanker had to be hastily sent up. Like Kvant 1, Kvant 2 had a set of gyrodynes. Being the first to be added, these two modules *had* to carry the gyrodynes, and because these provided *sufficient* manoeuvring capability, none of the later modules needed to have them. Also, Kvant 1 *had* to be in place before Kvant 2, so that its gyrodynes would be available when training the computer to handle the asymmetric configuration without using thrusters. Although modular, the complex was a closely integrated structure. Advancing Kristall (as the third module became known, on account of its furnaces) meant that it had to be *temporarily* placed on the lower port in order to balance the complex; once the module that it had displaced became available it was to be moved onto the side port for which it had been designed.

Each module extended its solar panels as soon as it reached orbit in order to provide power during the rendezvous. The panels on the modules for the upper and lower ports had frames similar to those on the base block. The modules for the side ports had a type of panel that could be retracted – to preclude their interfering with the base block's panels when the module was swung to the side. These retractable panels had been made detachable so that they could be transferred to another part of the complex. Prior to Kristall's relocation, its panels were to be moved to Kvant 1, which had none. Spacewalkers affixed frames to the sides of Kvant 1, mounted drive motors to accommodate the panels, and affixed a crane to one side of the base block in order to swing the first panel to the rear of the complex. But when it was decided to postpone further expansion until Kristall proved its commercial viability, Kristall had to stay on the lower port to maintain the complex in a symmetric configuration, and it was decided to leave its panels in place in the meantime. The cosmonauts were not short of external work, however. During a marathon series of spacewalks they erected the Sofora girder on Kvant 1 and placed the roll-control thruster block on its end, oriented to fire in the same plane as the base block's own roll-control thrusters. This structure's existence indicated not only considerable forethought, but also the degree to which constructing the complex was reliant on spacewalkers. And it was ironic that this construction centred on Kvant 1, the interloper. It was not until the fortunes of the programme began to improve – with the advent of the Shuttle–Mir programme – that the go-ahead was given to finish the other modules. Even then, the first of Kristall's panels was not moved to Kvant 1 until immediately before Kristall *had* to be moved. In fact, it took *five years* to finish the process of mounting solar panels on Kvant 1. This exercise demonstrates the interaction between long-term and short-term planning in the programme. The configuration of the modules was closely allied to the locations that they were to occupy on the complex; a redeployable solar panel was developed, a crane to transfer it was devised, supporting frames were fabricated, motors were installed, and cables were strung to deliver the power – all of which activities indicated long-term planning. On the other hand, the job was carried out rather haphazardly, with the individual tasks being passed from one crew to the next as the schedule slipped repeatedly. By assembling Mir much more slowly than originally envisaged, it was left perpetually short of power. Nevertheless, the crews showed remarkable ingenuity in adapting their programme to make the best possible productive use of their time. This indicated one important difference between Mir and its Salyut predecessors. Whereas previously a crew would not be sent up until their facilities were in place, continuously inhabiting Mir meant that its crews had to absorb delays, fix broken equipment, and do their best with what was available. In this respect, life on Mir was more Earth-like. And frustrating although it must have been for all concerned, relentlessly pursuing *long-term* objectives within the constraints of short-term expediency must have produced a valuable database for operational planning. It must be borne in mind, however, that we would have known little of this if it were not for Mikhail Gorbachev's *glasnost* initiative, for otherwise events would have been portrayed as having gone to plan.

The notion of completing the Mir complex within two years, using it for three years, and then replacing it with something much more elaborate, was clearly overoptimistic. Kvant 1 was not launched until one year after the base block. There was then a wait of nearly three years for Kvant 2, with Kristall following in six months. Then there was a five-year gap to Spektr, and another year for Priroda. As a result, the base block was 10 years old by the time the complex was finished, and the issue was how to keep it operating.

Table 10.2 Space station launches and re-entries

Spacecraft	Launched		Re-entered
Salyut 1	19 Apr 1971	0340	11 Oct 1971
Salyut 2	3 Apr 1973	~0300	28 May 1973
Cosmos 557	11 May 1973	~0300	22 May 1973
Salyut 3	25 Jun 1974	0138	24 Jan 1975
Salyut 4	26 Dec 1974	0715	3 Feb 1977
Salyut 5	22 Jun 1976	2014	8 Aug 1977
Salyut 6	29 Sep 1977	0950	29 Jul 1982
Salyut 7	19 Apr 1982	2245	2 Feb 1991
Mir	20 Feb 1986	0028	23 Mar 2001
Kvant 1	31 Mar 1987	0406	23 Mar 2001
Kvant 2	26 Nov 1989	1601	23 Mar 2001
Kristall	31 May 1990	1433	23 Mar 2001
Spektr	20 May 1995	0733	23 Mar 2001
Priroda	23 Apr 1996	1548	23 Mar 2001

Dates are Moscow Time.

CREWING

Commanders and flight engineers each received comprehensive training prior to being assigned to a crew. The training programme for guests included aspects of flying the Soyuz ferry relevant to handling emergencies, and lasted up to two years (although it could be compressed for people already fluent in Russian). The system evolved, but cosmonauts might begin by supporting a mission, and then be assigned to backup a specific mission without any real expectation of flying it, with the understanding of being first in line for the next comparable mission. Early on, all-rookie crews were sometimes formed, but after a number of docking failures a rule was introduced that each crew must include a veteran, and then, as confidence in the spacecraft increased, this rule was relaxed. Although crews were generally formed a year or so in advance, changes in the assignments were common, and substitutions for illness and accidents were not unknown. Valeri Ryumin, for example, although he had recently completed six months on Salyut 6, stood in for Valentin Lebedev when the latter injured a knee a few weeks before he was due to fly. Some crews were formed at very short notice to meet operational exigencies, an excellent example being the pairing of Vladimir Dzhanibekov and Viktor Savinykh at four months' notice to rescue Salyut 7 in 1985. Crewing changes tended to ripple on. For example, after Soyuz 25's failure to dock with Salyut 6, Alexander Ivanchenkov stood down from the next crew so that Georgi Grechko could inspect the docking system. As a result, when Vladimir Kovalyonok (the Soyuz 25 commander) moved up to take the next flight, he was teamed up with Ivanchenkov. As a result, the Romanenko–Grechko team, formed after the problem developed, had just two months to train for the flight that finally broke the 84-day Skylab endurance record. The final expedition to Salyut 6 was authorised only after a maintenance crew refurbished the station, which gave the Soyuz-T 4 crew three months to prepare. On the other hand, even a fully trained guest could wait years for a flight. Mohammed Faris of Syria, for example, was repeatedly rescheduled because of Salyut 7's problems, and was eventually sent to Mir. In contrast, political factors meant that Abdul Ahad Mohmand, the Afghan trainee, had to fly after six months of training (which was feasible because he was fluent in Russian) and, in fact, his hosts were already on Mir when he entered training. The ripple effect from this was that flight engineer Alexander Kaleri lost his seat and Sergei Krikalev was obliged to serve a double tour. Sometimes fate intervened at the preflight medical, and the prime and backup crews were switched. Valeri Kubasov showed symptoms of a lung infection, and the backup crew flew Soyuz 11 to Salyut 1. When Gennadi Manakov was found to have a heart irregularity, Valeri Korzun and Alexander Kaleri had 10 days' notice to prepare for a six-month tour on Mir. Surprisingly few cosmonauts have fallen ill in space. Although Romanenko developed a severe tooth-ache, he completed his tour on Salyut 6. When Vladimir Vasyutin developed a prostate infection, his crew was recalled from Salyut 7. When Alexander Laveikin showed a heart irregularity, he was replaced on Mir by Alexander Alexandrov.

While it is important that the members of a crew be psychologically compatible, too much can be made of this. It is true that fighter pilots tend to be competitive, but they are also highly professional. Cosmonauts understand their systems, know what is expected of them, and accept the deprivations in return for the opportunity to fly. Excluding antagonistic competitors, the process of selecting crews inevitably reduced to choosing 'likes' or 'dislikes'. People with *similar* traits were able to substitute for one another in their interrelationships without effort. They worked well together for short periods, but often became irritated with one another in the longer term. During

211 days on Salyut 7, Anatoli Berezovoi and Valentin Lebedev gradually diverged. They were *too* alike to get on well. They did not actually argue, they simply avoided interacting. It was better to choose people with *complementary* traits because, being aware of the fact that they had different backgrounds, expertise and outlooks, they tended to coordinate, acknowledge one another's point of view, and come up with a wider range of solutions to a problem. Hence the hasty pairings of Romanenko with Grechko, and Kovalyonok with Ivanchenkov were very successful. One factor that assisted in making disparates into an effective team was matching single-mindedness with analysis of alternative options. It was also often the case that the flight engineer was considerably older than the commander, who was usually a military officer and often on his first flight. Salyut was meant to support three cosmonauts, but the loss of the Soyuz 11 crew resulted in modifications to the ferry that (in the short term, at least) reduced its capacity to two. The introduction of the second-generation station meant that crews could host visitors. Although the Intercosmos missions were flown to satisfy Salyut 6's *operational* requirements, rather than for their own sake, they allowed welcome opportunities for veterans like Valeri Bykovsky, Viktor Gorbatko, Valeri Kubasov and Nikolai Rukavishnikov to fly again. The inability of Soyuz 33 to dock disrupted the schedule for a year, with the result that some guests trained with a succession of likely commanders. Although it would have been feasible, no in-orbit handovers were attempted with Salyut 6. The Soyuz-T reinstated the third seat, and Salyut 7's environmental system was capable of accommodating a crew of three for lengthy periods and six people for short periods. Psychologists initially feared that these expanded crews would fragment, with two excluding the third, but this did not occur. The first crew to make an extended mission included Oleg Atkov, but as a doctor he was already an *outsider* – because his rôle was to observe his colleagues. In the next case, the commander (Vasyutin) fell ill and his colleagues rallied to him. Mir's first three-person resident crew again involved a doctor (Poliakov). When the third seat of the ferry was used to enable guests to accompany a handover, the fact that the guest was an outsider by virtue of being a foreigner was masked by the sheer number of people present and the shortness of the visit. Cultural isolation became a factor only when guests began to fly full tours of duty. Nevertheless, although Norman Thagard felt isolated, his successors integrated rather better.

The task of assigning compatible crews was complicated when people started to fly individual missions, coming and going opportunistically, and on several occasions a ferry returned with three cosmonauts who had been launched separately. When Alexandrov replaced Laveikin he returned with Romanenko, whom he had joined, and Levchenko, who had been making a short visit. Similarly, when Krikalev was obliged to extend his tour, he returned with Volkov (his new commander) and Flade, who was a foreign guest. To illustrate the sometimes *ad hoc* nature of the crewing strategy, consider the case of Valeri Poliakov. Although Atkov had spent 236 days on Salyut 7 with two colleagues in order to monitor their physical adaptation (with authority to order them home if he thought it expedient), it was clearly impractical to have a doctor on Mir full time, and it was decided to send up one towards the end of a record-breaking flight to examine the residents prior to their return. Poliakov had planned to assess Romanenko towards the end of his 326-

308 In retrospect

Table 10.3 International space endurance records

Spacecraft	Launch	Date	Days
Yuri Gagarin	Vostok 1	Apr 1961	0.07
Gherman Titov	Vostok 2	Aug 1961	1.05
Andrian Nikolayev	Vostok 3	Aug 1962	3.93
Valeri Bykovsky	Vostok 5	June 1963	4.97
Gordon Cooper Pete Conrad	Gemini 5	Aug 1965	7.92
Frank Borman Jim Lovell	Gemini 7	Dec 1965	13.75
Andrian Nikolayev Vitali Sevastyanov	Soyuz 9	June 1970	17.71
Georgi Dobrovolsky Viktor Patsayev Vladislav Volkov	Salyut 1	June 1971	23.76
Pete Conrad Paul Weitz Joe Kerwin	Skylab	May 1973	28.04
Al Bean Jack Lousma Owen Garriott	Skylab	July 1973	59.49
Jerry Carr Bill Pogue Ed Gibson	Skylab	Nov 1973	84.04
Yuri Romanenko Georgi Grechko	Salyut 6	Dec 1977	96.42
Vladimir Kovalyonok Alexander Ivanchenkov	Salyut 6	June 1978	139.60
Vladimir Lyakhov Valeri Ryumin	Salyut 6	Feb 1979	175.06
Leonid Popov Valeri Ryumin	Salyut 6	Apr 1980	184.84
Anatoli Berezovoi Valentin Lebedev	Salyut 7	May 1982	211.38
Leonid Kizim Vladimir Solovyov Oleg Atkov	Salyut 7	Feb 1984	236.95
Yuri Romanenko	Mir	Feb 1987	326.48
Vladimir Titov Musa Manarov	Mir	Dec 1987	365.95
Valeri Poliakov	Mir	Jan 1994	437.75

day mission, but the third seat on Soyuz-TM 4 was requisitioned for a Buran pilot. Titov and Manarov (who took over) hoped to extend the record to a year, so towards the end of *their* mission Poliakov was assigned a seat on Soyuz-TM 6, which also carried an Afghan visitor. By remaining with the residents, Poliakov committed himself to staying on with the *next* resident crew, as a Frenchman was to accompany the forthcoming handover. Furthermore, because his new colleagues (Volkov and Krikalev) were *not* to host any visitors and another Buran pilot was to accompany the following handover, Poliakov would have to serve with Sevastyanov and Afanasayev *as well*. Unfortunately for Poliakov's hope of setting an 18-month record, the decision to vacate Mir until the new modules were ready meant that he had to return to Earth after 240 days.

Given the need to investigate how the human body adapts to weightlessness, it was not surprising that doctors were inducted into the cosmonaut corps. The first spacecraft to carry a multiple crew was Voskhod 1 in 1964. Pilot Vladimir Komarov was accompanied by Boris Yegorov, a physician, and Konstantin Feoktistov, one of Korolev's senior engineers. Almost all multiple-person crews (other than the Almaz crews, which were drawn exclusively from military pilots) had a military commander and a civilian flight engineer. Whereas NASA advertised for astronauts, the Korolev Bureau selected flight engineers from within its own ranks. The Buran pilots formed a separate group and when Igor Volk (the first to fly) visited Salyut 7, his hosts were eager to find out his background. Although a number of women were selected as cosmonauts, few actually flew. The plan to send a three-woman crew to Salyut 7 just before decommissioning it was scrubbed when Svetlana Savitskaya fell pregnant (at that time, each crew had to have an experienced member, and she was the only woman available). A number of scientists and journalists trained, but never flew. As a result, the predominantly male corps of pilots and engineers had to perform work that went to scientists on Skylab. Another difference between cosmonauts and their American counterparts was that, as spacecraft designers, the flight engineers migrated between passive and active rôles, and even served as flight directors. The only direct employment of astronauts in NASA's control room was as the communications link to a crew in space. The training regimes were strikingly different too. Whereas Shuttle training relied on state-of-the-art computer-based self-tuition, Mir crews employed the traditional lecture format with classroom exercises and formal examinations. The Russians relied on word of mouth rather than voluminous documentation, preferring to call in the designer of a piece of apparatus if a cosmonaut needed to know about it. In the Russian system, knowledge was informal. Both made heavy use of simulators and underwater training, of course, so there was also much in common. At first it was the European Space Agency's astronauts that cross-trained for Mir who experienced this difference, but following the agreement to cooperate, astronauts and cosmonauts began training together. Ironically, whereas NASA recruited new astronauts to work on the International Space Station, the cosmonaut corps was cut to a cadre, many of whom had flown several tours. Of the Buran group, who were all highly experienced test pilots, the leader, Igor Volk, was grounded for medical reasons, and others (including Anatoli Levchenko, Alexander Shchukin and Rimantas Stankiavicus, all of whom had familiarisation flights on Soyuz) died or were killed in air crashes and on the project's cancellation in the post-Soviet rationalisation the group was disbanded.

Mir passed through several phases of crewing strategy. The first crew could not take a guest with them because there was no returning crew to take the visitor home. In any case, since they had a demanding mission involving both commissioning Mir and mothballing Salyut 7, Soyuz-T 15 needed all the propellant that it could carry. If the planned handover had taken place, it might have been accompanied by the Syrian who was to have visited Salyut 7. In the event, even though it had been announced when Mir was launched that it was to be *continuously* inhabited, once it became clear that the first of the science modules could not be launched on time, it was decided to vacate the station. Consequently, the second crew could not take up a visitor either. However, in addition to achieving two handovers, the next five flights each carried a guest. After the first crew had commissioned Kvant 1, the emphasis became taking the endurance record to a year in space. Thereafter, Mir was left vacant once again while the next module was prepared. No visits were hosted during Soyuz-TM 8 to 10, while the front of the complex was expanded with the Kvant 2 and Kristall modules. Mid-tour visits ceased, but almost every handover was accompanied by a fee-paying researcher and the handover was lengthened to allow visitors up to a month onboard. Meanwhile, the endurance record was decoupled from routine operations (on his 14-month record-breaking mission, Poliakov flew with successive residents who flew regular tours). Later still, fee-paying researchers were allowed to make full tours, and NASA booked a continuous slot for a succession of astronauts who were delivered and retrieved by the Shuttle. Other visitors were limited to handovers until NASA's booking expired. By the time that Mir became an international facility, crewing had become a dynamic process involving a multitude of organisations from different countries.

WEIGHTLESSNESS

The basic objective of the space station programme was to determine whether the human body could survive long-term exposure to weightlessness, and the subsequent return to Earth. It is difficult now to appreciate the uncertainty that faced the first Salyut crew, and the shock of discovering them dead in their couches on returning to Earth. It was clear that they had survived three weeks of weightlessness without any debilitating effects, but it was far from certain that (if the technical malfunction had not taken their lives) they would have survived the return to Earth. Ironically, much longer flights later revealed that a three-week flight was just about the *worst* duration to choose, because no sooner had the body adapted to the space environment than it was forced to readapt to gravity from a state of relative weakness.

In general it takes a few days for the vestibular disorientation to abate, and about a week for the pooling of blood in the upper torso and puffiness in the face to wear off, during which time the body increases urination to shed what it takes to be excess body fluid. During the first week or so, cosmonauts take plenty of water to prevent dehydration. As the body adjusts, the capacity of the cardiovascular system reduces and the heart migrates up into the chest cavity. Physical exertion is avoided during this initial phase of adaptation, but thereafter a rigorous daily exercise regime is followed to compensate for muscular atrophy. By the end of the month, the body is in a state compatible with weightlessness. With cosmonauts making repeat flights, it was found that the body adapted more readily. A short familiarisation flight prior to a long mission seemed to have real benefit. After the rule was introduced that at least one member of each crew had to have previous flight experience, it was found that an experienced cosmonaut could teach a rookie how to adapt to weightlessness, thereby speeding up the process and making for more productive missions. As the duration of missions increased beyond four months, it was noted that there was an initial period during which every crew was highly motivated and very productive. In some cases this gave way to a lull in which motivation faltered until the flight was nearing its end, at which point the pace picked up. Sometimes only one member of a crew suffered that relatively unproductive phase, during which the other worked extra hard to make up. This 'gap' depended on the duration. For example, while Yuri Romanenko was very productive all through three months on Salyut 6, he became increasingly depressed in the latter half of an 11-month stay on Mir. Vladimir Titov and Musa Manarov, on the other hand, who followed him, spent a year onboard without ill effects. Once in-orbit handovers became routine, the task of pushing the endurance record was given to physicians who served with a succession of hosts flying relatively short 'tours of duty'. Romanenko had recommended four months, but the decision to exploit the in-orbit service life of the Soyuz ferry led to six months becoming the norm.

One factor that greatly affected a crew's long-term performance was the length of their day. Although the familiar idea of a 'solar day' means little in the context of a 92-minute orbit, the human body is linked by dint of its evolution on the planet to such a daily rhythm. However, the precession of their orbit imposed by the Earth's equatorial bulge meant that crews of the early Salyuts tried to follow a 'day' making optimal use of the times that they were in communication with the control centre. Unfortunately, week after week of awakening half an hour earlier each day proved debilitating, and the cosmonauts became progressively more exhausted. The remedy was to adopt a 24-hour day on Moscow Time. This was interrupted only for events dictated by orbital mechanics, such as a spacecraft docking, and each 'late night' was followed by a 'lie in' to regain lost sleep. In fact, it was found that whereas newly arrived resident crews slept a full eight hours, their requirement gradually diminished to about five hours. Crews on brief visits often tended to skip sleep to pursue their work, with the result that they burned out after a week. Resident crews, on the other hand, tried to maintain a five-day week. If they chose, they could work through their weekends, but books and videos were available for relaxation.[2]

When the in-orbit life of the Soyuz was only three months, a ferry required to be

---

[2] After so many years of occupancy by so many crews with different tastes, Mir had a real library of books and videos; John Blaha, for example, took some *Star Trek* tapes with him.

replaced several times on a record-breaking flight, and this operational requirement allowed the Intercosmos Organisation to invite its members to send researchers on brief visits to Salyut 6. Each such crew expanded the database on the early phase of adaptation. A wide variety of tests were devised to elicit data (both subjective and objective) on all aspects of space flight. These studies were further developed by the fee-paying visitors to Mir. The tests focused on vestibular, hormonal, chromosomal and immunological changes, the capacity of the cardiovascular system, the heart's rhythm, structure and migration within the chest, the composition and distribution of body fluid, the capacity of the respiratory system, and bone loss and muscular atrophy. Studies were also made of posture, skin sensitivity, sources of physical irritation, changes to the senses of hearing, taste and visual acuity, aspects of brain activity and cognitive function. Psychological tests were developed to monitor self-assessment of working efficacy and relationships within a crew and between a crew and the ground, and psychologists monitored the video downlink to independently assess the state of mind of each member of a crew. The cosmonauts were, therefore, 'laboratory rats' as well as pioneering engineers and research technicians, and they endured a great deal for biomedical science. For example, the procedure for sampling bone marrow from some visitors prior to and following a mission was unpleasant. At a more mundane level, Norman Thagard noted that the need to log his food intake was a disincentive to eating. Although cosmonauts accepted the exercise regime (and on Earth regarded it as a tonic) it was not a very pleasant experience in space because, in the absence of gravity, sweat clung to the body and pooled in the body's concavities, particularly over the sternum. In the cabin's artificial environment there was no breeze to cool the body and dry it. An all-over rub after exercise using a towel was hardly adequate, hence the efforts over the years to devise an effective shower.

In addition to the strict daily exercise regime during the flight to keep muscles in trim, extra measures were taken in the final few weeks of a flight. A drug was taken to stimulate the body and to develop stamina. Negative-pressure leggings were used to draw blood into the legs to increase the capacity of the cardiovascular system and put an increased load on the heart, and, as the body accumulated fluids in response, saline solution was drunk to overcome dehydration. A cocktail of vitamins and mineral supplements to balance the electrolytes rounded off the pre-return procedure. Contrary to initial concerns, the 3–4 g load imposed by re-entry did not pose a serious threat even after a lengthy period in space. The fact that the crew of a Soyuz descent module took this load flat against their backs, with their feet elevated above the torso, helped greatly. Although most cosmonauts had to be assisted from the capsule, after a few minutes resting in the recliners nearby they were able to stagger to the medical tent. Despite the rigorous exercise regime in space, it generally required several days for the vestibular system to recover, but with the sense of balance and coordination reinstated it was possible to walk (albeit on weak legs and with a distinctive gait) for as long as stamina permitted. In contrast, arm muscles tended to grow stronger due to moving cargo about. Although most cosmonauts tended to lose weight in space, a few actually gained several kilograms. In the first week after a six-month tour, positive pressure was applied to the lower body by wearing the type of *g*-suit leggings used by pilots of high-performance jets in order to prevent blackout from orthostatic intolerance as blood drained from the head, and to ease the adaptation of the heart to the increased load of pumping against gravity. One effective way of easing the readaptation to Earth was to swim because, just as immersion simulated training for spacewalking, it masked the force of gravity. In fact, it was not uncommon for a newly returned cosmonaut to try to float out of bed on awakening, and to set objects in mid-air in the expectation that they would remain there. Such minor irritations apart, even the weakest individual was fully recovered in a few weeks. After his first mission, Georgi Grechko complained of chest pains for several days, but was fine the second time; it seemed that the first time his internal organs were slow to 'fall' back into place. Valeri Ryumin recovered much faster from his second six-month tour, which he volunteered for within months of returning from his first. The vital point though, was that the recovery time had been *decoupled* from flight duration. A six-month tour was medically and psychologically practical.

Given this certainty, it might legitimately be asked why NASA devoted so much effort to studying adaptation. Poliakov had made a significant start on his 14-month marathon, but NASA wished to track the process in much greater detail using its sophisticated apparatus to measure bone density, muscle mass and blood chemistry, and dovetail this into pre- and post-flight monitoring. The worst effect of exposure to prolonged weightlessness was bone demineralisation, because it seemed not to reach a plateau. Calcium leached into the bloodstream and accumulated in the renal glands, so kidney stones could ultimately be an issue. The effect on the skeleton was similar to osteoporosis, but was more pronounced. On Earth a sufferer typically lost bone mass at 2 per cent per annum, but in space the loss rate was an order of magnitude greater. In fact, the effect was not uniform, because large bones such as the femur tended to lose most. Norman Thagard developed what cosmonauts referred to as 'chicken legs' after losing nearly 10 per cent of the mass of his femur in five months. Bone loss was not a simple correlation with duration, however. One cosmonaut lost 8 per cent after six months, which was the same as was lost by the final Skylab crew in three months. To complicate the situation further, whereas one cosmonaut lost fully 20 per cent in five months, another lost 15 per cent in seven months. The effect also depended upon the initial state of the bone. Shannon Lucid, for example, had relatively dense bones and she was within acceptable bounds even after six months. The effect *was* reversible, but the rate of recovery was slow. Nevertheless, exposure to weightlessness did not seem to trigger runaway life-threatening biological processes following return to Earth.

THE FRUSTRATIONS

One of the frustrations of living in a weightless environment was that unless an item was firmly strapped down it floated freely. Elasticated straps were run across every flat surface to grip small items in place temporarily; unused items were wrapped in nets and bundled in spare nooks. Although the fan-driven air flow tended to carry loose items to the grilles, small things were often lost. Yuri Romanenko lost his watch, and Jean-Loup Chrétien lost a roll of exposed film. Thomas Reiter could not find a centrifuge blood separator for two months! One night, Anatoli Berezovoi and Valentin Lebedev stayed up late looking for the locking mechanism for the docking drogue, and eventually found it behind the cone, just out of sight. This illustrates the folly of trying to impose a rigid schedule on activities in space.

Improvements to the telemetry system meant that the doctors could monitor the biomedical sensors on spacewalks, but this was a two-edged sword – it showed that Alexander Laveikin suffered a heart irregularity while stressed, but was insufficient to form a diagnosis and so he was recalled; although follow-up tests established that his condition was not life-threatening, he never flew again. Although the cosmonauts put up with biomedical testing, they found it frustrating because unpacking the sensors, pastes, cables, cuffs and belts, and plugging in and verifying the monitoring systems took a long time. And then they had to wait until they were in radio range so as to be able to transmit telemetry. Furthermore, if the tests could not be completed within a pass, they had to break off and, remaining wired up, await the next orbit. Finally, the apparatus had to be disconnected and packed away again. It was a major item on the schedule.

While everybody involved accepted the need for teamwork, there was occasional friction between a crew and the people on the ground. At times the consequences of different motivating factors emerged. For example, it was particularly irritating to be hustled to adhere to the exercise schedule at the expense of an experiment on which they were working. On one occasion, having set up a camera, they were told to break off even though this would have meant not being able to take the intended pictures. On several occasions, cosmonauts urged the ground not to feel obliged to talk simply because the radio link was open. This feedback was of interest to the psychologists, of course. As one crew packed to return home, a cosmonaut observed that he would make faster progress without being reminded of what he had to do. However, there were times when a crew was not disturbed unnecessarily, such as during ablutions and at meal times. The fact that the radio link was not continuous was welcomed by most crews. Although Shannon Lucid acknowledged the need for teamwork, she relished the self-reliance imposed by being out of contact for long periods due to Mir's poor communications. Often, the only sound on the downlink was laughter, as the crew enjoyed their own company. On the other hand, the two-way video was of considerable psychological benefit on a long flight because it enabled the cosmonauts to interact with their families. The arrival of a resupply ship was always a highlight because of the fresh food that it brought. No matter how nourishing the prepackaged food, the cosmonauts soon grew bored with it (in part, this was due to the change in the sense of taste in space). The switch from complete-meal packs to variety packs from which individuals could pick-and-mix was a welcome development. Much of the food was pretty normal; it just had to resist flaking (items such as bread and cake were cut into bite-sized chunks and moistened so as not to crumble). Nevertheless, everyone craved fresh food. On occasion, cosmonauts ate the onion bulbs sent to be planted in a cultivator for an experiment. To make amends, one cosmonaut took a big plastic cucumber with him and claimed that it had grown while the station had been vacant! It turned out that watching a plant grow in space was therapeutic, even for cosmonauts with no previous horticultural bent. Simply gawking out of the porthole at the Earth was always a popular pastime. It was not unheard of for a cosmonaut to find a companion *asleep* at the window. The view out of the porthole was surpassed only by that from a suit visor during a spacewalk.

Imagine the frustration of riding a rocket into orbit in the expectation of spending several months on a space station, and then having to return to Earth 48 hours later due to a failure to dock! Even successfully boarding a station was no guarantee of a smooth flight. Laveikin suffered the disappointment of his recall, and Valeri Poliakov had his first attempt to extend the endurance record curtailed when Mir was vacated. On the other hand, Sergei Krikalev was obliged to serve a double tour. It must have been frustrating to train to commission a module, only to have it not appear; to have to cancel a spacewalk because the crane handle had worked loose and drifted free; to order a piece of kit and, on unpacking it, find that it was not what was requested; to discover that the food packs had been rifled by hungry members of the launch team; to have a return-capsule lost; or worst of all, to work on the assigned programme for day after day without knowing if anything worthwhile was being achieved. Despite the danger and the frustrations, however, there was no shortage of volunteers.

THE WORK

With so many automated satellites in use, it was fair to ask the value of maintaining a human presence in low orbit. What could cosmonauts see that the sophisticated sensors could not? While vision was initially degraded upon entry to weightlessness, it recovered and, eventually, cosmonauts found that they could distinguish extremely subtle variations of colour which, to their disappointment upon seeing the results of their picture-taking, standard panchromatic film was unable to capture. Hence, visual observations of oceanic currents, thin oil above shoaling fish, and blooming plankton proved particularly valuable. In fact, the recovery of visual acuity was hastened if an experienced cosmonaut was able to offer tuition to draw the newcomer's attention to the fine detail. In addition, the cosmonauts were rather amazed to find that under oblique illumination they could *see* shallow sea floors and seamounts in open ocean. It turned out that sun-glint was very useful, as it enabled inundation to be studied in heavily vegetated areas. Even although there were meteorological satellites above, the cosmonauts were often asked for visual observations of hurricanes, glaciation and snow coverage, melt-water run-off and river capacity. Cosmonauts on long missions made particular note of seasonal variations. It turned out that topographic detail was best seen in the autumn (after the vegetation had thinned out and before the terrain was masked by snow) and so the search for faults likely to contain natural resources was best done in the autumn. Cosmonauts often noted that prospectors had started work at sites identified by their predecessors. Sometimes cosmonauts communicated directly with fishing fleets, to steer skippers to the most promising-looking shoals. Cosmonauts were often the first to report forest fires. More can be seen from low orbit than might be imagined. It is not generally appreciated that from low orbit the Earth does not simply look like a map – topographic relief is discernible. In fact, the perspective is comparable to viewing vertical structure in a 4-centimetre band from a height of two metres. Because the rate of change of perspective is faster (due to the speed of orbital flight) the eye is easily drawn to vertical detail, and since the view is considerably broader the impression is more striking than from a high-flying aircraft. The hollow 'eye' of a tropical storm is pronounced. The Himalayas are spectacular. A volcano pokes up from its surroundings. Not only is the pillar of a volcanic plume rising into the stratosphere striking, if it is then caught by a jet-stream it can be traced for thousands of kilometres. The depth perception provided by the binocular vision of the human eye is surpassed only by a stereoscopic camera.

Crew visual observations were of course augmented by apparatus for specialised studies. Multispectral sensors enabled the resources of the unexplored wilderness of Siberia to be surveyed, and the agriculture in the Ukraine (the Soviet Union's 'bread basket') to be assessed. Overhead imagery from the later Salyuts contributed to the national economy by selecting routes for hydrological and transportation projects, in particular the 3,500-kilometre railway being built in eastern Siberia, running from Lake Baikal to the River Amur on the Manchurian border.

Atmospheric studies using spectrometers made feasible the first global survey of the processes involved in heat transfer to space, and the distribution of water vapour and aerosols in the upper atmosphere. The data revealed a steady-state glow above the equator that had not been suspected. Transient flashes high above thunderstorms were also discovered, which are now known to be a form of lightning discharge to the ionosphere, rather than to the surface. Measurements of charged particles at orbital altitude suggested a correlation with tectonic activity, as if rock that is under stress generates a localised intensification of the Earth's magnetic field. Cosmonauts could peer obliquely *downwards* at aurorae, and trace the glowing bands for thousands of kilometres across the surface. They also made unique studies of noctilucent clouds, which turned out to be more common and more extensive than had been believed.

The materials-processing research covered organic as well as inorganic materials. A wide variety of alloys of substances that are immiscible on Earth were made and returned for study, as were different types of semiconductor. Despite the hope that installing a bank of furnaces would transform Mir into a commercial 'factory', it was difficult to show economic viability because this output did not feed an enterprise economy. The biological materials-processing, on the other hand, was of immediate benefit to the pharmaceutical industry. Electrophoresis is a highly effective way of separating active biological substances, and interferon and a variety of vaccines were manufactured. Polycrylamide gel was used to refine terrestrial products. Automated versions of both the furnaces and the electrophoresis units were later installed on automated satellites – once the cosmonauts had discovered the flaws in their original designs, since many items designed specifically to operate in weightlessness require some refinement before they work as intended. American experiments on Mir grew crystals of proteins and three-dimensional tissue cultures for study in terrestrial laboratories. The advantage of doing this on Mir was that crystals could grow for *months* rather than a fortnight, which was the longest time that a Space Shuttle could remain in orbit.

Mir's utility increased with each module added, and as visitors delivered more apparatus. Later visitors could rerun experiments to broaden the database. Although storing it all became an issue, having a range of apparatus in-house also enabled the residents (and later visitors) to adapt experiments in order to expand on earlier work. And, of course, the residents could fill in gaps in Earth-studies coverage lost to their visitors due to adverse weather. A large part of the value of Mir, therefore, was that its research was *ongoing* and *long term*. It was a long struggle, but cultivators were perfected to grow higher plants in space. Arabidopsis (a weed) was first to complete its cycle and yield seed. Later wheat (a staple) was 'harvested'. Experiments showed that chlorella thrived in space, and could be used in a closed-cycle aquarium. These results are encouraging for food production on future stations and (in the long term) for the vast space colonies envisaged by the visionary Princeton physicist Gerard K O'Neill in the 1970s. One key element of this *ultimate* habitat has yet to be studied, however: fish, frogs and quail chicks have all been successfully hatched in space, but (as far as is known) no human being has yet been conceived in space.

THE BURAN SHUTTLE

No study of the Mir space station can ignore the Buran shuttle, but not *too* much should be made of it because, in the final analysis, it proved to be unnecessary. The VKK (the acronym for atmospheric spacecraft) was ordered by the military in the late 1970s to counter the NASA Space Shuttle, the development of which was backed by the Department of Defense. As such, it was developed independently of the space station programme. Surprisingly, considering the secrecy surrounding its development, Buran's maiden launch in 1988 was shown live on television. Even more remarkably, it flew with automatic avionics. (It could not carry a crew until the development of the environmental systems was finished.) The prospect of operating Buran in conjunction with Mir prompted the installation of two androgynous ports on the Kristall module. Buran was to dock on the axis and mount an X-ray telescope on the side port. A module similar to Kvant 1 was once shown in a mock-up of the payload bay, so it is possible that Buran was to have delivered modules to further extend the front of the Mir complex after the cluster there was complete. It is also possible that the requirement to have the androgynous port in place was a factor in advancing Kristall's launch. The development of the YMK cosmonaut manoeuvring backpack that was sent up in Kvant 2 was to be tested for use by cosmonauts in Buran's payload bay. Although conceived independently, the two programmes were clearly seen as being symbiotic. Several mission profiles were considered for testing Buran. One scenario involved sending two Buran pilots up to Mir by Soyuz and then launching Buran for an automatic rendezvous and docking. The pilots would board Buran, undock, put it through its paces, and then, depending on how the vehicle performed, either return directly to Earth or redock and leave it to make an automatic return. In preparation, two members of the pilot group visited Mir in 1984 and 1987. Another such mission was planned for 1989, but was cancelled. These flights were to provide experience of weightlessness and to assess piloting skills immediately following return. However, financial realities in post-Soviet Russia prompted Buran's cancellation. In hindsight, it is clear that the programme consumed resources that could have been better spent. Ironically, when Kristall's androgynous port *did* accommodate a Shuttle, it was the American one.

MIR'S LEGACY

By the time of Mir's de-orbiting in March 2001, the assembly of its successor, the International Space Station, was well underway. It had a pair of Russian modules at its *core*. The Zvezda habitat was a copy of Mir's base block, but with fewer radial ports. Its front port hosted the Zarya module with a configuration that was based on Kristall, but with an additional radial port. The axial port led to Unity, NASA's first module. The first three-person crew flew up on a Soyuz, but thereafter the plan was to make handovers during Shuttle visits. Nevertheless, the Soyuz would be replaced at six-monthly intervals to serve as a 'lifeboat' to enable the residents to escape in an emergency. When the Shuttle fleet was grounded by the loss of Columbia in 2003, it was only the venerable Soyuz that enabled International Space Station operations to continue.

Mir's legacy – the core of the International Space Station.

Appendices

Appendix 1 Soyuz launches and recoveries

Spacecraft	Launched		Recovered		Days
Soyuz 1	23 Apr 1967	0335	24 Apr 1967	0623	1.12
Soyuz 2	25 Oct 1968	1200	28 Oct 1968	1100	2.96
Soyuz 3	26 Oct 1968	1134	30 Oct 1968	1025	3.95
Soyuz 4	14 Jan 1969	1030	17 Jan 1969	0951	2.97
Soyuz 5	15 Jan 1969	1004	18 Jan 1969	1058	3.04
Soyuz 6	11 Oct 1969	1410	16 Oct 1969	1253	4.95
Soyuz 7	12 Oct 1969	1345	17 Oct 1969	1225	4.95
Soyuz 8	13 Oct 1969	1320	18 Oct 1969	1211	4.95
Soyuz 9	1 Jun 1970	2200	19 Jun 1970	1450	17.71
Soyuz 10	23 Apr 1971	0254	25 Apr 1971	0240	1.99
Soyuz 11	6 Jun 1971	0755	30 Jun 1971	0216	23.76
Soyuz 12	27 Sep 1973	1518	29 Sep 1971	1434	1.97
Soyuz 13	18 Dec 1973	1455	26 Dec 1983	1150	7.87
Soyuz 14	3 Jul 1974	2151	19 Jul 1974	1521	15.73
Soyuz 15	26 Aug 1974	2258	28 Aug 1974	2310	2.00
Soyuz 16	2 Dec 1974	1240	8 Dec 1974	1104	5.93
Soyuz 17	11 Jan 1975	0043	9 Feb 1975	1403	29.55
Soyuz 18	24 May 1975	1758	26 Jul 1975	1718	62.97
Soyuz 19	15 Jul 1975	1520	21 Jul 1975	1351	5.94
Soyuz 20	17 Nov 1975	1737	16 Feb 1976	0542	90.50
Soyuz 21	6 Jul 1976	1509	24 Aug 1976	2133	49.27
Soyuz 22	15 Sep 1976	1248	23 Sep 1976	2242	8.41
Soyuz 23	14 Oct 1976	2040	16 Oct 1976	2046	2.00
Soyuz 24	7 Feb 1977	1912	25 Feb 1977	1236	17.73
Soyuz 25	9 Oct 1977	0540	11 Oct 1977	0626	2.03
Soyuz 26	10 Dec 1977	0419	16 Jan 1978	1422	37.42
Soyuz 27	10 Jan 1978	1526	16 Mar 1978	1419	64.95
Soyuz 28	2 Mar 1978	1828	10 Mar 1978	1645	7.93
Soyuz 29	15 Jun 1978	2317	3 Sep 1978	1440	79.64
Soyuz 30	27 Jun 1978	1827	5 Jul 1978	1631	7.92

Appendix 1 continued

Spacecraft	Launched		Recovered		Days
Soyuz 31	26 Aug 1978	1751	2 Nov 1978	1405	67.84
Soyuz 32	25 Feb 1979	1454	13 Jun 1979	1918	108.18
Soyuz 33	10 Apr 1979	2034	12 Apr 1979	1935	1.96
Soyuz 34	6 Jun 1979	2113	19 Aug 1979	1530	73.76
Soyuz-T 1	16 Dec 1979	1530	26 Mar 1980	0050	100.38
Soyuz 35	9 Apr 1980	1638	3 Jun 1980	1807	55.06
Soyuz 36	26 May 1980	2121	31 Jul 1980	1815	65.87
Soyuz-T 2	5 Jun 1980	1719	9 Jun 1980	1541	3.93
Soyuz 37	23 Jul 1980	2133	11 Oct 1980	1250	79.63
Soyuz 38	18 Sep 1980	2211	26 Sep 1980	1854	7.86
Soyuz-T 3	27 Nov 1980	1718	10 Dec 1980	1226	12.79
Soyuz-T 4	12 Mar 1981	2200	26 May 1981	1638	74.77
Soyuz 39	22 Mar 1981	1759	30 Mar 1981	1442	7.86
Soyuz 40	14 May 1981	2017	22 May 1981	1658	7.86
Soyuz-T 5	13 May 1982	1358	27 Aug 1982	1904	106.21
Soyuz-T 6	24 Jun 1982	2029	2 Jul 1982	1821	7.91
Soyuz-T 7	19 Aug 1982	2112	10 Dec 1982	2203	113.03
Soyuz-T 8	20 Apr 1983	1711	22 Apr 1983	1729	2.01
Soyuz-T 9	27 Jun 1983	1312	23 Nov 1983	2258	149.41
Soyuz-T 10	8 Feb 1984	1507	11 Apr 1984	1750	63.11
Soyuz-T 11	3 Apr 1984	1709	2 Oct 1984	1357	181.86
Soyuz-T 12	17 Jul 1984	2141	29 Jul 1984	1655	11.80
Soyuz-T 13	6 Jun 1985	1040	26 Sep 1985	1352	112.13
Soyuz-T 14	17 Sep 1985	1639	21 Nov 1985	1331	64.87
Soyuz-T 15	13 Mar 1986	1533	16 Jul 1986	1634	125.04
Soyuz-TM 1	21 May 1986	1222	30 May 1986	1049	8.93
Soyuz-TM 2	6 Feb 1987	0038	30 Jul 1987	0404	174.15
Soyuz-TM 3	22 Jul 1987	0559	29 Dec 1987	1216	160.26
Soyuz-TM 4	21 Dec 1987	1418	17 Jun 1988	1413	179.00
Soyuz-TM 5	7 Jun 1988	1545	7 Sep 1988	0450	91.54
Soyuz-TM 6	29 Aug 1988	0823	21 Dec 1988	1257	114.16
Soyuz-TM 7	26 Nov 1988	1850	27 Apr 1989	0629	151.48
Soyuz-TM 8	6 Sep 1989	0138	19 Feb 1990	0736	166.25
Soyuz-TM 9	11 Feb 1990	0916	9 Aug 1990	1135	179.10
Soyuz-TM 10	1 Aug 1990	1332	10 Dec 1990	0908	130.83
Soyuz-TM 11	2 Dec 1990	1113	26 May 1991	1404	175.12
Soyuz-TM 12	18 May 1991	1650	10 Oct 1991	0712	144.52
Soyuz-TM 13	2 Oct 1991	0859	25 Mar 1992	1151	175.08
Soyuz-TM 14	17 Mar 1992	1354	10 Aug 1992	0504	145.63
Soyuz-TM 15	27 Jul 1992	1009	1 Feb 1993	0648	188.86
Soyuz-TM 16	24 Jan 1993	0858	22 Jul 1993	1042	179.08
Soyuz-TM 17	1 Jul 1993	1833	14 Jan 1994	1119	196.70
Soyuz-TM 18	8 Jan 1994	1305	9 Jul 1994	1428	181.94
Soyuz-TM 19	1 Jul 1994	1625	4 Nov 1994	1420	125.92
Soyuz-TM 20	4 Oct 1994	0142	22 Mar 1995	0704	169.22

Appendix 1 *continued*

Spacecraft	Launched		Recovered		Days
Soyuz-TM 21	14 Mar 1995	0911	11 Sep 1995	1052	181.07
Soyuz-TM 22	3 Sep 1995	1300	29 Feb 1996	1342	179.03
Soyuz-TM 23	21 Feb 1996	1534	2 Sep 1996	1141	192.84
Soyuz-TM 24	17 Aug 1996	1715	2 Mar 1997	0944	196.73
Soyuz-TM 25	10 Feb 1997	1709	14 Aug 1997	1617	184.92
Soyuz-TM 26	5 Aug 1997	1936	19 Feb 1998	1210	197.73
Soyuz-TM 27	29 Jan 1998	1933	25 Aug 1998	0925	207.54
Soyuz-TM 28	13 Aug 1998	1343	28 Feb 1999	0515	198.69
Soyuz-TM 29	20 Feb 1999	0718	28 Aug 1999	0435	188.84
Soyuz-TM 30	4 Apr 2000	0902	16 Jun 2000	0444	72.82
Soyuz-TM 31	31 Oct 2000	1053	6 May 2001	0942	186.91
Soyuz-TM 32	28 Apr 2001	1137	31 Oct 2001	0800	185.89

Dates are Moscow Time.

Appendix 2 Progress launches

Spacecraft	Launched	MT
Progress 1	20 Jan 1978	1125
Progress 2	7 Jul 1978	1426
Progress 3	8 Aug 1978	0103
Progress 4	4 Oct 1978	0209
Progress 5	12 Mar 1979	0847
Progress 6	13 May 1979	0717
Progress 7	28 Jun 1979	1225
Progress 8	27 Mar 1980	2153
Progress 9	27 Apr 1980	0924
Progress 10	29 Jun 1980	0741
Progress 11	28 Sep 1980	1810
Progress 12	24 Jan 1981	1718
Progress 13	23 May 1982	0957
Progress 14	10 Jul 1982	1358
Progress 15	18 Sep 1982	0859
Progress 16	31 Oct 1982	1420
Progress 17	17 Aug 1983	1608
Progress 18	20 Oct 1983	1259
Progress 19	21 Feb 1984	0946
Progress 20	15 Apr 1984	1213
Progress 21	8 May 1984	0247
Progress 22	28 May 1984	1813
Progress 23	14 Aug 1984	1028
Progress 24	21 Jun 1985	0440
Cosmos 1669	19 Jul 1985	1705
Progress 25	19 Mar 1986	1308

Appendix 2 *continued*

Spacecraft	Launched	MT
Progress 26	23 Apr 1986	2340
Progress 27	16 Jan 1987	0906
Progress 28	3 Mar 1987	1414
Progress 29	21 Apr 1987	1914
Progress 30	19 May 1987	0802
Progress 31	4 Aug 1987	0044
Progress 32	24 Sep 1987	0344
Progress 33	21 Nov 1987	0247
Progress 34	21 Jan 1988	0152
Progress 35	24 Mar 1988	0005
Progress 36	13 May 1988	0330
Progress 37	19 Jul 1988	0113
Progress 38	10 Sep 1988	0334
Progress 39	25 Dec 1988	0712
Progress 40	10 Feb 1989	1154
Progress 41	16 Mar 1989	2154
Progress-M 1	23 Aug 1989	0710
Progress-M 2	20 Dec 1989	0631
Progress-M 3	1 Mar 1990	0211
Progress 42	6 May 1990	0044
Progress-M 4	15 Aug 1990	0801
Progress-M 5	27 Sep 1990	1437
Progress-M 6	14 Jan 1991	1750
Progress-M 7	19 Mar 1991	1606
Progress-M 8	30 May 1991	1204
Progress-M 9	21 Aug 1991	0250
Progress-M 10	17 Oct 1991	0305
Progress-M 11	25 Jan 1992	1050
Progress-M 12	20 Apr 1992	0129
Progress-M 13	30 Jun 1992	2043
Progress-M 14	16 Aug 1992	0219
Progress-M 15	27 Oct 1992	2020
Progress-M 16	21 Feb 1993	2132
Progress-M 17	31 Mar 1993	0834
Progress-M 18	22 May 1993	1042
Progress-M 19	11 Aug 1993	0223
Progress-M 20	12 Oct 1993	0035
Progress-M 21	28 Jan 1994	0512
Progress-M 22	22 Mar 1994	0754
Progress-M 23	22 May 1994	0730
Progress-M 24	25 Aug 1994	1825
Progress-M 25	11 Nov 1994	1022
Progress-M 26	15 Feb 1995	1948
Progress-M 27	9 Apr 1995	2334
Progress-M 28	20 Jul 1995	0704

Appendix 2 *continued*

Spacecraft	Launched	MT
Progress-M 29	8 Oct 1995	2152
Progress-M 30	18 Dec 1995	1731
Progress-M 31	5 May 1996	1104
Progress-M 32	1 Aug 1996	0000
Progress-M 33	20 Nov 1996	0221
Progress-M 34	6 Apr 1997	2004
Progress-M 35	5 Jul 1997	0812
Progress-M 36	5 Oct 1997	1809
Progress-M 37	20 Dec 1997	1145
Progress-M 38	15 Mar 1998	0246
Progress-M 39	15 May 1998	0212
Progress-M 40	25 Oct 1998	0715
Progress-M 41	2 Apr 1999	1528
Progress-M 42	16 Jul 1999	2038
Progress-M1 1	1 Feb 2000	0947
Progress-M1 2	26 Apr 2000	0008
Progress-M1 3	6 Aug 2000	2226
Progress-M 43	17 Oct 2000	0037
Progress-M1 4	16 Nov 2000	0432
Progress-M1 5	24 Jan 2001	0429

Glossary

Aelita-1
A multipurpose biomedical kit on Salyut 7 (to supersede the Polynom-2M). Specifically designed for use in space, it could measure a wide range of parameters, including cardiovascular, cerebral, circulation and blood flow. Unfortunately, it overheated upon being inadvertently left on overnight, but was eventually repaired. Cosmonauts generally found full medical check ups frustrating because of the time it took to unpack all the sensors, pastes, cables, cuffs and belts, assemble and test the monitoring equipment, then wait to fly within radio range so that the telemetry could be transmitted to the doctors on the ground. Unless the data could be completed within one pass, they would have to break off until the next pass. Finally, the equipment had to be disconnected and packed away again.

Aerosol
An experiment on Salyut 7. It used a spectrometer to study the upper layers of the atmosphere at visible and infrared wavelengths. These observations were made by pointing the sensors directly down at the Earth, so that it precisely followed the ground track, rather than by pointing it off to the horizon to monitor the absorption of sunlight at sunrise and sunset. This had the advantage of localising the measurement, and also permitted continuous monitoring over extended periods.

Afamia
An experiment on Mir using the Kristallisator furnace to grow gallium-antimonide monocrystals.

Agidel
An experimental shaver tested on Salyut 7. It incorporated a chamber to collect stubble, but in practice this proved to be inadequate, so a larger vacuum cleaner was also used.

AIAA
American Institute for Aeronautics and Astronautics.

Ainur
An electrophoresis apparatus on the Kristall module used to purify proteins. It was said to be capable of processing 100 kilograms of material per year.

328 Glossary

Akustika
: An experiment on Mir to measure the background noise at various points within the complex. The hum from the environmental system was generally around 80 decibels.

Alice-1
: A French experiment on Mir that investigated the transport of heat and mass in gas-liquid systems, and phase-change phenomena at their critical points (when the properties of the gaseous and liquid phases were identical).

Alice-2
: A French 64-kilogram furnace that accurately controlled the temperature of a fluid. It used a CCD camera to investigate the fine-scale phenomena of the critical-point phase transition. Its data and video output were stored for return to Earth. It was a refinement of the Alice-1 experiment.

Alissa
: A French experiment on the Priroda module. It used a lidar at 5270 Angstroms aimed straight down to study the atmosphere. In addition to the vertical structure of clouds, it could detect tropospheric aerosols (and hence complement the stratospheric capability of Ozon-M). Unfortunately, because its four-laser probe consumed 3 kW its use had to be limited. It had initially been meant for Salyut 7, so had been designated l'Atmosphere Lidar Sur Salyut (hence Alissa).

Almaz (Diamond)
: The reconnaissance platform designed by the Chelomei Bureau, the shell of which was modified by the Korolev Bureau to serve as Salyut.

Altai-1
: An experiment on Salyut 6. The Splav furnace was used to study the diffusion and mass transfer properties of lead and tin (metals that are soluble in their liquid state) and the effects of convective flows occurring across the thermal gradient.

Altai-2
: An experiment on Salyut 6. A crystal of vanadium pentoxide (an active semiconductor widely used in the production of thermistors) was grown in the Splav furnace to test whether very homogenous crystals could be created in microgravity conditions.

Altyn
: A biotechnology experiment on Mir which involved a genetic study of plant (wheat) cells.

Amak-3
: A blood analyser on Salyut 4.

AMEE
: A NASA experiment. The Advanced Materials Exposure Experiment was deployed outside Mir. Its optical properties monitor could be rotated in different directions to enable a video camera and a spectrometer to examine each sample and provide real-time data on the thermal, optical and mechanical degradation of the exposed materials.

Amplituda
: An experiment on Salyut 6 to determine the structural stability of a three-

spacecraft complex. Like the Resonance experiment, this involved a crewman jumping on the KTF in a rhythm defined by a timing signal from the ground, while a transducer measured the effects.

ANBRE

This analogue biomedical recorder was a skin-tight suit supplied to Mir by the European Space Agency, to study the dynamics of posture by using the ELITE apparatus to measure the motion of the major limbs in weightlessness.

Anna-3

A telescopic spectrometer on Salyut 1. It used a Cerenkov counter to measure the flux of gamma-rays having energies in excess of 100 MeV, for astrophysical research. It had a 1-degree pointing accuracy.

Anthropometry

An experiment on Salyut 6 to determine changes in muscular mass and bone structure (specifically the loss of calcium) during the initial phase of adaptation to weightlessness.

Antibiotik

An experiment on Salyut 7. Absorbent pads were used to sample bacteria and microflora from different parts of the body. Once these had developed in the Cytos-2 incubator, they were subjected to various antibiotics to determine their responses. The biomedical data was transmitted to Earth on the telemetry link.

AO-1

The star tracker in the main compartment of Salyut 7. The cosmonauts could orientate the station by reference to the stars using the sextant and star tracker manually. The two instruments had first been coaligned so that when a star was centred in one it was in the centre of the field of view of the other. Sightings were taken as the station passed through the Earth's shadow. The star tracker could be used to determine the orientation of the station, then the sextant used to hold it steady. This was a backup capability in case the Delta/Kaskad system failed.

APA

A NASA biomedical experiment on Mir. Anticipatory Postural Activity was to observe posture and muscle coordination during various activities.

APAS

An Androgynous Peripheral Assembly System. The term 'peripheral' applies because its initial capture mechanism is mounted peripherally (unlike the axially projecting probe of the standard Soyuz unit, which, by its nature, requires a male/female combination; it is the absence of this restriction which makes the peripheral unit androgynous). The system devised for Apollo-Soyuz had the three guide plates canted outwards. Since this historic link-up took place in 1975, this is now know as the APAS-75 configuration. The port on the Kristall module of the Mir complex, intended to accommodate the Buran shuttle, had inward-canted guide plates, and was known as the APAS-89 configuration. This assembly was bought from Energiya by NASA and attached to the ODS to enable Atlantis to dock at the Kristall module. As the two sets of guide plates meshed, the capture latches they carried engaged to achieve the soft docking. Springs within the mechanism damped out the residual relative motions as the two spacecraft jostled one another. After fifteen minutes in this state, the Shuttle fired its thrusters to force the extended guide ring against that on Kristall to ensure proper alignment, and then the twelve primary latches around the periphery of two collars were commanded to extend and engage, and the ring was retracted (this released the capture latches) to establish the hard docking which formed a rigid connection and a hermetically sealed tunnel within the mechanism.

Apollo
The spacecraft built by NASA to fly three astronauts into lunar orbit and return them to Earth. Far larger than Gemini, it comprised a blunt-cone command module and a cylindrical service module. It had considerably greater manoeuvring and navigational capability than the contemporary Soyuz.

Arabidopsis
A wallcress plant, commonly regarded as a weed, widely used in genetic experiments because it has the advantage of rapid growth (barely 40 days per generation), and a comparatively simple genome (only about twenty genes) which is well understood.

Aral-91
A programme to study the movement, concentration, composition, temperature and wind speed of dust and aerosols blowing from the recently exposed bed of the Aral Sea.

Arfa
An experiment deployed outside Kvant 1 to investigate the ionosphere and magnetosphere. Its data was used to test the correlation between charged particles at orbital altitude and seismic activity in the Earth's crust.

Argon 16
The computer introduced on Salyut 4 to handle navigation, flight dynamics and attitude control. It also comprehensively monitored and displayed the status of its systems to the crew. It was afterwards installed on Soyuz-T. The Americans pointed out that this technology was comparable with that used by Gemini, and greatly inferior to that used by Apollo. Nevertheless, it was a significant advance for the ferry, because it enabled it to manoeuvre independently of ground tracking and computer commands.

Argument
A biomedical ultrasound scanner on Salyut 7 which produced a sectional image of the heart on a screen. It was an adaptation of a device used in emergency clinics in the Soviet Union, and was similar to, but not as easy to use as, the Echograph.

ARIS
A Boeing device intended for the International Space Station. The Active-Rack Isolation System used a mechanical suspension system to damp out micro-accelerations to facilitate microgravity experiments on the 'noisy' environment of an inhabited platform.

ARIZ
An X-ray spectrometer mounted on the ASPG-M platform.

Armadeus
A French experiment on Mir which involved the erection of a 28-kilogram model

of a solar panel, to test the articulated deployment mechanism. The structure comprised four motorised winding blades that incorporated Carpentier joints to eliminate friction. It was performed inside Mir. The deployment was filmed so that the deployment kinematics could be analysed later.

ASPG-M

A 225-kilogram Czech-built scan platform (similar to those carried on the Vega planetary probes). It was mounted on Kvant 2, and could be pointed by remote control (either by the flight controllers or the crew) to make Earth observations without requiring the complex to be reorientated. It carried a variety of apparatus (including the ITS-7D infrared spectrometer, the ARIZ X-ray spectrometer, the MKS-M2 multispectral optical spectrometer and the Gamma-2 videospectral television cameras), and had a pointing accuracy of better than 0.5 minutes of arc.

Astra-1

A mass spectrometer to analyse the gaseous environment around Salyut 7. It revealed a tenuous stream of matter trailing the station, composed of air vented from the small scientific airlocks and the exhaust products from the attitude control thrusters. After a spacewalk, it observed a stream of gas trailing behind the station resulting from the venting of the airlock.

Astra-2

A spectrometer on the Spektr module to measure the constituents of the gaseous environment at orbital altitude.

Astro

An experiment on Salyut 6. The Emission apparatus was used to study the cosmic ray background in the near-Earth environment.

Astro

A CCD star tracker built by Jena Optronik of Germany. One 80-kilogram unit was affixed to each side of the unpressurised compartment of Kvant 1 during a spacewalk in order to assist in orientating the enlarged complex.

ASU

A toilet on Mir; there were two – one in the base block and one in the Kvant 2 module.

Atlantika-89

A programme carried out jointly with Cuba to investigate a suspected correlation between ozone and the formation of hurricanes. Mir measured the amount and distribution of ozone.

Atlantis

A NASA Space Shuttle.

Atlet (Athlete)

A load-inducing suit which incorporated elasticated straps which linked a waist corset to the shoes, with braces running over the shoulders, to make the muscles continuously work to straighten out. It was tested on Salyut 4, but abandoned in favour of the Penguin suit.

Audimir

An Austrian experiment on Mir to study the auditory system.

Glossary

Audio
An East German experiment on Salyut 6 to investigate frequency characteristics of sound in the station. It tested the ability to distinguish subtle nuances of sound in weightlessness. The way sound propagates is dependent on air pressure; in partial pressure it does not propagate as far and its tone is altered. Audiograms made with the Elbe apparatus to determine auditory threshold over a range of frequencies were compared with tests made before and after the flight to determine whether hearing changed. The results revealed that the hearing threshold is lower in space.

Azolla
A biological experiment on Salyut 6, carried out with a fast-growing nitrogen-rich water fern (appropriately, azolla pinnata was used) commonly used as fertiliser in rice-growing areas, to see if it held promise for use in a closed-cycle hydroponics system on a future station to properly cycle chemicals in its air.

Balance
An experiment on Salyut 6 to determine the changes in the body's water and mineral balance by measuring water intake, urine excretion, body mass and concentrations of a range of substances in the blood.

Balaton
A Hungarian apparatus (named after a lake in Hungary) on Salyut 6, designed to determine the intellectual and motor performance of a cosmonaut. It was held in the palm of the subject's left hand in such a way that the pulse could be measured from one finger while sensors recorded galvanic skin resistance (that is, the electrical conductivity) of two other fingers as an indication of the level of perspiration. This measured the subject's effort in responding to a series of mental exercises involving patterns of audio tones and lights. Each exercise had a number of solutions of different levels of difficulty, and as the cosmonaut passed from one level to the next the skill factor required was increased. Throughout, the cosmonaut was subjected to a series of disruptive rhythms and tones played over his headset.

Balkan
A Bulgarian experiment on Salyut 6. It was a study of the Earth with the Spektr-15K spectrometer.

Balkan-1
A lidar on the Spektr module to determine the altitudes of clouds.

Ballisto
An experiment on Salyut 7 which monitored chest movements to measure the accelerations of the body caused by the heart's pumping action, and changes in the shape and location of the heart from the body's adaptation to the absence of gravity. The experiment was continued on Mir.

Bania
The shower unit on Salyut 7. It took about two hours to set up, use, clean, and stow away, as it involved a complicated procedure which began with pumping out the contents of the ASU urine collector from the small EDV container, which is

the temporary receptacle for fluid waste, into the much larger EDV tank in which it is stored. This would enable the smaller container to be used to take dirty water extracted from the shower unit. The cosmonauts then filled the two Kolos containers in its ceiling, and then pulled the curtain down from its mount in the ceiling of the station and turned on the air filtration unit in the floor of the unit which would extract odours which might overtax the environmental unit if they were allowed to leave the enclosure. Once the water in one of the Kolos tanks had been heated, the first man could don his air tube and goggles which would prevent him from drowning and keep the water out of his eyes, and begin his shower. The Kolos containers, however, held only 5 litres each of hot and cold water, so he did not have very long to get himself clean. It took at least five minutes for the enclosure to fill with a water, air and steam mixture, which was made to flow downwards by pumping air through the enclosure. A sponge soaked in highly abrasive katamine (quarternary amines) was used to literally scrape off the dirt which had accumulated on their bodies during a month in space. Clearly, taking a shower in space was not just a matter of stepping into the compartment and turning on the water, and as a result, the cosmonauts were limited to monthly showers. Unfortunately, there was no room in the station to keep it in place. Once, when pumping out the EDV tank, a plug came free and a mixture of dirty water and urine sprayed out, and when this had been cleaned up the cosmonauts really did need the shower that they had started out to erect.

Batyr
A biomedical experiment on Mir to evaluate the effect of breathing exercises as a way of easing the initial phase of adaptation.

BAZK
A star camera on Salyut 7.

BCAT
A NASA experiment performed in the Glovebox on Mir. The Binary-Colloid Alloy Test was to observe the crystallisation of alloys of colloids. Samples using different relative concentrations were processed for 24 hours to establish their behaviour, and then one was left to grow for three months.

Bealuca
A Hungarian experiment on Salyut 6 which smelted copper-aluminium in the Splav furnace.

Berolina
A series of East German experiments in crystal growth using the Splav and Kristall furnaces on Salyut 6. Of the two experiments using the Splav furnace, one boiled beryllium-thorium to produce homogeneous glass of far higher quality than possible under the influence of gravity, and the other used a special quartz matrix to produce crystals of a bismuth-antimony semiconductor. Experiments in the Kristall furnace grew lead-telluride crystals for the first time using the sublimation technique which vaporised them by heating one end of the ampoule, and seeded a crystal at the other cooler end. Kristall was also used to grow a crystal of bismuth-antimony by stretching the seed. It was processed with the material sandwiched between two plates, within an ampoule. This yielded a tree-

structured crystal which was about five times larger than that which can be produced on Earth. It was done this way to compare the result with that produced by the Splav furnace.

Beta

A multifunction medical unit on Salyut 4. It complemented the electrocardiogram functions of the Polynom-2M, but it also measured lung capacity and recorded seismograms to determine the rhythm of the heart and the force of blood pumping.

Big Bird

An umbrella name for several classes of US Department of Defense's photographic reconnaissance satellites, the earliest of which was introduced in the early 1970s, contemporary with the Almaz.

Biobloc-3

A French experiment on Salyut 7 which studied the effects of cosmic rays on biological materials by sandwiching samples of artima salina cysts and tobacco seeds between heavy-ion detectors.

Biodose

A French experiment on Mir which measured the biological effects of cosmic rays, to assess long-term effects of exposure.

Biogravistat

A plant growth apparatus on Salyut 6. It comprised two disks, one fixed and the other rotating to simulate gravity, one as the experiment and the other as the control. It was a small centrifuge similar to that which had been flown on the Cosmos 782 Biosat, which generated partial gravity to offer plants that had been unable to grow in weightlessness a more conducive environment. At first it excessively aerated the roots, but once this was rectified it was found that the seeds in the gravity chamber grew at twice the rate of those in the stationary chamber, and the roots aligned themselves in the direction of the imparted force. Once the seeds had germinated, they were transferred to other apparatus for further development. Barley was grown in it. Mushrooms grew well, but developed oddly curled stems. Tests demonstrated that plants were able to react to a force as low as one ten-thousandth of normal gravity.

Biokat

An aquarium first flown on Soyuz 19, during the Apollo-Soyuz mission.

Biokhim

A biomedical device on Salyut 7 used to measure blood electrolytes.

Biokryst

An electrophoresis apparatus on Mir.

Biosfera (Biosphere)

An experiment on Salyut 6 to investigate the state of the Earth's environment. It involved making visual observations of the Earth's oceans which would be compared with a specially created chromatic atlas to determine optical properties of the atmosphere under different conditions. This would yield information which would improve photographic techniques by evaluating the ability of different films, filters and exposures to record different types of detail. It would also determine the conditions under which different phenomena were best imaged. Because this was a long-term project, observations would be made by successive crews. Eventually, it was to establish global levels of marine and atmospheric pollution.

Biostoykost
An apparatus on Mir for making polymers.

Bioterm-1,2,3,4
A series of thermostatic plant growth experiments flown on successive Salyuts.

Biryuza
An apparatus on Mir used to study "the dynamics of physio-chemical processes", and "the processes of growing crystals, solutions and molten baths in microgravity".

Black Sea-84
A photographic and spectrographic study of the Black Sea region, organised by Intercosmos as the first phase of a multi-year programme. Data from Salyut 7 was correlated with that from the Cosmos 1500 oceanographic and Meteor-Priroda multispectral Earth-resources satellites, aircraft and research ships. It was a survey of the hydrophysical and biological characteristics of the surface of the Black Sea.

Bodyfluids
An Austrian experiment on Mir that studied the composition and distribution of blood and bodily fluids in weightlessness by measuring the speed of sound in blood to determine the dynamics of transient fluid motions after stimulation.

Bosra
An experiment on Mir to gather data with which to improve mathematical modelling of the upper layers of the atmosphere and ionosphere.

BPAS
A NASA experiment on Mir. The Belt-Pack Amplifier System consisted of a harness carrying sensors to measure muscle stimulation and work output during exercise on an ergonometer. The data on bone and muscle density was compared with pre/post-flight scanner imagery.

Braslet (Bracelet)
Elasticated rings introduced on Salyut 7. They were worn around the thighs during the initial phase of adaptation to restrict the migration of blood from the legs to the upper torso, trapping blood in the legs. Wearing them for up to an hour, several times a day, during the first few days had much the same effect as spending time in the Tchibis unit.

Brillomir
An Austrian experiment on Mir designed to measure critical fluctuations during the decomposition of binary liquid mixtures in the absence of gravity.

BST-1M
The primary instrument on Salyut 6. This 650-kilogram instrument was contained in the massive conical mount used by the OST-1 on Salyut 4. It was a telescope with a 1.5-metre-diameter mirror for infrared, ultraviolet and submillimetre observations from 50 micrometres to 2 millimetres. Although it operated in the

vacuum of space its sensors had to be cooled to –269°C; this was done using an improved version of the closed-cycle cryogenic unit first tested on Salyut 4. Helium was produced using a compressor, two gas-refrigerating machines and intermediate heat exchangers. It passed through an expanding throttle valve to lower its temperature. Although it consumed 1.5 kW, once the sensor had been cooled to its operating temperature it could be maintained at that level for extended periods. The telescope could be used only when the station was in the Earth's shadow, and for the rest of the time it had to be protected by a cover. It had a pointing accuracy of 1 arcmin. Infrared observations revealed small-scale variations in the submillimetre flux of the atmosphere in the locations in which cyclonic activity originated.

BTS
A NASA experiment on Mir. The Biotechnology System was to grow cell cultures and protein crystal. Mammalian (cow) cartilage cells were suspended in a liquid growth-medium in a bioreactor vessel, delivered within a 'powered transfer' package. It exploited microgravity to grow a 3-D tissue culture (in contrast to 2-D on the surface of a nutrient in a Petri dish) to study cell attachment patterns and interactions. The growth was recorded on video, and samples were extracted at regular intervals and then frozen for later analysis; it was a long-term experiment which ran for three months. The bioreactor kept the sample at the right temperature, fed it nutrient, removed waste and produced a log of parameters. It was found that cells of cartilage grew faster in space than on Earth. The result was frozen for subsequent analysis.

Buket
A telescopic gamma-ray spectrometer on the Kristall module used to measure the energy spectrum and spatial characteristics of cosmic radiation.

Buran (Snowstorm)
The Soviet space shuttle. The new vehicle was called the Vozdushno-Kosmichesky Korabl (VKK) which was best translated as 'atmospheric spacecraft'. Its one and only flight, which was conducted automatically, was in November 1988. It was cancelled in the early 1990s.

C-1,2
The sextants in the forward transfer compartment of Salyut 7.

Calibration
An experiment on Salyut 7 to measure the effect of small accelerations on the Magma-F furnace to develop a mathematical model of its behaviour in microgravity, where thermo-convective agitation is absent.

Caribe
A Cuban experiment on Salyut 6. It involved smelting five ampoules in the Splav and Kristall furnaces. Splav was used to make alloys of gallium arsenide with aluminium and of tin-telluride with germanium-telluride. Kristall was used to produce an alloy of germanium and indium, and of zinc-indium-sulphide (ZIS). In addition to creating semiconductors, these studied nucleation and the mechanisms of crystal growth.

CCD
: An electronic camera technology which uses a charge-coupled detector integrated circuit to form the image. It has the advantage that its efficiency (its ability to register an impinging photon) is about 80 per cent, but is limited both by the spatial resolution of the cells (the pixels) in a detector array and by the dimensions of the array. It has two other key advantages: it yields its output electronically, and in a form ideal for image processing.

CDS-1
: The Crimean Diffraction Spectrometer built into the OST-1 on Salyut 4. Its high-precision optical elements could resolve 600 lines per millimetre and yield an ultraviolet spectrum in the range 800–1,300 Angstroms, producing an average resolution of under 2 Angstroms.

Cerenkov counter
: An instrument to detect the secondary radiation (Cerenkov radiation) caused by a very-high-speed charged particle passing through a gaseous medium.

Cervical Shock Absorber
: A Mongolian-designed experiment on Salyut 6. It applied pressure to the cervical part of the spinal column and restricted head movement, and was able to simulate local loads encountered under normal gravity. It was hoped that this would inhibit the onset of the motion sickness commonly suffered during the initial phase of the process of adapting to weightlessness. It was worn continuously for the first three days, except when sleeping.

CFE
: A NASA experiment on Mir. The Candle Flame Experiment was a combustion experiment performed in the Glovebox so that its cameras could record the event for later analysis. It was later augmented by the FFFT. In all, these trials consumed 80 candles and eight samples of solid fuel. A flame in microgravity forms a sphere rather than the familiar pear shape, and it soon smothers itself because there is no convection to draw off the carbon dioxide that it creates. A forced air flow will keep it burning.

Chelomei Bureau
: The rocket design bureau (OKB-52) established by Vladimir Chelomei.

Chlorella
: A Czech-designed experiment on Salyut 6 which investigated the growth of an algae culture in a nutrient medium. The algae used was, obviously, chlorella. This is a rapid-growth seaweed which might play an important part in future station operations because it absorbs carbon dioxide and gives off oxygen.

CHR
: A German experiment on Mir supplied by the Genetics Institute of Essen University to investigate chromosomal aberration in the lymphocytes. It involved taking samples before and after the flight.

Circe
: A French radiation dosimeter on Mir which used a low-pressure tissue-equivalent gas proportional counter; it proved very effective. It was used to monitor ionising radiation, and gamma-ray and neutron dosages in the station.

338 Glossary

Long after the joint mission was over, measurements continued to be taken twice a day.

Claznoye
A biomedical device on Salyut 7 used to measure blood flow in the eye, and the movement of the blind spot.

Climate
An experiment on Mir which used the EFO-1 to study the optical density of the upper atmosphere as part of a long-term investigation into air pollution.

CMT
Cadmium-mercury-telluride.

CNES
The French Space Agency.

Cogimir
An Austrian experiment on Mir that analysed cognitive functions during the process of adaptation to weightlessness.

Cognilab
A French neurosensory experiment on Mir. It involved strapping into an instrumented chair which measured the body's response to muscular stimulation under different conditions.

Columbus
The European Space Agency's laboratory module intended to be attached to the International Space Station.

Comet
A French micrometeoroid detector deployed outside Salyut 7. Its two chambers could be operated by remote control from within the station. The first was to be exposed to Comet Giacobini-Zinner, the second to Halley's comet. In each case it was hoped that it would collect dust ejected by the comet during its passage through the inner Solar System.

Coordination
An experiment on Salyut 6 to investigate the effects of weightlessness on the body's voluntary motor functions, using a mechanical hand/eye psychomotor test.

Cortex
An experiment on Salyut 6. It monitored the brain's electrical activity in the absence of gravity by taking electroencephalograms (EEG) to determine responses to various stimuli. The in-flight EEGs would be compared with pre/post-flight data. The data was collected with an instrumented helmet and was stored on a small tape recorder.

Cosmic radiation
The radiation extant in space. It comprises three components: solar wind, solar flares and cosmic rays, all of which consist primarily of charged atomic nuclei with differing speeds and energy.

Cosmic rays
The most energetic component of cosmic radiation. These heavily-ionised atoms (some as heavy as iron) have been accelerated by unknown processes in the Galaxy, far beyond the Solar System, to speeds approaching that of light, so they

arrive with energies in excess of 1 billion electron volts, sufficient to enable them to pass straight through the radiation shielding used to protect spacecraft. They pose a serious long-term threat to space missions, both to equipment and to their crews. The astronauts on the Apollo lunar missions reported 'seeing' sporadic cosmic rays that passed through their eyes. The transit of a heavy nucleus through a crystalline lattice will leave a microscopic track, so such materials were flown on Mir specifically to assess the cosmic ray flux. Other experiments exploited the fact that at narrowly-sublight invacuo speeds, a cosmic ray will stimulate Cerenkov secondary radiation when passing through a gaseous medium.

CSA

The Canadian Space Agency.

CSK-1,2,3,4

The Kristallisator furnace.

Cytos

A joint Soviet-French experiment on Salyut 6. A variety of micro-organisms were delivered in a Bioterm at 8°C to inhibit their development. The French experiment used paramecium protozoa and the Soviet experiment used proteidae. These were put in the Cytos apparatus, which was maintained at 25°C to facilitate their development, and after 12 hours were reinhibited by being returned to the Bioterm for return to Earth for analysis. Over a four-day period, eight consecutive generations of micro-organism were grown. This experiment was designed to determine the effects of microgravity and radiation on the process of cell division (the basis of the life process) to study the kinetics of cell division. This was part of an investigation of the long-term ability of organisms to live in space. The French team, which had designed and built the Cytos apparatus specifically for this experiment, was led by Professor Hubert Planel of the National Space Research Centre at the University of Toulouse, and the experiment was conducted in collaboration with Professor Yuri Nefedov of the Soviet Institute of Medicobiological Research. The test showed that bacteria behaved as they did on Earth, but that simple micro-organisms thrived on the nutrients provided.

Cytos-2

A French incubator on Salyut 7. It was used to investigate the structure and functioning of bacteria cells in weightlessness and the effect of a variety of antibiotics against bacteria.

Cytos-3

A French incubator on Salyut 7.

Danko

A cassette outside Mir. It exposed construction materials, and produced electronic data that could be downloaded periodically.

DARA

Reunified Germany's Aviation and Space Research Institute.

DCAM

A NASA experiment sent up to Mir. The Diffusion-controlled Crystallisation Apparatus for Microgravity used a semi-permeable membrane to grow protein monocrystal over a long period. The growth process was filmed, and the results were returned to Earth for study.

Deformatsiya (Deformation)
An experiment on Salyut 6 that used optical sensors to measure the distortion of the complex's structure due to solar heating. The complex was orientated and maintained in a solar-stabilised attitude which created a 300°C temperature differential. Because this attitude baked one side and froze the other, it could not be held for long without risking damaging its systems, but the results indicated that even several hours of such exposure distorted the complex's primary axis by no more than 0.1 degree.

Delta
A self-contained navigational system tested by Salyut 4. It was a significant advance over previous stations, which had required data on the path to be measured on every revolution by a network of tracking stations. The new system used a set of Sun sensors to keep track of the station's position in its orbit by monitoring the rising and setting of the Sun on the Earth's horizon in order to determine the period of each orbit. It incorporated the Argon 16 computer, which combined this with the readings from the radio altimeter to compute the station's orbital parameters, monitor its progress around the globe and compute the times during which it would be within communications range of the various ground stations. It could pinpoint the station's location over the surface of the Earth to within 3 kilometres, and its altitude to within a few hundred metres. It greatly simplified the ground-based support systems and reduced the loads on the cosmonauts. It was tested on Salyut 4 and then made operational on Salyut 6 and Salyut 7.

Diagnost
A Hungarian experiment which began during the preparations for launch to Salyut 6 and was then repeated immediately after landing, to study the effects of space flight on the cardiovascular, respiratory and vestibular systems, and on the hearing threshold.

Diagramma
This experiment on Mir used a magnetic discharge transducer on a short boom projecting from the scientific airlock to measure the physical characteristics of the atmosphere at orbital altitude (measuring the aerodynamic flow around the complex would help evaluate aerodynamic drag).

Diffusia (Diffusion)
An experiment on Salyut 5 to produce a more homogeneous alloy of dibenzyl and toluene than can be achieved under normal gravity.

Diffusion
A French experiment on Salyut 7. It employed the Magma-F furnace to study the dissolution of a polycrystalline solid alloy in its own liquid when in a state of thermodynamic equilibrium, to develop a mathematical model of the solidification process. It investigated capillary forces during crystallisation of alloys at different temperatures.

Discovery
A NASA Space Shuttle.

Glossary

Diusa
An apparatus on Salyut 7. Capable of measuring minute air pressure variations, it was used to test for suspected cabin leaks.

DM
The Docking Module mated to the axial port on the Kristall module to provide extra clearance for Shuttles from Mir's projecting solar panels, to provide an extra airlock chamber, and to store cargo. At 4.6 metres long, 2.2 metres in diameter, with a mass of only 4.2 tonnes, it was essentially an extremely stretched Soyuz orbital module fitted with an APAS port at each end.

DOM
A German experiment on Mir which evaluated five different dosimeters as methods of measuring radiation within the complex.

DOS
An acronym for permanent orbital station. This designation was assigned to the scientific Salyut adapted by the Korolev Bureau from the Almaz design:

DOS 1	Salyut 1	April 1971
DOS 2	<failed rocket>	July 1972
DOS 3	Cosmos 557	May 1973
DOS 4	Salyut 4	Dec 1974
DOS 5	Salyut 6	Sept 1977
DOS 6	Salyut 7	Apr 1982

Although DOS 1 to 4 differed in detail (such as solar panels), they were all basically the same configuration with a single docking port at the front. DOS 5 and 6 represented a significant redesign to bring back features of the OPS (notably the rear docking unit, the peripherally-mounted engines) and this then served as the basis of the Mir base block, which in turn led to the Service Module of the International Space Station.

Dose
A Hungarian experiment on Salyut 6 to measure radiation doses. A number of sensors were worn by the cosmonauts, and others were distributed around the station. Each day they were analysed using the Pille thermoluminescent device to monitor the incremental accumulation of radiation.

Dosimir
An Austrian experiment on Mir which tested a TLD dosimeter.

Dosug
An experiment on Mir that attempted to formally evaluate the influence of music, video and games on a cosmonaut's morale whilst off duty.

Doza-B
An experiment on Mir which involved placing biological samples and radiation sensors at specific places in the complex to investigate exposure levels.

Drosophilae
A fruit fly, widely used in genetic studies because it has the advantage of a short reproductive cycle and a well-understood genome.

DS-1
: A French biomedical instrument on Salyut 7. It was a doppler device to measure blood flow rate.

Duga
: A Bulgarian electrophotometer on Salyut 6, attached to one of the portholes and used while in the Earth's shadow, pointed by a cosmonaut using an optical sight. It detected, intensified and measured natural emissions at 6,300, 5,577, 4,278 and 6,563 Angstroms (all in the optical band) to study a range of upper atmospheric phenomena including the vertical structure of aurorae, the luminous red ionospheric arcs created by severe magnetic storms and stable luminous arcs in the plasmasphere. Such studies were part of an ongoing programme, so were conducted whenever the opportunity arose. Unfortunately, a fault in the construction of its image converter meant that it displayed an inverted image (a new image converter was later flown up and installed). It revealed that the atmosphere above the equator glowed. This was most noticeable in the red oxygen line, at 6,300 Angstroms, whose intensity varied by a factor of 10. A similar, but less intensive, glow was detected at 5,577 Angstroms. However, because it was not detected at 4,278 or 6,563 Angstroms (generated by electron and proton emissions respectively), it was clear that the observed glow could not have been produced by intense flows of such charged particles but by some other mechanism.

Dynalab
: A French experiment on Mir to measure the propagation of vibrations through the complex (a variation on the Resonance theme).

Eceq
: A French experiment on Mir which measured the flux of cosmic rays, and its effects on electronic equipment in the complex.

Echantillon (Sample)
: A French cassette which exposed materials samples to the space environment. The Comes segment exposed samples of paints, reflectors, adhesives, filament-reinforced composites and optical materials; Mapol exposed samples of polymeric materials thought suitable for creating inflatable structures in space; MCAL exposed materials to determine the evolution of absorptivity and emissivity, and also incorporated a pair of dust detectors, one active and one passive (called DIC and DMC respectively), to collect cosmic dust for subsequent examination. A 16-kilogram 0.75-metre square box, it was affixed to the outside of Mir for a year.

Echograph
: A French biomedical instrument on Salyut 7; an ultrasound scanner to make a sectional image of the heart. It was later transferred to Mir.

Echograph-II
: An improved ultrasonic body scanner on Mir, to measure blood flow in deep vessels (in particular the main truncus and venous return) as well as the capacity of the heart and other internal organs.

Echography
: A French experiment on Salyut 7 which involved using the Echograph ultrasound

scanner and the DS-1 doppler device to measure the cavity dimensions, mycocardiac thickness, artery blood content and the speed of blood supply to the vessels. The hypothesis was that, in the absence of gravity, the heart moved up into the chest. The fact that Chrétien initially had some difficulty locating his heart lent support to this supposition. He also used the DS-1 to determine the distribution of blood in the cranial artieries.

EDLS

A NASA technology experiment performed on Mir. The Enhanced Dynamic Load Sensors involved a push-off pad to measure the force used to move within the complex, and an instrumented foot-restraint to monitor movements whilst nominally working 'in place'. The empirical data was to be used in fitting out the modules of the International Space Station.

EFO-1

A Czech electrophotometer used onboard Salyut 7 to make measurements of the upper atmosphere. It was used to monitor stars as they dropped down to the horizon, measuring the flickering light as it passed through the layers of the Earth's atmosphere. (An intensive programme of such observations was to be made to locate and study layers of aerosols in the upper atmosphere). Usually for these observations, the station was orientated with its axis perpendicular to the velocity vector and parallel to the horizon, and then placed in a slow roll timed to keep the apparatus in the porthole facing the horizon to monitor a succession of stars rising or setting through the layers of the atmosphere. Occasionally, however, the station was put in gravity-gradient mode, and stabilised so that the porthole faced ahead (depending on whether microgravity materials processing was underway at the same time). It was transferred to Mir. Its main contribution was to the Climate experiment.

EFU-Robot

An electrophoresis unit on Salyut 7. It was a pilot-scale form of an automatic 'factory' designed to produce biological preparations which could be useful to the food, health and agricultural agencies. It was transferred to Mir.

Elbe

An audiometer. It produced an audiogram which measured the hearing threshold across a range of sound frequencies.

Electron volt, eV, keV, MeV

A unit for measuring the energy of a particle. It corresponds to the energy required to displace a free electron across a potential difference of one volt within an electric field; that is 1.6×10^{-19} Joules.

Electrotopography

An experiment on Salyut 7 performed in one of the scientific airlocks. Deformations on the surface of selected materials exposed to space for different periods were studied. The aim was to explore by electromagnetic means the surface of samples exposed to space. Each sample was put between two metal plates, with photographic film next to the sample. When a charge of several thousand volts was applied across the plates, the electric field was contoured around microscopic defects in the sample. It could detect variations as small as

344 Glossary

10^{-11} metres in the shape of a sample. However, a problem developed on the first trial. As soon as the small scientific airlock was opened, the temperature in it fell rapidly to $-150°C$, as did the apparatus. If the airlock was opened immediately to retrieve the sample, the water vapour in the station's atmosphere condensed on it and ruined the experiment. Yet if the airlock was closed, and the temperature was allowed to rise slowly over a period of hours before it was opened, the sample acquired an electric charge that distorted it, again ruining the test. To overcome this, the cosmonauts orientated the station so that the airlock faced the Sun towards the end of the exposure so that the sunlight would warm it up, and if the timing was right the airlock could be closed at just the time when it was at a comfortable $20°C$. The trial was rerun a number of times using more sensitive film and longer exposure periods (in some cases, 40 hours).

Electrotopograph-7K
An apparatus on Mir used to measure surface distortions of advanced plastics and high-temperature superconductors exposed to space in the scientific airlock for various times.

Elektron
This produced oxygen by electrolysing a 30 per cent potassium hydroxide (KOH) solution, to maintain the required composition of the gaseous environment within the Mir complex without requiring so much liquefied air to be delivered by Progress ferries. It was connected by flexible tubes which were to be distributed throughout the complex. There were twelve electrolysis cells in the unit, which were cooled by the base block's primary coolant loop. It consumed about 4 kilogram of water per day; whenever possible water recycled from urine was employed. The hydrogen released was vented.

Eleutheroccus
A drug taken over a prolonged period (a daily dose of 4 ml) towards the end of a long spaceflight to act as a tonic, to stimulate the body to work harder, and thereby increase long-term stamina to help cosmonauts prepare to adapt to gravity.

ELITE
A European Space Agency experiment that used four infrared television cameras to monitor the subject's posture in weightlessness.

Elma-1
A French smelting experiment on Salyut 6 using the Kristall-2 and Splav furnaces. Aluminium-copper, aluminium-tin, aluminium-lead and lead-tin were processed to study the processes of diffusion during melting and subsequent cooling of metal alloys. Magnetic alloys of magnesium, cobalt and other metals which do not mix under a strong gravity field were produced, as were monocrystals of germanium and vanadium oxide. The process of diffusion during melting and subsequent cooling of alloys of tin-lead and aluminium-copper was studied. A sample of gadolinium-cobalt (a magnetic material which has computing applications) was also produced.

Elma-2
A French experiment on Salyut 7. It involved creating an alloy of aluminium and indium (which are immiscible in Earth's gravity).

Emission
A Mongolian experiment on Salyut 6. It involved using dielectric detectors to monitor the strength and composition of cosmic rays in the 10 MeV range. The apparatus was inserted in the scientific airlock, and exposed to space for three days. The tracks bored through the detector substance would be examined after the detector had been returned to Earth. A similar detector (also Mongolian-built) had previously been flown on the unmanned Intercosmos 6 satellite, and had revealed that energetic nuclei with charges up to 28 units constituted over 90 per cent of the incident radiation. Another detector was exposed within the station to measure how well the station's hull was able to block radiation.

Emissiya (Emission)
An experiment on Salyut 4 involving a photometer and a number of spectrometers at the back of the station which scanned the Earth's horizon to measure the luminescence intensity of the red atomic oxygen spectral line to study processes taking place at a height of between 250 and 270 kilometres, the most Sun-sensitive part of the ionosphere where the electrons in the Earth's magnetic field interacted with the upper atmosphere. The photometer had been built by the Academy of Sciences. When combined with data gathered by geophysical stations on the ground, the results would assist forecasting of short-term changes in the upper part of the atmosphere and could ultimately facilitate reliable forecasting of global climate change.

Endeavour
A NASA Space Shuttle.

Energiya (Energy)
A heavy-lift rocket in the Saturn V class. Its first launch carried the Polyus platform, and its second the Buran space shuttle. Dr Boris Gubanov, the chief designer of the Energiya Bureau, which was responsible for its development, said that it was to be used to launch the building blocks of Mir 2.

Energomash
The manufacturer of the engines for the *Semyorka* rocket.

Epitaxy
A process for growing the crystalline structure on the surface of an existing crystal in such a way that the two lattices are aligned.

Epsilon
Contained in Kvant 2, this was an experiment to assess the thermal protection of the complex.

Eotvos
A Hungarian experiment on Salyut 6 which used the Kristall furnace to produce semiconductors of gallium arsenide, indium antimonide and gallium antimonide, and a crystal of an alloy of gallium arsenide and chromium using the moving solvent method.

Era (Air)
A French technology experiment involving a 240-kilogram self-deploying articulated framework that was erected outside Mir. The experiment comprised a 0.6-metre-diameter compressed stack of 1-metre-long carbon fibre rods, a

support platform, a video unit and a control panel. After the platform had been affixed to the anchor, the stack was affixed to an arm on the platform set at an angle of 45 degrees to keep it away from the surface of the complex. Once a 50-pin umbilical had been connected, the articulated pin-jointed rods were supposed to spring open automatically over a four-second period to create 24 identical prisms forming a thick hexagon some 4 metres wide, but it did not do so. It was nudged a few times to try to shake it loose, but it refused to deploy. Rather than jettison it, the cosmonauts persisted and it eventually deployed. When fully unfolded, a set of accelerometers were attached to it to measure its vibration modes. It was subsequently concluded that the structure must have been locked in its folded position by water vapour which had frozen when exposed to vacuum. After the experiment was complete, the structure was jettisoned. This experiment was to evaluate one of the options under consideration by CNES to erect antennas in space, as part of its long-term planning for its Hermes mini-shuttle development programme.

Ercos
A French cassette left on Mir for six months to determine the effect on computer memory chips of exposure to cosmic rays.

Erdem
An experiment on Salyut 6 to study Mongolia, a vast, inaccessible and largely unexplored country; in this case, it was self-evident that space-based observation was the most cost-effective means of surveying. The KATE-140 and MKF-6M cameras (the latter coordinated with similarly equipped aircraft) were used to survey the Earth resources. The data was used to locate geological fault lines, compile a map of soil conditions, identify subterranean water deposits, and map and assess crops and forests. Appropriate Mongolian sites were photographed using hand-held cameras (for the on-going Biosfera experiment). These observations, together with spectroscopic data, were intended to provide an assessment of pasturelands, map the boundaries between fertile and arid terrain, assess water capacity of rivers, glaciers and snowfields, assess atmospheric pollution, and locate ancient geological ring structures (usually the result of meteor impacts).

Erdenet
A materials experiment on Salyut 6 which used the food heater to dissolve copper sulphite in water and then recrystallise it; the process was filmed. This was performed to investigate the processes of diffusion and redistribution of impurities in a microgravity environment. This test was undertaken as part of a study of exceptionally high quality cleaning materials.

ERI
An experiment on Mir to test methods of depositing galvanic coatings.

ERTS
The pioneering Earth-Resources Technology Satellite which demonstrated the multispectral imaging system later used by NASA's Landsat series.

ESA
The European Space Agency

ESEF
: This European Science Exposure Facility comprised four cassettes. Three were passive traps to accumulate ambient particulate debris, and the fourth was an active system that recorded the speed, mass and trajectory of each impact. Umbilicals were connected so that the clam-shell covers of the traps could be opened by remote control from within Mir. Each was to be opened for a specific period. One of the dust collectors was opened in October 1995 to trap debris from the Draconid meteor stream associated with Comet Giacobini-Zinner. They were closed while spacecraft were manoeuvring nearby, to preclude contamination. Analysis of the active experiment revealed that the complex passed through a stream of debris every ten hours or so, at which time it suffered 5,000 microscopic impacts in an interval of only one minute.

Etalon
: An exposure cassette fitted on Salyut 7 prior to launch, and retrieved in space. It contained a variety of different optical coatings, some of which had bubbled and peeled off.

Euphrates
: A Syrian experiment on Mir which involved visual, photographic and spectrographic surveys of Syrian territory to study atmospheric pollution, water deposits and mineral resources, and imaging by the KATE-140 camera to identify ancient sites.

Extinctia (Extinction)
: A joint Soviet-Czech experiment on Salyut 6 to investigate the micrometeoroid dust layer which exists at between 80 and 100 kilometres altitude by measuring the changes in the apparent brightness of stars as they dropped down behind the Earth's dark horizon.

FAI
: The Federation Aeronautique Internationale. Its rules required that a record be exceeded by 10 per cent in order to be considered officially broken.

Faza
: The AFM-2 multispectral radiometer on the Spektr module sampled in the 0.4–2.2 micrometre range to study the transition zones between Earth's surface and the atmosphere and the upper atmosphere and space.

FEK-7
: A cosmic ray detector which used photographic film to study nuclei of transuranic elements. It was hoped that it would prove the existence of Dirac monopoles, but as yet these hypothetical particles remain elusive. Such detectors were initially flown on the circumlunar Zond spacecraft and then on Salyut 1.

Fem
: An Earth-resources experiment on Mir which used the MKF-6MA in Kvant 2 to survey Austrian territory.

Feniks
: A spectrometer on the Spektr module to study the Earth's surface.

Ferma-1 (Beam)
A 20-kilogram structure comprising a set of lattice-and-pin frames of aluminium-titanium alloy. When fully erected, outside Salyut 7, the 15-metre-long girder had a square section, about half a metre on a side. It was developed by the Institute of Electrical Welding, in Kiev. It was a prototype of a structure being evaluated for later use in space. There were three deployment options: manual, semi-automatic and automatic. It was deployed automatically, and then retracted. A flat plate was attached to the top and several experiment packages mounted on this before it was extended again, this time manually. One instrument was the 3 mW LED laser which illuminated a sensor set up in a porthole to measure the stability of the girder. The data would help predict the behaviour of larger framework structures. Then with an improved version of the URI multipurpose toolkit, some of the pins were beam-welded to lock the girder. Later the entire package was disconnected and jettisoned. Later versions for Mir were named Rapana and Strombus.

Ferrit
A cassette outside Mir to investigate the change in structure of ferro-magnetic materials.

FFFT
A NASA experiment on Mir. The Forced-Flow Flamespread Test involved passing an airflow over a flame to study growth of the flame under such circumstances, flammability of samples of cellulose and polyethylene, and the ignition process.

Fialka-F
An apparatus on Mir to study the ultraviolet radiation in near-Earth space.

Filin-2 (Eagle-Owl)
A telescopic X-ray spectrometer on Salyut 4, mounted on the outside of the station. It was sensitive to wavelengths in the range 1–60 Angstroms. The telescope used by the spectrometer was boresighted with a 6-centimetre-diameter sight with a 1-degree field of view. It was essentially autonomous, so when it detected a signal all the operator had to do was use the sight to establish the location of the source with respect to the starfield. Since it could relay its results to Earth, it continued to be used while the station was unoccupied.

FM-107
A Fourier mass spectrometer on Salyut 7.

Fon
This studied the gaseous environment in the immediate vicinity of the station.

Foton (Photon)
A class of automated satellite for microgravity materials-processing, typically involving apparatus tested by cosmonauts (the Splav-2 and Zona furnaces, and the Kashtan electrophoresis unit have all been used). The first satellite was launched in April 1988. The Vostok spacecraft was used in order to be able to return the results to Earth, but because this did not incorporate solar panels, materials-processing apparatus had to be powered by storage batteries, which limited both payload and duration.

Freedom
: The orbital complex proposed by NASA to fulfil President Reagan's order to design and assemble a space station in low orbit.

Functionality
: An experiment on Salyut 6 to assess sensory and motor reactions to a range of acoustic and visual irritants introduced whilst performing a monitored mental task.

Gallar
: A furnace on Kvant 1, derived from the Korund-1M. It produced a variety of semiconductors, and some smelts ran for a week or more.

Gamma-1
: A multi-function biomedical test kit on Mir. Amongst other things, it monitored cardiovascular activity.

Gamma-2
: A multispectral television camera cluster mounted on the ASPG-M platform, to enable researchers on the ground to make remote-control observations of the Earth. It was used to assess the pollution levels in industrial zones, with the long-term goal of mapping the spread of pollution from large cities. It was also used on favourable ground passes to assess the ecological state of water basins, heavily forested areas and agricultural land to monitor seasonal variations in crop growth.

Gel
: An electrophoresis experiment on Salyut 7. It made biological gels (a gel is a twin-colloid) with a purity 100 per cent greater than could be achieved on Earth. It made a polycrylamide gel, which could then be used to render the synthesis of biologically active materials on Earth more efficient. A similar apparatus was used on Mir.

Gemini
: The two-seat spacecraft produced by NASA to develop rendezvous and docking, spacewalking, and long-duration experience in Earth orbit, during 1965 and 1966, preparatory to Apollo. The first US spacewalk was on Gemini 4, the first rendezvous was by Gemini 6 and 7, the first docking with an Agena target was by Gemini 8, and the longest flight (14 days in December 1965) was by Gemini 7, with Frank Borman and James Lovell. The Air Force adapted the design (in a configuration known, appropriately, as Blue Gemini) to operate with the MOL.

Genom (Genome)
: A biotechnology experiment on Salyut 7, involving the use of an electrophoresis unit to separate a DNA solution into different strands. The process was lit by an ultraviolet lamp and filmed.

Geoex-86
: Organised by Intercosmos; a remote sensing programme taking place within East Germany, for which imagery from Mir was correlated with that from aircraft, ground teams and Cosmos 1602.

GFZ-1
: GeoForschungsZentrum (GFZ) was a 22-centimetre-diameter, 20-kilogram

sphere incorporating 60 small laser retroreflectors. A geodetic satellite built by Kaiser-Threde in Germany and ejected from Mir, it was to provide detailed data on the Earth's gravitational field. Because its orbit was lower than all other geodetic satellites, it was hoped that over its two-year life it would permit the distribution of mass within the planet to be mapped.

Glasar-1

A telescope for ultraviolet spectrography in the wavelength range 1,150–1,350 Angstroms, with a resolution of 2 Angstroms. It had a 40-centimetre-diameter primary mirror, and a 1.3-degree field of view. Although it used an electronic image intensifier, exposures of up to 10 minutes were needed to record faint stars (about magnitude 17) on film. There was a small airlock in the rear transfer compartment of Kvant 1 to enable the cosmonauts to reload it with film. Exposed film was returned to Earth for processing. It was used to survey bright quasars, active galactic nuclei and stellar associations. It was developed jointly by the Byurakan Astrophysical Observatory in Armenia and the Swiss.

Glasar-2

An ultraviolet telescopic spectrograph on the Kristall module.

Globus (Globe)

The main control panel of Salyut 4; a navigational display that indicated the station's position as it travelled around the world.

Glovebox (GBX)

A NASA chamber on the Priroda module for experiments requiring physical isolation. It was fitted with television cameras to record visible results.

Glucometer

A biomedical device on Salyut 7 used to monitor carbohydrate exchange in cells. It showed that this process operated at a greatly reduced rate in microgravity, suggesting that cellular functioning might be disrupted during very long flights.

Granat

A spectrometer on the Kristall module used to study the spatial characteristics and energy spectrum of radiation.

Greben

An experiment on the Priroda module; an altimeter to measure mean sea level directly below the complex's path to an accuracy of 10 centimetres.

Greenhouse

A NASA experiment conducted in the Svet cultivator to study plant reproduction, metabolism and biochemistry. It was first used to grow a strain of dwarf wheat. Shoots were extracted at regular intervals and frozen for return to Earth. They were 'harvested' when they produced seed, and some of these seeds were planted to make a second generation. It was the first time that a staple completed the growth cycle, and this success had important ramifications for food production and atmospheric processing on future stations. For this experiment, the Svet was enhanced by instrumentation from Utah State University to monitor light, temperature, air pressure, the level of carbon dioxide, water vapour, and substrate moisture.

Grif
A gamma-ray detector on the Spektr module to measure emissions from the complex, generated by its passage through the Earth's magnetic field.

Gyrodynes
This attitude-control system was tested on Salyut 3, used operationally on Salyut 5 and then built into Salyut 6, Salyut 7 and Mir's Kvant 1 and Kvant 2 modules. It used a set of electrically-driven flywheels on magnetic bearings, and converted electricity from the solar panels into torque to create an inertial attitude-control system that could reorientate the station without consuming propellant. It proved to be very responsive to commands. Although the system consumed power (each wheel was rated at 90 W) the propellant saving enhanced the station's sustainability. The two Mir modules each had six flywheels (each 165 kilograms) spinning in pairs, one pair on each cartesian axis. Once the desired orientation had been attained, the flywheels damped out vibrations from other equipment, to hold the station stable.

Gyunesh-84 (Sun)
Organised by Intercosmos, this was a photographic and spectrographic study of Azerbaijan. One of the targets was a remote-sensing test site. Data from Salyut 7 was correlated with that gathered by aircraft and ground teams. It produced maps of the most economical grazing pastures for livestock, assessed the extent of infestation in forest regions, and identified subsurface water deposits.

Halong (a bay on the Vietnamese coastline)
An experiment on Salyut 6. It used the Kristall furnace to produce an alloy of bismuth, tellurium and selenium and an alloy of bismuth-antimony-telluride (BAT), and to grow a monocrystal of gallium phosphide semiconductor.

Hautey
An experiment on Salyut 6 to investigate how yeast cells (a monocellular microorganism which grows extremely rapidly) divided as part of an ongoing study of intracellular processes.

Hermes
A French-sponsored project to develop a mini-shuttle for delivering crews to space stations and for servicing satellites in low orbit. In 1993 it was cancelled for financial reasons.

HEXE
The High-Energy X-ray Experiment incorporated a set of four Phoswich scintillators on Kvant 1, sensitive to X-rays in the 15–200 keV range, and with a 1.6 degree square field of view.

HTLPE
The High-Temperature Liquid-Phase Experiment sent to Mir by NASA.

Hohmann transfer orbit
The most energy-efficient trajectory to move between any two orbits.

Horizon-Dawn
An experiment on Salyut 6. It used the Spektr-15K spectrometer to monitor sunlight scattered by transmission through the upper atmosphere at low angles of

incidence immediately prior to dawn, and immediately after sunset, to measure the diffusion of various molecules and aerosols at an altitude of 100 kilometres.

HPM

A German experiment on Mir to study hormonal changes, involving the collection of blood, saliva and urine samples.

HSA

The temperature and humidity control system on Salyut 7. Once, when its NOK-3 condenser pump failed and Lebedev opened the panel to examine it, he found an enormous blob of water floating inside the compartment. After water had precipitated on the cold, porous surface of the dryer, it was supposed to be pumped to a tank in the regeneration system to be recycled, but the pump had failed and the water had accumulated. When the pump was disassembled it was found that the bearing had broken. Ironically, a month earlier, Lebedev, his hearing keenly attuned to the sounds of the various apparatus within the walls of the station, had warned the flight controllers that the pump's bearing was running rough, but they had assured him that there was no problem.

HSD

A German experiment on Mir devised by the University of Berlin. It employed the Tchibis suit to determine tissue layer thickness and compliance.

Igla (Needle)

A rendezvous radar transponder introduced by Salyut 6. As previously, it required the station to reorientate itself to face an approaching spacecraft. Once the spacecraft's final transfer orbit brought it within 25 kilometres of a station, Igla was activated. It provided data on range, closing rate, line-of-sight angular velocity, and perpendicular deviation from the straight-in approach vector. It automatically controlled the approach until it was 200 metres out, then paused so that the crew could perform a final check. At the 200-metre pause, it rolled the vehicle to align for the final phase. If everything was as it should be, it was standard practice for the cosmonauts to let a ferry perform an automated docking.

Ikar

An experiment on the Priroda module that was built jointly by Russia and Bulgaria. It comprised three sets of radiometers sampling at six wavelengths in the 0.3–6.0-centimetre microwave range. The Ikar-N radiometers looked straight down at the ground track, the Ikar-D scanning radiometers had an oblique view and Ikar-P offered a panoramic view. The oblique-looking radiometers viewed a swathe of track up to 750 kilometres wide.

Illuminator

This experiment recorded the degradation of the glass of Salyut 6's portholes due to exposure to space. The accumulation of particles on their external surface was recorded by taking photographs with the Pentacon-6M camera. Changes in their transmission characteristics were measured using the Spektr-15K spectrometer. During a spacewalk, one of the cosmonauts wiped the glass of one of the portholes with his glove to try to collect some of this dust for analysis (this smear was still clearly noticeable a year later), but noted that it seemed to be embedded in the glass rather than accumulated on its surface. The cosmonauts regularly examined the glass for any signs of impacts by micrometeoroids. The portholes had a double layer of glass, each layer 14 millimetres thick. A 4-millimetre-diameter crater was found in one porthole, and the micrometeoroid particle appeared to be embedded in the glass.

Illusion

A French experiment on Mir which used the Physalie-M apparatus to investigate how the sensory and motor systems adapt, in a continuation of a previous experiment.

Imitator

An experiment on Salyut 6 which measured the thermal profile of the Kristall furnace in its operating environment. This, together with the examination of the malfunctioning Kristall-2 unit, enabled the designers to develop an optimised smelting procedure that would yield extremely homogeneous crystals.

Immunity

An experiment on Salyut 6 to investigate changes in the proteins and minerals directly related to the body's immune system by determining antibody and immunoglobin levels in blood.

Immunology

A French experiment on Mir which studied the characteristics of the immune system.

Impulse

An apparatus on Salyut 4 to assess the state of the vestibular apparatus by measuring the threshold of sensitivity of the vestibular system to different stimuli (brain impulses which create the illusion of a yawing or banking motion in the absence of gravity). It was also on Salyut 5.

Inquiry

A psychological test.

Inkubator-1

An apparatus on Salyut 6 to study embryo development. Quail eggs developed much more slowly than those on the ground.

Inkubator-2

A Czech-supplied apparatus on Kvant 2 to further the study of embryonic growth. The quails that hatched after two weeks were the first Earth-based creatures to be born in space. After another two weeks, it was evident that the chicks were not developing properly; they were extremely frail, and appeared to be unable to feed themselves from the food tray in their habitat. Despite this, Dr Ganna Maleshko, of the Institute of Medical Biology in Moscow, reported that they had demonstrated that the embryo had developed normally; this was a significant step forward in the study of adaptation to weightlessness. Such experiments were a vital precursor to the development of closed-environment spaceborne habitats, because fowl could eventually contribute a valuable source of food. Subsequent investigations studied the post-hatching development process.

Inspektor
   A German manoeuvrable sub-satellite sent to Mir for trials.
Intercosmos
   A research organisation founded by the Soviet Academy of Sciences. In 1976 a number of Socialist countries signed an accord with the Soviet Union each to fly a cosmonaut to a Salyut station under this banner.
Interferon
   An experiment on Salyut 6 to test the possibility of manufacturing interferon (a chemical produced by human cells which interferes with the development of viruses) on a commercial scale in space. One part of this experiment involved injecting a vial of human white corpuscles into an interferon-producing substance to determine whether the rate of production was increased in microgravity. A related blood analysis test determined whether weightlessness affected the generation of interferon within the body. Another test evaluated whether existing interferon pharmaceutical preparations, which had been delivered both in the liquid state and in lyophillised gel, were rendered more or less effective against viruses by being exposed to weightlessness. Subsequently, attempts were made to to determine the capacity of human lymphocytes to synthesise interferon.
Iskra-2 (Spark)
   A satellite released from Salyut 7. This was the first time that a manned station in Earth orbit had released a satellite (although satellites had been left in lunar orbit by Apollo missions). The 28-kilogram package had been built by students of the Moscow Institute for Aviation. Its hexagonal structure incorporated a simple repeater designed to relay communications between radio hams across the Soviet Union. Its surface was embossed with the emblems of the countries participating in the Intercosmos project. It had no propulsion, and was simply ejected from the scientific airlock by a spring and left to drift away from the station. Prior to deployment, the cosmonauts tested its battery and transmitters. It deployed its two side antennas immediately it cleared the airlock. Its orbit rapidly decayed and it re-entered the atmosphere about two months later.
Iskra-3
   The second amateur radio satellite deployed from Salyut 7. Its orbit decayed within a month.
Isparitel (Evaporator)
   A 24-kilogram electron-beam gun designed by the Ukrainian Academy of Science's Institute of Electrical Welding in Kiev, used in the Vaporiser experiment to deposit thin coatings. It was tested on Salyut 6 and Salyut 7, and was set up in the scientific airlock. Exposures could range between 1 second and 10 minutes, depending on the thickness required. At its finest setting, it could deposit a layer of just a few micrometres; at its thickest about 1 millimetre. After some teething problems, this low-voltage electron beam was used to melt tiny granules of either aluminium or silver so that they could be vaporised and sprayed to condense on small disks of carbon and titanium, to determine whether such coatings would be a practical way of protecting stations from the deteriorating effects of the space

environment. This test followed directly from the successful respraying of the main mirror of the OST-1 on Salyut 4.

Isparitel-M
Apparatus on Salyut 7 to extend the Vaporiser experiment and further investigate the deposition of thin-layer coatings.

Istok
A long-duration exposure cassette containing connectors and bolts.

Istok-1
An experiment on the Priroda module, built jointly by Russia, Poland, Romania and the Czech Republic to study the oceans. Its infrared radiometer sampled in 64 channels between 1.6 and 3.6 micrometres. It incorporated a television camera sampling in the range 0.4–0.75 micrometres.

ISX
A German experiment on Mir. It involved wearing knee restraints while performing calf exercises to evaluate the effect of isometric exercises on muscles, blood pressure and heart rate.

ITS-7D
An infrared spectrometer mounted on the ASPG-M platform. It sampled in the 4–16 micrometre range, and was used to study the transition zones between the Earth's surface and the lower atmosphere and the upper atmosphere and space.

ITS-K
A telescopic infrared spectrometer carried on Salyut 4 and Salyut 5. It could be used to observe the sky, but its primary objective was the Earth's atmosphere. The infrared sensors carried by previous stations had used conventional compressor-based cooling units, which had used large amounts of electricity and been unreliable. In this case, however, a far more advanced cryogenic system was employed which used an 'ice coating' of solid nitrogen. This had been designed by the Kharkov Physical-Technical Institute of Low Temperatures, in the Ukraine. Not only did this draw far less power, it could be run at $-223°C$ for long times. The telescope employed a 30-centimetre-diameter mirror and projected a 20-arcmin field of view onto the slit of the spectrometer. It incorporated a fluorite prism which selected radiation in the range 1–7 micrometres to yield a spectrum with a resolution of 600 lines per millimetre. With the slit parallel to the horizon at sunrise and sunset it was used to measure the temperature of the layers of the upper atmosphere and to determine the density and distribution of water vapour (using the 2.7 micrometre absorption feature). It was so sensitive that it could detect minute changes in the concentrations of water vapour in the upper atmosphere. To make such observations the spectrometer was aligned exactly on the centre of the Sun's disk, and measurements were taken as the Sun rose or set behind the horizon (the slit of the spectrometer was aligned parallel to the horizon). It was a fleeting opportunity, but the automatic attitude control system was able to add data regularly as the station orbited the Earth every 92 minutes. At such times, it was possible to measure sunlight passing through a 'layer' of air up to 1,000 kilometres long. The infrared spectrometer measured water vapour absorption. The general level of understanding of the conditions and processes

operating in the upper atmosphere were so basic that until the matter could be resolved, two competing models were to be tested by the observations. One model assumed a dry atmosphere, and the other assumed a moist atmosphere. The difference in the levels of possible moisture concentrations according to these two models was of the order of between 200–300 per cent. Although the data from Salyut 4 argued in favour of the dry model, the observations were only preliminary, applying only to a narrow range of locations and times, and later studies established that the temperature of the upper atmosphere ranged between 400°C and 1,700°C because of heating resulting from the absorption of most of the Sun's ultraviolet radiation. Such data were important for achieving an understanding of global weather systems and long-term climatic changes. It was concluded that nitrogen oxide, formed in the disintegration of the flow of corpuscular particles from the solar wind, was the main source of infrared emission from the upper layers of the atmosphere. It also helped to define the energy spectrum of the Sun across the infrared range (absorption by water vapour in the lower atmosphere prevents this being done from the ground), refined the value of the solar constant (the total amount of energy emitted by the Sun) and enabled the number of carbon monoxide molecules in the solar corona to be determined (it is thought that this plays a major part in the process which heats the corona to one million degrees).

Izvestia (News)
A newspaper in the Soviet Union.

Kaliningrad
The flight control centre (known as TsUP, pronounced "soup") in a north-eastern Moscow suburb, superseding Yevpatoria. It was first used in 1973, and recently named the Sergei Korolev Control Centre.

KAP-350
An advanced topographic mapping camera in Kvant 2.

Kapillyar (Capillary)
An experiment on Salyut 6 with the Kristall furnace. A crystal of germanium was grown by using capillary action within a molybdenum matrix. One novel test grew a disk-shaped monocrystal of silicon to evaluate the feasibility of producing solar cells in space during the construction of large power systems in orbit.

Kardiokassett
An experiment on Salyut 6 to evaluate a device to measure cardiovascular parameters for the right-hand part of the heart.

Kardiolider (Cardioleader)
A Polish apparatus on Salyut 6 designed to record cardiovascular reactions via one chest sensor which monitored cardiac strain and another which monitored the speed of the ergonometer. By matching the level of exertion to the performance of the heart a cosmonaut could be given warning if the exercise was excessive, and this would enable an optimum programme of exercises to be developed for future flights.

Kaskad (Cascade)
An automated attitude control system tested by Salyut 4. Orientation was

measured with respect to infrared horizon sensors, which established the local vertical, and the Neytral velocity-vector sensor which determined the angle of the station with respect to its direction of travel. It greatly assisted observational work (previous crews had spent up to 30 per cent of their time on routine tasks related to orientating earlier stations), and also resulted in "a considerable reduction in fuel consumption". It was made operational on Salyut 6. When refined for Salyut 7, it could orientate the station to within 1 degree (an order of magnitude better than before), and was considered sufficiently reliable to permit the station to be manoeuvred without crew supervision.

Kasyun
An experiment on Mir which made use of the Kristallisator furnace to smelt an aluminium-nickel alloy.

KATE-140
A 140-millimetre focal length large-format topographic mapping camera. From a 350-kilometre orbit, pointed straight down at the ground, its 85-degree field of view could photograph an area of 450 by 450 kilometres with a resolution of 50 metres. It operated in both the visual and infrared bands. The film cassette held 600 pictures. It was a stereoscopic topographical mapping camera, and it could give individual frames or create an extended strip-image. It could be operated either by the crew or by remote control from the ground. It was carried on Salyut 6 and Salyut 7, and although the one on Salyut 7 was transferred to Mir, a second was soon sent up.

KATE-500
A multispectral camera flown on Salyut 4. It had 500-millimetre focal length.

Kazbek-Y
The couches in the Soyuz ferry. Each cosmonaut has a personalised contoured couch liner. When a cosmonaut moves from one ferry to another, the liner and the Sokol pressure suit are transferred.

KFA-1000
A film camera with a ground resolution of 5 metres.

KFV
A German experiment on Mir that used the applied potential tomography (APT) apparatus and the Tchibis suit to investigate changes in the distribution and flow of body fluids. The APT measured the distribution of fluids in different parts of the body.

KGA-1
A holographic camera flown up to Salyut 6 to further study the deterioration of the portholes. This 5-kilogram device used a helium-neon laser to create a series of holograms. It was tested by recording a dissolving salt crystal, to reveal how the density of the crystal was distributed through dissolution without convective flaws. The imaging process was extremely sensitive to vibration interference; on Earth, it required an enormous supporting structure to eliminate interference. The laser system had been developed by the Physical-Technical Institute in Leningrad. In the future, holographic cameras would enable engine performance and other processes which are otherwise difficult to monitor to be investigated.

KGA-2
A holographic camera on Salyut 7.

Khrunichev
The manufacturer of the Soyuz, Progress, Almaz, Salyut, Mir and TKS-based spacecraft.

Kinesigraph
A French experiment on Mir to make stereoscopic images with which to investigate the restitution of movement of the corporal segment.

KL
Two television cameras (KL-103 and KL-140) on the ASPG-M scan platform, used to indicate to the remote operator what the primary instruments were looking at.

Klimet-Rubidium
A materials experiment on Mir which used the Kristallisator furnace using a mixture of rubidium, silver and iodine to test a concept for extremely lightweight batteries and condensers. Because this experiment consumed so much power, it was performed while the cosmonauts slept.

Kolosok
An experiment on Mir to investigate aerosol structure in zero gravity.

Koltso (Ring)
A walk-around communications system in Salyut 6 which cosmonauts could operate without having to return to the control panel to speak to the control centre.

Kometa
An experiment on Salyut 7. It involved making observations designed to improve a mathematical model for reproducing the colour of the sea as viewed through the atmosphere so that more accurate colour-based measurements would be able to be made in the future.

Komplast
A cassette outside Salyut 7. It contained non-composite materials whose physical properties were to be tested after prolonged exposure to the space environment.

Komza
A Swiss-Russian experiment on the Spektr module to sample gas in the immediate vicinity of the station. It employed interchangeable cassettes.

Kondor
A Canadian experiment on Mir designed to develop a means of controlling the radiation within the complex.

Kontrol
An experiment to test the structural stability of the docking collar on which Kvant 2 was mounted.

Konus (Cone)
The conical drogue in the standard probe/drogue docking system. The unit hinged into the station's docking compartment. The 2.2-metre-diameter spheroidal multiple docking adapter on Mir was too cramped to accommodate five drogues, so, since it was not operationally necessary for them all to be so equipped, only

two were installed, and the other ports had flat covers. One drogue was kept on the axial port, where vehicles docked, and the other was placed as necessary to facilitate the movement of modules onto the radial ports. (In some cases, moving the drogue required depressurising the compartment, and this was referred to as an 'internal spacewalk'.)

Korabl-Sputnik
Any spacecraft incorporating a cabin for a crew.

Korolev Bureau
The rocket design bureau (OKB-1) established by Sergei Korolev.

Korolev Control Centre
The control centre at Kaliningrad, renamed after Sergei Korolev by President Boris Yeltsin.

Korund
A 200-kilogram electric furnace on Salyut 7. This incorporated a revolving sample holder which could be loaded with up to a dozen ampoules. Although each sample was processed singly, it could be set up to operate either automatically or by remote control from the ground. This meant that a departing crew could prepare it so that it could run after they had gone (when they would no longer be around to vibrate the station). It would also be possible to set up such a semi-automatic system in a free-flying crew-tended module. In contrast to the earlier Splav apparatus, in which the sample was fixed and the thermal field was varied to control the growth of a crystal, and Kristall, in which the sample was moved while the thermal field was held constant, the Korund furnace was capable of moving the sample (in an ampoule 25 millimetres in diameter and 30 centimetres in length) at speeds varying from a few millimetres a day to a few centimetres a minute, and could vary the temperature between $20°C$ and $1,270°C$ at rates of $0.1°C$ to $10°C$ per minute, to produce a monocrystal of up to 1.5 kilograms. With a fully loaded rotating table of samples, it was able to create 18 kilograms of monocrystal without human supervision. It was hoped that the second generation technology would demonstrate the feasibility of an orbital manufacturing facility for extremely pure semiconductors which could be used either on an unmanned spacecraft or a later station. Unfortunately, it switched itself off as soon as it attained its operating temperature. The cosmonauts realised that the thermal coefficients of the heaters had been defined by ground tests, where thermal convection had drawn off energy, but in microgravity, where convection does not develop, the elements had heated up much more rapidly, leading the control system to believe that it was overheating. Once reconfigured with more appropriate coefficients it was found to be usable, so they started a crystal-production run to test it. Then the outer compartment overheated (to $100°C$) due to a design flaw. Because it was such an important apparatus, however, the cosmonauts figured out a way to repair it, but even so, it could not be used for more than a few hours at a time. Later, the mechanism that moved the ampoules within the thermal chamber jammed and its motor burnt out.

Korund-1M
A 136-kilogram semi-industrial-scale furnace on Mir, for crystal growth. It was an

improved form of the apparatus tested on Salyut 7. The samples were processed individually for periods ranging between six and 150 hours. Once set up, it was completely automatic, but because it consumed 1 kW its use had to be limited.

KR-5

A spectrometer on the Spektr module.

Krasnaya Zvezda (Red Star)

The newspaper of the armed forces of the Soviet Union.

Krater

A furnace (strictly speaking a melting zone) on the Kristall module used to study crystallisation of a high critical-temperature superconductor, to produce epitaxial silicon and monocrystals of exotic alloy (such as barium oxide, yttrium oxide and copper oxide) and semiconductors (most notably gallium arsenide). Many smelts ran for ten days.

Kristall (Crystal)

The 'technology' expansion module of the Mir complex. It had a pressurised volume of 60 cubic metres and two solar panels with a total area of 72 square metres which provided 9 kW. Its solar panels were retractable, and could be extended out to a maximum 36-metre span. Its primary scientific payload was a bank of furnaces, but it also contained Earth-observation apparatus.

Kristall

An experiment on Salyut 5 to study how crystals grow in microgravity conditions. Seed crystals were placed in containers with a water solution of potash and alum. Once the crystals had matured, they were extracted for return to Earth. It was hoped to produce much larger monocrystals than was possible on Earth. The first experiment ran for three weeks. In the second, a dyeing agent was added to the solution of potash alum to study the diffusion of the mixture into the developing crystalline structure.

Kristall-1

An electric furnace used to create monocrystals of semiconductors. Unlike Splav, which it superseded, it could be operated within the station (set up in the rear transfer compartment). An improved version of the Splav, it used zone-melting to crystallise semiconductors. Whereas Splav created a temperature gradient to facilitate crystallisation, Kristall employed uniform heating and exposed a sample to a steady-state thermal zone at a temperature in the range 400-1,200°C. It could operate in four different ways: the first method produced monocrystals from the gaseous phase by sublimation, which evaporated the sample and then transported the gas to the cooling zone so that it could settle on and enhance the seed; a second produced films of monocrystal using chemical gas-transportation; a third produced monocrystals using a 'moving solvent' to obtain a high-temperature solution; and the fourth method employed a seeding technique in which as the temperature was slowly reduced in the crystallisation zone one side of the sample was cooled while the other side was kept hot. This yielded a seed on the cold side of the sample, and as it was slowly stretched across the zone, incremental crystallisation produced a single, highly regular crystal. Also, because the temperature could be controlled more accurately than in Splav, the crystals were considerably more homogeneous. The samples were carried in capsules which were 10 millimetres in diameter and 175 millimetres long, and could be moved through the heating chamber at rates between 0.188 and 0.376 millimetres per minute. Although this was the first time that this apparatus had been flown, when it was tested it produced a monocrystal of gallium arsenide employing the high-temperature solution technique (this semiconductor could be used in the construction of highly efficient solar collectors). It was tested on Salyut 6, and worked well, but broke down after successfully performing 40 smelting operations.

Kristall-2

Flown up to Salyut 6 to replace the earlier model.

Kristall-3

Flown up to Salyut 6.

Kristallisator

A Czech-built automated semiconductor furnace on Salyut 7. It could maintain heat and pressure constant at temperatures up to 1,000°C, to grow crystals in smelts lasting from several hours to several days. The unit on Salyut 7 was transferred to Mir, and then used for a series of experiments to investigate the process of crystallisation in silver-germanium and lead chloride-silver chloride, which are eutectic alloys (that is, they have an extremely low freezing point).

KRT-10

A 10-metre-wide radio telescope. The 350-kilogram unit comprised the antenna and a device to attach it to the rear docking unit of Salyut 6 in 1979. It was a hexagonal structure of rods deployed by springs. By using shorter rods in the centre and longer rods at the edge, it formed a parabolic dish. A fine mesh over the surface formed the reflector, and a 5-metre-tall triangular mount supported the five radiometers (four horns in the 12-centimetre band and a spiral antenna operating in the 72-centimetre band). It was attached to the rear docking assembly, projecting, furled, within the Progress ferry, so that it was exposed when the ferry withdrew, at which point it was commanded to deploy. This action was televised by the ferry. As it unfurled, it completely hid the view of the station behind. A video tape had been delivered showing the crew how to operate it. They monitored its reaction to a series of pointing commands. On alternate days (to permit the batteries to be recharged) over the ensuing weeks, it was to be used to carry out a series of pioneering observations. It was to be used both to observe the Earth and celestial sources. When aimed down, it could reveal the structure of meteorological phenomena, map sea state, and measure water salinity and soil humidity, but the 12-centimetre band was limited to 7,000-metre resolution. For the astrophysical part of its programme, it was to work in conjunction with the new 70-metre ground-based radio telescope of the long-range space communications centre in the Crimea (together creating an interferometer with a diameter as wide as the Earth). It gave mixed results, but it was seen as being just an early test of the use of an orbital telescope, and so the very act of trying to use it generated valuable design feedback. For the deep sky observations it had been used in two modes: in one mode the station had been stabilised so that the telescope gave continuous coverage of a given source, and in the other mode it was set in a rotation which had been orientated so that the telescope would scan the Milky Way, so that it could map sources. When aimed at the Earth, it demonstrated that such an instrument would be able to provide useful geological and oceanic data. It was also used to monitor an eruption by Mount Etna. The dish was then jettisoned to clear the rear docking hatch, but, as it slowly drifted away, it began to wobble and became entangled in the apparatus which projected from the back of the station. The mesh could be seen clearly by the television camera on the docking unit. At the suggestion of ground controllers, the station's main engines were fired to try to blast the dish free and leave it behind, but this failed to dislodge it. The next suggestion was that the station should be pitched up and down to shake it free, but this did not work either. Eventually, Vladimir Lyakhov and Valeri Ryumin spacewalked to release it. Although the task involved only four quick snips with the pliers, it was a potentially dangerous operation because Ryumin had to crawl down between the dish and the rear of the station. As a general rule, mission planners regard any item not specifically built to be manipulated by spacewalking cosmonauts as a potential death trap, and Ryumin was reminded of this when he cut a strand of the entangled mesh away and discovered that he had set the heavy structure rotating towards him. Finally, after he had completely cut it free, he pushed it away from the station.

KSS-2
A photographic spectrograph carried on Soyuz 13 and Salyut 4. This 2-channel fast-operating spectrometer was aimed at the centre of the solar disk, its slit parallel to the Earth's horizon, at the day and twilight horizons, to measure the strength and distribution of atmospheric pollutants and the vertical distribution of water vapour in the atmosphere. It could also be used to measure the reflectance spectrum of the Earth's surface under direct insolation and distinguish soils with different moisture content.

KTF
The Kompleksny Trenazher Fisichesky was an integrated physical trainer based around an electrically driven moving belt treadmill almost a metre long by half a metre wide. It was used in conjunction with a customised harness using adjustable elasticated cords which attached to the framework of the track in order to provide a hold-down load of around 50 kilograms. KTF-1 was installed on Salyut 1.

Kurs (Course)
The advanced rendezvous radar transponder introduced by Mir, which did not need the entire complex to continually reorientate itself to face an approaching spacecraft; the spacecraft need only be within the transponder's wide field of view. It operated in the S-Band.

Kursk-85
A survey of the Kursk Oblast region. Salyut 7's MKF-6M and KATE-140 data were correlated with the Meteor-Priroda satellite and aircraft and ground teams.

Kvant 1 (Quantum)
An expansion module for the Mir complex. It was delivered by a TKS-based tug and docked at the rear of the base block. It had a pressurised volume of 40 cubic

metres and contained the Vozdukh and Elektron regenerative life support systems. It had no power-generating capability of its own to run its astrophysical observatory.

Kvant 2

An expansion module for the Mir complex. It had a pressurised volume of 60 cubic metres and two solar panels with a total area of 50 square metres which provided 7 kW. It incorporated a large airlock, a Bania, a second ASU, a second Elektron life-support system, and the Vika oxygen production system (to replenish the airlock). Its primary scientific payload was for Earth observation (primarily the ASPG M and the MKF-6MA), but it also contained astrophysical apparatus.

Kyulong

A Vietnamese experiment on Salyut 6. It used the MKF-6M and KATE-140 cameras to carry out Earth-resources observations of the Vietnamese peninsula, to make the first comprehensive survey of its geological, botanical, mineral and coastal resources to yield information on tidal flooding, hydrological features of the Central Plateau and silting in the Mekong and Red River deltas, and assess the effects of defoliants sprayed during the war, and used the Spektr-15K spectrometer to assess the health of crops.

Labrint

An experiment on Mir which studied the interaction between the visual and vestibular systems in the early phase of adaptation to weightlessness.

Levka (Lion's Cub)

An experiment performed on Soyuz 13 which measured blood flow to the brain as it adapted to weightlessness. It involved a chest expander which imparted a force of 15 kilograms when stretched. A cosmonaut was to flex this at a rate of 30 times a minute while electrodes measured the way in which the flow of blood to the brain developed.

Levkoi (Gillyflower)

An apparatus on Salyut 4 which measured specific features of blood flow within the brain. Also on Salyut 5.

Lidar

A radar which uses a laser pulse rather than a radar pulse.

Ljappa

A mechanical arm carried by the modules that docked at the front of the Mir complex. It transferred the modules from the axial port, where they docked, to a permanent position on a lateral port. This involved performing a 45-degree rotation, so the axial collar was offset by that amount to ensure that the modules would be properly aligned once finally docked.

Logion

An Austrian experiment on Mir to study the characteristics of a liquid metal ion source to evaluate whether it could be used to remove the electrostatic charge that accumulates on a satellite in orbit. It was done to try to find a way of making a magnetospheric research satellite more sensitive, by discharging it so that its own charge would not interfere with extremely fine measurements.

Lotos
   An apparatus on Salyut 6 to mould shapes of structural elements from polyurethane foam.
Luch (Beam)
   A SDRN requiring three geostationary satellites to relay signals (voice, telemetry and real-time experiment data) between Mir and Kaliningrad to provide continuous communications coverage. However, the network was never completed, and more often than not the satellites were leased to relay commercial television.
Lungmon
   An Austrian experiment on Mir which tested a new electrical heart and lung monitoring unit.
LV-1
   The optical viewfinder on Salyut 7. It provided a 6-times magnified image, so was often used to observe features on the Earth in somewhat higher resolution than was possible with the naked eye.
Lyulin
   An experiment on Mir which used a microprocessor (Zora) to test psychological and physiological analysis of their reactions, in particular their reaction time.
Lyulin
   A dosimeter incorporating a microcomputer processor used to make measurements of the flux of ionising radiation, providing a high-resolution temporal profile around the complex's orbit.

Magma-F
   An electric furnace on Salyut 7. It was an improved form of the Kristall furnace used on Salyut 6. An accelerometer and a magnetic sensor had been added to measure vibrations affecting the melting and crystallisation processes, as vibrations tended to disrupt the process. It could accommodate quartz ampoules up to 20 millimetres in diameter and 200 millimetres long, could heat samples to 900°C and continuously monitor its temperature at 14 points, and send this data to Earth as part of the telemetry stream. To eliminate perturbations caused by the cosmonauts moving about inside the station, it was found to give better results if it was unbolted and then left to float freely. At one point, cadmium theioselinide (a material with important laser and communications applications) was produced.
Magnetobiostat
   A plant apparatus on Salyut 6. Popov and Ryumin had made a nationwide appeal for suggestions for better ways of growing plants in the absence of gravity; the use of a magnetic field had been thought an excellent idea, so the necessary apparatus had been put together. Arabidopsis, crepis and ginseng grew successfully. A similar unit was installed on Mir.
MAK-1 (Poppy)
   A small (16-kilogram) satellite equipped to study the upper atmosphere and radio its data direct to Earth. It was ejected from Mir's scientific airlock. It was

meant to unfurl a wide dish antenna immediately, but did not do so, probably because its battery was flat. It had been intended to carry out a three-day programme. As the first in the series, it had two objectives: firstly, to verify the electrical, temperature regulation and telemetry systems, and secondly, to run the Fokus payload to study phenomena arising from electrons and plasmas at orbital altitude.

MAK-2

A small satellite ejected from Mir's scientific airlock to investigate the physical characteristics of the ionosphere.

Malakhit

A greenhouse apparatus on Salyut 6. It used a synthetic soil incorporating capsules which released ion-exchange fertiliser at a controlled rate. It was used to investigate the growth of orchids which, because these were adapted to a dry atmosphere, it was hoped might prove more able to survive in the station's artificial environment. Unfortunately, even though the orchids had already started to flower when they were delivered, they deteriorated immediately and the flowers disintegrated, but once back on Earth they reflowered. Orchids planted in space developed roots which were long enough to protrude from the apparatus, but they did not flower. In a similar unit on Salyut 7, tomatoes, coriander, celantro, radishes, borage and cucumbers were all planted. Unfortunately, the borage and radish did not grow at all; the celantro grew to a height of 7 centimetres and then died; only the cucumbers thrived.

Marina-2

A gamma-ray telescopic spectrometer on the Kristall module.

Mariya

A magnetic spectrometer to study the mechanisms by which flows of high-energy particles, X-rays and gamma rays are generated in near-Earth space. It was first used on Salyut 7, and an improved form was installed on Mir. Its data was used to evaluate a correlation between concentrations of charged particles streaming in the Earth's magnetic field in space above areas of the Earth's surface subjected to seismic activity. Subsequent terrestrial studies revealed that magnetic fields are indeed associated with tectonically active rock.

Mariya-2

A magnetic spectrometer on the Kristall module.

Mars

A visual test performed on Salyut 7 to assess the ability to distinguish subtle colour variations. (It was planned to employ colour charts to calibrate observations for some experiments).

Maskat

An experiment on Mir which tested the ability of given compounds to enrich genetic material.

Medilab

A French experiment on Mir which involved taking a series of blood and urine samples to test for hormonal changes.

Medusa
: A cassette attached to Salyut 6 to expose materials to the space environment. It involved a number of biological samples, half of which were attached to the outside of the station, whilst the rest were inside. The external cassettes held quartz ampoules with various mixtures of amino acids and other 'building blocks' of life. Mounting them outside exposed them to the full effect of solar and cosmic radiation in the station's orbital path. These would eventually be retrieved and compared with containers of biopolymers (microcultures which make up any living organism) kept within the station. The experiment, related to 'the problem of the origin of life in the Universe,' was designed to identify any functional changes which an elementary living culture undergoes as a result of being exposed to the ambient space environment over a prolonged period. Analysis demonstrated that exposure to sunlight caused components of nucleic acids to develop into substances similar to nucleotides; a step in the process towards creating nucleic acids. A similar biopolymer cassette was placed outside Salyut 7 to further study the effect of the radiation environment on the chemical processes fundamental to the creation of life.

MEEP
: The four exposure cassettes of NASA's Mir Environmental Effects Package were affixed to the DM by spacewalking astronauts. Two of the cassettes sampled ambient particulate matter (one was passive, but the active one recorded the time, size and trajectory of all particles hitting it). The other two exposed construction materials (including insulation, paints, and optical glass coatings) to be used on the International Space Station.

MEFCE
: A NASA experiment on Mir. The Mir Electric Field Characterisation Experiment was set up to sample ambient emissions in the 400 MHz to 18 GHz radio range.

Membrane
: A biomedical experiment on Salyut 7 to study calcium loss in bone by measuring the exchange of calcium in cell membranes. It revealed that taking an anti-oxidant preparation could significantly reduce this calcium metabolism.

Mera
: A long-range radio-technical system on Salyut 7, to assist approaching spacecraft. On its first test, although it was switched on at a distance of 250 kilometres, the scanner did not lock on until the separation was down to 30 kilometres. Nevertheless, it led the ferry in until the Igla could take over.

Merkur (Mercury)
: The three-seat descent module developed for Almaz and its associated TKS ferry. It had an Apollo-style blunt-cone configuration. There was a hatch in the side for entry, and a hatch through the heat shield in the base to gain access to the parent vehicle behind. After separation, thrusters on its nose orientated it for its de-orbit burn; releasing the tower exposed the parachute compartment at its base. Several capsules were launched (in 1976, 1978 and 1979) to verify the re-entry control system, and those released by the TKS were successfully recovered (one of which

returned cargo from Salyut 7; it could return 500 kilograms of material, which was ten times the capacity of the Soyuz returning with a crew of two), but it was never flown by cosmonauts. The empty capsule had a mass of 2.5 tonnes.

Meteor
A class of meteorological satellite.

Meteor-Priroda (Nature)
A satellite launched in June 1980. Its instrumentation included a ten-channel multispectral camera (Fragment) which transmitted 2,000-kilometre long swath-images 30 kilometres wide, with a resolution varying between 30 metres and 800 metres. During the early 1980s, Earth observations by the Salyut crews were often correlated with this data.

MGAS
A NASA biomedical monitor on Mir to assess metabolic response to exercise. The Metabolic-Gas Analyser System measured inspired and expired air to yield data on protein metabolism. This was employed in conjunction with the NASA ergonometer. It had previously been flown on a life sciences Spacelab mission.

Microaccelerator
A French experiment on Mir which used a video camera to measure microscopic accelerations.

MIDAS
A NASA experiment on Mir. Materials In Devices As Superconductors was to investigate the electrical properties of high-temperature superconductors.

Migmas
An Austrian experiment on Mir which tested a mass spectrometer intended for use in microgravity. A similar device was under development for the European Space Agency's Columbus module as a materials analysis tool.

MIK
The building in which *Semyorka* rockets are integrated with Soyuz spacecraft and loaded on the rail transporter.

Mikrovib
An Austrian experiment on Mir which studied skin sensitivity by analysing the spontaneous and stimulated microvibrations of the body's surface in the absence of gravity.

MIM
The Microgravity Isolation Mount on Mir supplied by Canada. It employed a magnetic field to damp out ambient microaccelerations, and isolated apparatus from microvibrations in the range 0.01–100 Hz. It was placed in NASA's Glovebox so that residual vibrations transmitted onto the surface of a liquid could be videotaped for its initial trial. SAMS simultaneously recorded the ambient vibrations to assess the MIM's ability to filter them out. It was hoped that such an active filter would be able to offer a significantly better vibration environment for microgravity research.

Minidoza
An experiment on Salyut 6 to measure the flux and energy spectrum of cosmic rays.

368 Glossary

MIPS
  The Mir Interface-to-Payloads System used a portable computer to enable NASA's apparatus on Mir to download data via the station's telemetry link.

Mir (Peace, New World, Community)
  The 'base block' of the Mir orbital complex; a habitat module with permanent facilities for a crew of two, providing power generation and regulation, navigation and attitude control. It comprised three main compartments (the multiple docking adapter, the main compartment, and the rear transfer compartment), around which was the unpressurised engine compartment. It had a pressurised volume of 90 cubic metres, two solar panels with a total area of 76 square metres providing 9 kW, and a third panel of 22 square metres which was subsequently installed to bring the overall output to 11 kW. Over a 10-year period, it was expanded by the addition of five modules (Kvant 1, Kvant 2, Kristall, Spektr and Priroda).

Mir 2
  A follow-on to Mir, envisaged as an elaborate orbital complex comparable with NASA's Freedom space station concept.

MIRAS
  The French-Belgian Mir InfraRed Atmospheric Spectrometer was affixed to a platform at the far end of the Spektr module. It scanned the absorption lines in the atmosphere's infrared reflection spectrum, to identify its trace constituents, to map their distribution, and to monitor the evolution of the atmosphere over a period of at least a year, in order to help understand the interaction between solar illumination and the atmosphere. Its data (which was stored in memory and downloaded daily) was integrated with that from satellites and aircraft. An earlier version of this instrument had been used on the ATLAS-1 Space Shuttle mission in 1992.

Mirgen
  An Austrian experiment on Mir which used blood analysis to evaluate the effect of space flight on genetic material.

Mirror
  A mirror-beam furnace on Mir. It employed two small lamps and a special optical system to smelt small samples at temperatures up to 1,000°C. It was to be used to experiment with different alloy materials, and tested procedures for furnaces in the Kristall module.

MISDE
  A NASA test on Mir. The Mir Structural Dynamics Experiment used set of 31 portable and 22 fixed accelerometers and strain gauges were installed throughout the complex to measure the transient stresses resulting from manoeuvres and docking operations, and from thermal effects due to repeatedly flying in and out of the Earth's shadow, to further characterise the microgravity environment.

MKF-6
  A multispectral camera, flown on Soyuz 22 in 1976 to test it prior to installation on Salyut 6. The 200-kilogram package was mounted on the front of the orbital module instead of on the docking unit. It had been built by Carl Zeiss, at its factory in Jena in East Germany. From the 250-kilometre orbit, the camera was

able to image an area 110 by 150 kilometres with 15-metre ground resolution. By testing the camera in this way it would be possible to identify any improvements which would be necessary before it was deployed operationally. When not in use, the camera's optics were protected by covers. According to Dr Heinz Kautzleben, Director of the Central Institute for Earth Physics in Berlin, the test version of the camera was capable of functioning in space for only a week. It produced 'photographs', each of which comprised six separate frames taken through filters selecting narrow wavelength bands between 0.46 and 0.86 micrometres (four visual and two in the infrared spectrum). A set of six film cassettes was needed to complete each photographic sequence across the full spectral range. After being developed on Earth, these images would be examined using a four-channel Zeiss MSP-4 multispectral projector. Unfortunately, because each cassette weighed 13 kilograms only two could be returned in the cramped Soyuz descent module, so the use of the camera had to be restricted in this test. Each cassette contained film for up to 1,200 images, however, so it would provide a comprehensive test of its capabilities. It was aimed vertically at the ground. Some observations were to be correlated with data collected simultaneously by Antonov-30 aircraft and ground teams in selected regions of the Soviet Union for the Raduga Earth-resources programme, to help interpret the imagery. Orbital multispectral photography offered a far more economic way of prospecting for scarce resources in the Earth's crust, but for its data to be interpreted its sensors had to be calibrated, so similar observations were made from aircraft flying at around 7,500 metres, and all of this overhead imagery had to be combined with ground observations of soil and atmospheric properties. An orbital platform can survey territory at a far greater rate than can a high-flying aircraft. In ten minutes an orbital camera can photograph 1 million square kilometres, a task which would take several years for an aircraft to complete.

MKF-6M

An improved form of the MKF-6, flown on Salyut 6. It was similar to the unit which had been tested on Soyuz 22, but upgraded to function for several years. At Salyut 6's higher altitude (350 kilometres) each photograph covered an area of 165 by 220 kilometres with a resolution of 20 metres. There was a 60 per cent overlap between adjacent pictures in a sequence, so stereoscopic views could be generated. Simultaneous photographs were taken in six spectral bands (specifically, 0.46–0.50, 0.52–0.56, 0.58–0.62, 0.64–0.68, 0.70–0.74 and 0.78–0.86 micrometres) on film in cassettes holding 1,200 frames. Because photographic film deteriorated rapidly in the increased radiation environment at orbital altitude, new cassettes of film would have to be flown up, and exposed film would have to be returned before it fogged over, so a series of ferries would have to visit the station on a regular basis. One of its major tasks was to survey the planned route of the Trans–Siberian railway, then being built from Baikal to Amur. During a four-month tour, one crew took 18,000 images (it would have taken two years of surveying from aircraft, or a lifetime of fieldwork, to perform the same survey on Earth). At the request of the Ministry of Agriculture, it regularly photographed the Salsky test site near Rostov, in the

Ukraine. This was one of four agricultural areas which had been set up to calibrate multispectral cameras on orbital stations (the other sites were at Voronezh, near Lake Baikal, and in Kirghizia). These had been planted with specific cereals, vegetables and grasses to assess the ability of such a camera to distinguish between various plants under different conditions. A similar apparatus was installed on Salyut 7.

MKF-6MA

Contained in Kvant 2, this multispectral camera was an improved version of MKF-6M.

MKS

An East German spectrometer flown on the Intercosmos 21 satellite.

MKS-M

An East German spectrometer-camera carried on Salyut 7. It could be pointed with greater freedom than the other fixed cameras (namely the MKF-6M and KATE-140), and could be fitted with different spectrometers for studying the atmosphere (A) or the biosphere (B). The -AS combination sensed 6 wavelengths in the 0.7–0.8 micrometre range. The -BS combination sensed 12 wavelengths in the 0.4–0.9 micrometre range, selected to detect chlorophyll.

MKS-M2

A photoelectric spectrometer mounted on the ASPG-M platform on Kvant 2. It measured visible and near-infrared wavelengths, and was used to study the transition zone between the Earth's surface and the lower atmosphere. It was remotely operated and its results were transmitted to Earth.

MMK-1

A number of micrometeoroid detection panels with a total area of 4 square metres were preset on the outside of Salyut 4. Each panel incorporated two layers of metal separated by a thin layer of insulation and formed a capacitor. When a micrometeoroid particle hit the outer layer it deformed it and generated a brief electric current which could be counted. The current indicated the size of the particle and its penetrating power. It could detect a particle as small as a billionth of a gramme.

MOL

The US Air Force developed a Manned Orbiting Laboratory to serve as a reconnaissance platform. It was to be launched on a Titan-IIIM with a Blue Gemini mounted on top. In orbit, the crew would open a hatch through the heat shield and enter the 4-metre-long, 3-metre-wide pressurised module. It was to have carried crew-operated telescopic cameras. The maximum mission capability was one month. There was no provision for replenishment or crew exchange, so the platform was to be discarded when the crew returned to Earth in their spacecraft. Serious preliminary studies began in 1964, firmed up in 1965 and fixed in 1967. A target date of 1968 was set for an unmanned test of the entire system, and 1969 for the initial operational flight. Development was slow, however, and only the test of the modified spacecraft had been made to verify the integrity of the hatch during re-entry when the project was cancelled in 1969, partly for financial reasons, but mainly because it had become clear that the reconnaissance systems on the automated Big Bird spy satellites would offer an equivalent capability and be both cheaper and more flexible.

Molniya (Lightning)
A class of communications satellite placed in highly elliptical orbits ranging between 460 and 39,350 kilometres at 65 degrees so that they would spend most of their time far above the Soviet Union, serving as relays.

MOMS-2P
A German experiment on Mir; an electro-optical multispectral imager. It had previously been flown on the Spacelab-D2 shuttle mission, and could be operated in various modes, including high resolution surface-relief topography. It was mounted on a scan platform on the Priroda module, and carried its own GPS-based package so that the position of the complex (within about 5 metres) and its orientation (to within 10 arcsec) could be assigned to each image.

Monimir
An Austrian experiment on Mir which analysed postural reflexes to assist in the development of a computerised neurological analyser. It used the Motomir apparatus to investigate the movement of the subject's head and arms.

Monitoring
A test by acoustic sensors to assess the structural stability of the docking port on which Kvant 2 was mounted; if the hermetic integrity was lost, the entire complex would depressurise.

Morova
A Czech-designed materials processing experiment on Salyut 6. It was performed in the Splav furnace over a two-day period, and was designed to evaluate the possibility of producing rare metals and alloys in microgravity. Small quartz ampules containing a variety of silver-lead chlorides and copper-lead chlorides had been prepared by the Czechoslovak Academy of Sciences. These were processed to produce extremely pure crystals and semiconductors. Other samples combined glass and metal to create new materials with "certain electro-optical properties." Hardened melts of crystalline and glass-forming materials were far more homogeneous than those which could be produced on Earth. In one case, lead chloride and copper chloride was held at 500°C for 20 hours and then cooled at 10°C per hour. Although this produced lead crystals which were both bigger and more homogeneous than possible on Earth, some displayed physical deformities (such as a helical surface) caused by microaccelerations due to the operation of other equipment on the station.

Motomir
An Austrian experiment on Mir which involved neurophysiological analysis of human motorics. It used a large four-element ergometer to measure the force and velocity characteristics of the subject's arms and legs during defined movements.

Moz
An experiment on the Priroda module developed jointly by Russia and Germany. Its multispectral spectrometer sensed reflected solar insolation using 17 channels at visible and infrared wavelengths between 0.4 centimetre and 1.0 centimetre, looking straight down to study the oceans.

372 Glossary

MSRE
: A NASA experiment. The Mir Sample Return Experiment was for micrometeoroid research. It was placed outside Kvant 2.

MSS
: A photoelectric spectrometer on Salyut 7.

MSS-2
: A spectrometer on Salyut 7. It had a 7-degree field of view and observed the 0.4–0.8 micrometre range with a resolution of 1,200 lines per millimetre. Although its wide field was suitable for studying terrain features on the ground track, or the ocean, it was inappropriate for making horizon observations. It turned out that using the KATE-140 to image the site of the spectra was cumbersome.

MSS-2M
: An improved form of the MSS-2 that both showed its output visually to the cosmonauts and also incorporated a boresighted camera to enable them to record the site of an interesting spectrum.

MSU-E
: A high-resolution (45-metre) optical scanner on the Priroda module. Its CCD sensed three wavelengths between 0.5 and 1.0 micrometre.

MSU-KS
: A medium-resolution optical scanner on the Priroda module that sampled five wavelengths between 0.5 and 12.5 micrometres.

Multiplikator (Multiplier)
: A biological experiment on Salyut 6 to study the growth rate of micro-organisms.

Nanovesi
: An experiment on Salyut 6. Samples of structural materials and optical coatings were placed in the scientific airlock to expose them to the space environment to assess their degradation.

NASA
: The US National Aeronautics and Space Administration.

NASDA
: The National Space Development Agency of Japan.

Nausicca-1
: A French experiment on Mir which measured radiation in the complex, and correlated this with the complex's movement in its orbit.

Neptune
: The main control panel of the Soyuz-TM ferry.

Neptune
: A visual test performed on Salyut 7 to study the degradation in visual acuity during the early phase of adaptation to weightlessness.

Neva-5
: A radiometer on the Spektr module.

Neytral-Vektor
: An ion sensor which determined its orientation with respect to its direction of flight through the ambient 'ionic wind'.

Glossary 373

Niva
  A videotape system on Salyut 7 which enabled the cosmonauts to record visual observations while out of communications range, and then relay them to specialists on the ground once back in range.

NSF
  The US National Science Foundation

Oasis
  A hydroponic apparatus on Salyut 1 for higher plants. The seeds were placed in a nutrient solution under a strong fluorescent lamp. Cameras photographed the chamber at regular intervals to monitor the rates of growth. Khibiny (chinese) cabbage, flax, onion, marrow-stem kale and crepis (actually the hawk's beard strain, an unobtrusive weed producing tiny white flowers) were all tried. Crepis was a common subject for studies of plant genetics. Flax was selected because it has a strong tissue structure which should be unnecessary in space. It was important to find out whether kale would grow and, if so, whether its nutritional value was altered by being grown in the absence of gravity, because it could be eaten and might well serve as the basis for hydroponic gardens in future orbital stations. The final form of the apparatus had lamps, fans to create an airflow, a water pump and layers of woven material emulating soil to aerate the roots. Peas finally grew in the Salyut 7 Oasis apparatus, which used an ion-exchange resin substrate. They germinated, then grew 20 centimetres tall (so much that they touched the lamps in the compartment's ceiling, and their pots had to be lowered to give them room to grow). When they were 30 centimetres tall, the tallest had sprouted six branches, 23 leaves, and a thick tangle of tendrils. Although the leaves were well-extended, they were coated with mould; some white, some brown. The white root systems had expanded out from the artificial soil. When the tendrils started to twist together, they developed buds that slowly unrolled to form leaves. Wheat was also grown successfully. These experiments not only investigated the characteristics of the plants grown in weightlessness; they also evaluated their potential as a source of food on future stations.

Oasis-2
  A biological experiment carried on Soyuz 13 in 1973. It was a closed-cycle system involved in producing proteins. It comprised two interconnected cylinders. In one, water-oxidising bacteria was cultivated. This consumed hydrogen released by water electrolysis, to sustain growth. Oxygen was drawn off into the second cylinder containing urobacteria which absorbed the oxygen and released carbonic acid, which was fed back into the first cylinder to synthesise more biomass. In essence, the waste from one process was the raw material for the other. This symbiotic regenerative process increased the biomass by a factor of 30 in just two days, demonstrating that it would be feasible to produce protein from simple raw materials to drive a food production system on a future orbital station.

ODS
  The Orbiter Docking System was used by the Shuttle on visits to Mir. Set in a twin-triangular truss spanning the front of the payload bay, it was basically two

interconnected tubes, one leading from the mid-deck hatch to a hatch to the transfer tunnel to a Spacelab or a Spacehab module further aft, and the other running upwards through an airlock to the APAS at the top to mate with Mir.

ODU

The unified propellant system introduced by Salyut 6 in which both the main orbital manoeuvring engines and the attitude control thrusters used the same propellants (so that it could be more easily replenished than if each engine unit used different propellants). After being installed in the Progress cargo ferry, it was fitted to the Soyuz-T form of the crew ferry (and thereafter to the Soyuz-TM). On Mir, the twin-chamber engine produce a total of 600 kilograms of thrust for orbital manoeuvring. The chambers could be gimballed within a 5-degree cone to aim their impulse through a slightly offset centre of gravity.

OLIPSE

A NASA experiment on Mir. The Optizon Liquid Phase Sintering Experiment used the Optizon furnace in the Kristall module to process samples of metals.

Opros (Questionnaire)

A psychological test which involved filling out a comprehensive series of questions concerning the cosmonauts' eating and sleeping habits, level of physical fitness and posture, sense of smell, vision and hearing and other factors which would enable the psychologists to correlate their physical and mental health. Visiting cosmonauts would fill it out every day, the residents once a week.

OPS

Orbiting piloted station, the Chelomei Bureau's Almaz military reconnaissance platform configuration.

OPS 1	Salyut 2	April 1973
OPS 2	Salyut 3	June 1974
OPS 3	Salyut 5	June 1976

An improved reconnaissance platform was built for launch in 1979, with its crew flying the TKS rather than the Soyuz, but this was cancelled. An automated version for commercial purposes was subsequently developed.

| OPS 4 | Cosmos 1870 | July 1987 |
| OPS 5 | Almaz 1 | March 1991 |

Its primary instrument was an S-Band (10-centimetre) synthetic aperture radar that produced images of a 25-kilometre-wide swath to the side of the ground track. The radar had 25-metre resolution in the case of Cosmos 1870, and 15-metre resolution in the case of the automated Almaz. Because its data were transmitted to Earth there was no need for a film-return capsule. It was the same configuration as the manned platform, but instead of the rear-mounted solar panels it had larger panels on either side of the forward section. These had a total area of 86 square metres, and generated the 10 kW needed to operate the radar. Two radar antennas extended on mounts on the forward compartment, and extended along the length of the body. The attitude was controlled by gyrodynes. It later emerged that the first satellite was to have been tested in the early 1980s, but had been repeatedly postponed; the second was a year late, and the third was apparently cancelled for financial reasons.

Optizon-1

A non-crucible furnace in the Kristall module which heated a sample by zonal melting (that is, by focusing radiant energy from an electric light source) to produce semiconductors (notably silicon).

Optokinez

A biomedical apparatus on Salyut 7 used to investigate the relationship between the vestibular and visual systems during adaptation to weightlessness.

Optovert

An Austrian biomedical device on Mir to study the interaction of the visual and vestibular systems to measure eye-movement in response to optokinetic stimulation, and to assess its effect on motoric performance.

Orion-1

This was the primary instrument on Salyut 1. It comprised a complex and highly accurate optical and electronic system involving a 280-millimetre and a 50-millimetre mirror. One telescope and a spectrograph were mounted on the outside of the station, and another telescope, mounted inside the station, looked through a quartz screen. Being above the Earth's atmosphere, it was able to observe far into the ultraviolet. It recorded stellar spectra on film. It was sensitive in the 2,000–3,000 Angstrom band, with a resolution of about 4 Angstroms, and it had an automatic tracker to follow the desired star with an accuracy of 1 arcsec. It could be used only while the station was in the Earth's shadow. An airlock enabled the film to be replaced, and exposed film was returned to Earth for processing.

Orion-2

An ultraviolet telescope carried on Soyuz 13. This was mounted on the front of the orbital module in place of the docking unit. The cosmonauts had received extensive training in how to operate it by the staff of the Byurakan Observatory in Armenia. It was an improved form of the Orion-1 apparatus tested on Salyut 1, but in this case the entire unit was mounted outside the spacecraft. When not in use it was protected by a shroud. It employed a wide-field meniscus telescope which could simultaneously sample an area of about 20 degrees square, and the optical components were made of crystalline quartz. It was sensitive to radiation in the range 2,000–3,000 Angstroms. Designed by Grigor Gurzadyan of Armenia, it was mounted on a fully stabilised platform incorporating a dozen electric motors. Once the spacecraft had been orientated so that it pointed in the general direction of the target and had been provided with reference stars, it could automatically lock onto the desired coordinates and hold itself steady throughout the required exposure, which could last up to about 20 minutes. It used special film supplied by NASA. In all, the cosmonauts recorded 10,000 spectrograms of 3,000 stars in the constellations of Auriga, Perseus, Taurus, Gemini and Orion. Most of these stars were of 10th magnitude at visual wavelengths, but some were as faint as 12th magnitude.

Orlan-DM
: An improved spacesuit used on Mir superseding that carried by Salyut. It comprised an aluminium alloy body section with an integral helmet and backpack. It operated at about 6 psi (relatively high pressure compared to NASA standard), but its elasticated constant-volume arm and leg joints gave much greater flexibility and dexterity than the previous suit. The operator wore an inner garment containing tubes for water coolant. It was entered by a hatch in the rear (the backpack was hinged and swung aside). All equipment was accessible for maintenance simply by opening the rear access hatch (there were no access ports on the backpack). It required just 30 minutes prebreathing oxygen at 10 psi before dropping the pressure to 6 psi (this was done with the suit closed, and thus consumed the backpack's resources, so this period had to be kept as brief as possible). This new suit did not use a communications/power umbilical because its backpack supplied power and had transmitters for the voice link and telemetry. The oxygen tank, and the lithium hydroxide scrubber which removed carbon dioxide, could sustain seven hours' operation (the Salyut suit had been limited to five hours) in addition to several hours pre/post-activity in the airlock. Another feature, welcomed by the doctors, was the biomedical sensors that relayed data via the telemetry link. Like earlier suits, its systems were rated for about 10 excursions before requiring maintenance. (The original Orlan D suit bore the postfix 'D' to signify that it had been designed for use on the DOS stations.)

Orlan-DMA
: An improved Orlan-DM suit on Mir.

Orthostatism
: A French experiment on Mir that involved a study of orthostatic resistance using an Echograph to monitor the changes in cardiovascular capacity and venous circulation, and sampling to investigate hormonal changes due to the absence of gravity. A subsequent refinement of this study used the Echograph, Diuresis and Tissue, and made use of the Haut Schicht Dicke apparatus left by a German visitor.

OST-1
: A solar telescope, the main instrument on Salyut 4, housed in a tall, broad, conical mount projecting up from the floor of the main compartment. It employed a 25-centimetre-diameter mirror with a 2.5-metre focal length, and incorporated the CDS-1 diffraction spectrometer. The entire assembly had been designed and built by the Crimean Astrophysical Observatory specifically for the flight, and the crew had received extensive training from the staff of the observatory in how to use it. It had been used by remote control for two weeks before the crew had arrived, and at some point a sensor had been burnt out, disabling the pointing system and preventing further remote control use. Although the cosmonauts could not actually observe its mechanism to align the secondary mirror, they discovered that by listening to its servo mechanism and timing its travel they were able to estimate its position sufficiently accurately to align it so that sunlight reflected directly onto the main mirror. Although this meant that the station

would have to be rotated to scan the telescope over its target, at least the programme could now be resumed. Without their intervention the telescope would have remained unusable, and a significant portion of the scientific programme would have been lost. The mirror was resurfaced by the Zentis experiment, to extend its useful life. It was used to study dynamic processes on narrow cross-sections of the solar disk, including flocculi and prominences. In addition, at sunrise and sunset, its ultraviolet spectrometer could be used to measure the strength of ozone's absorption lines, to extend investigations by Earth-based instruments. It was a fleeting opportunity, but the automatic attitude control system was able to add data regularly, as the station orbited the Earth every 92 minutes. At such times, it was possible to measure sunlight passing through a sample of air 1,000 kilometres deep. The results helped assess the level of pollutants and, later, contributed to a long-term study of what is now universally referred to as 'ozone depletion.'

OVI

A German experiment on Mir. It was devised by the University of Mainz and used a pair of glasses equipped with stimulators and sensors to investigate the influence of vestibular asymmetry and the stimulation of the throat receptors on eye movement and the 'subjective horizontal' in the absence of gravity. It involved a series of psycho-physiological tests.

Oxymeter

A Czech experiment on Salyut 6 to determine the concentration of oxygen in skin tissue in the absence of gravity.

Ozon-M

An experiment on the Priroda module. Its spectrometer assessed the distribution of ozone and other aerosols in the upper atmosphere by measuring their absorption of sunlight at sunrise and sunset at four wavelengths between 0.12 to 0.26 micrometre.

Palma (Palm)

An experiment first conducted on Salyut 4, designed to evaluate reaction to external irritants. (Psychologists would eventually evaluate a large range of visual and acoustic irritants during tests conducted by the Intercosmos crews on later stations). It was also on Salyut 5.

Palmyra

An experiment on Mir. It mixed two substances to produce a synthetic material with a crystalline structure similar to that of human dental and bone tissue.

Parallax-Zagorka

An image intensifier for the Rozhen experiment. It was also used on its own to study the vertical distribution of luminescence in the polar, middle and equatorial latitudes and also the luminescence of the complex itself as it flew through the extremely rarefied atmosphere at its orbital altitude.

PAS

A NASA experiment on Mir. The Passive Accelerometer System measured the low-level but continuous accelerations on the complex (as opposed to transient shocks transmitted through the structure), to assess the extent to which factors such as air drag decelerated the complex in its orbit, and how differential gravity drew items inside towards the Earth. It involved placing a small metal ball within a reference frame and recording how it drifted over time. These measurements helped to characterise the effects influencing microgravity experiments. These data complemented those from the MISDE and SAMS instrumentation.

Payload Systems

The first American apparatus on Mir. Payload Systems Inc's commercial contract was to fly six experiments over several years. The first was a protein crystal growth experiment in which two enzymes (hen egg white lysozyme and D-amino transferase) were submitted to 112 tests by three crystallisation processes (batch, vapour diffusion and boundary-layer diffusion). It was returned after two months. Protein crystals had been grown by similar apparatus on the Space Shuttle, but on for about 10 days. Crystalline protein was valuable because it enabled the normally delicate material to be studied. The structure of a crystal grown in microgravity is extremely pure and free of distortions. Protein is by far the most important substance in the body. The hope was that such crystals could shed light on genetic defects such as those producing cancers.

PCN

A French camera on Salyut 7. It was a hand-held low-light camera incorporating a variety of filters. It was intended to be used to make observations of faint upper atmosphere phenomena such as luminescent clouds, and was used to photograph a variety of faint sky phenomena, including the zodiacal light, noctilucent clouds and lightning storms. It was transferred to Mir.

Pelican

A remote manipulator outside the Spektr module. This 2-metre long twin-segment arm, which had a gripper on its end, was to be used to extract small experimental packages from a scientific airlock and affix them to a number of external anchor points, each of which had a socket for a power, control and telemetry umbilical. It enabled exposure cassettes to be deployed and retrieved without requiring a cosmonaut to go outside. Energiya developed a standardised experiment package. With the package in place, four pallets could be unfolded individually, as required, to expose the samples.

Perception

A Cuban experiment on Salyut 6. It evaluated visual, tactile and muscular sensitivity using the Cuban-supplied Contact apparatus.

Phoswich

A scintillation detector consisting of two crystals, e.g. NaI(Tl) and CsI(Na), optically coupled and forming a PHOSphor sandWICH. The NaI(Tl) acts as an X-ray detector, while the CsI(Na) scintillator acts as an active shield. The scintillation light produced in a phoswich is viewed, though a light guide, by a photomultiplier tube.

Physalie

A French experiment on Mir to evaluate the coordination between the body's sensory and motor systems. This involved using a number of devices to record biomedical data whilst being filmed performing a variety of bodily movements.

Physiolob

A French experiment on Mir to study the rapid changes to the cardiovascular system which occur in the early phase of adaptation to weightlessness, particularly how the body senses blood pressure and regulates its flow. This involved donning a harness supporting various sensors.

Phyton

A cultivator on Salyut 6 for growing higher plants, developed by the Ukrainian Institute of Molecular Biology. It incorporated a very powerful lamp, a nutrient medium, and a filter to extract air impurities which, although no danger to the crew, might inhibit the growth of the plant. It was designed so that the roots grew 'down' in the dark, while the shoots grew 'up' towards the lamp. Seeds of onions, cucumbers, tomatoes, garlic and carrots were placed into compartments of a black panel containing a nutrient solution. The objective of the experiment was to grow the plant in the hope that it would flower and, hopefully, produce a seed of its own. This would demonstrate the complete growth cycle from seed, grown plant, and back to seed in weightlessness and thus achieve the goal of almost a decade of effort. It was supplemented by data gained from other experiments investigating the development of individual cells. Unfortunately, the arabidopsis plant grew so rapidly that it choked on its own waste products. The onions very quickly grew seed stalks (this was unexpected because on Earth onions do not produce seeds in so short a period). Another type of higher plant was a strain of wheat, characterised by a very short stalk and a very heavy yield of grain; although this grew, it did so more slowly than expected, and, as a result, it had only just reached its most complex stage, and was in the process of producing ripe grain, when it was necessary to cut short the experiment and return to Earth, so it was not possible to proceed to the second generation in space. In the case of fungi, which have a very good geotropic reaction (that is, they react to gravity), the fruit bodies formed were totally disorientated unless they were provided with a light source, in which case they would grow towards it, and so reacquire a measure of orientation.

Phyton-2

A plant-growth apparatus on Salyut 6. It incorporated three light sources and interchangeable plant pods containing an ion-exchange nutrient. The plants were given measured quantities of water by an automatic sequencer and ventilated. It subsequently turned out that shoots developed better if the apparatus was put near a porthole so that they could be exposed to sunlight rather than the artificial light. Despite all efforts, however, the peas and wheat still died early in their development cycle.

Phyton-3

A plant-growth apparatus on Salyut 6. After many years of experiments, in 1980 a higher plant flowered in space (this was a key step towards the goal of completing the growth cycle, from seed to seed). This was an arabidopsis. Grown from its bud, it flowered four days after the control experiment on Earth. The next objective was to have one complete the cycle by producing seeds. Arabidopsis in the apparatus on Salyut 7 podded. The largest plant was 65 millimetres tall, had 12 green leaves, three tiny pink flowers about 2 millimetres across and a pod 3 millimetres long. When the pods ripened, they burst open to reveal the seeds within. It was great news for the biologists, who were keen to receive seeds from a plant grown in space.

PIE

A NASA experiment on Mir. The Particle Impact Experiment was for micrometeoroid research. It was placed outside Kvant 2.

Pille (Moth)

A thermoluminescent ionising radiation dosimeter. Being portable, it was employed to measure the dosage rates at different parts of the Mir complex. It was also used to assess the increased dosage resulting from passing through the South Atlantic Anomaly. It measured doses in the 10-millirad to 10-rad range.

PILOT

A standard NASA computer program used to test piloting skills (and thereby retain them) during shuttle missions (in particular to practice landings just before returning to Earth), which Thagard used on Mir to evaluate how his performance changed over such a long period.

Pingvin (Penguin)

A lightweight work-suit with elasticated straps sewn into its fabric to impose compressional loads on specific muscle groups, to counter the atrophying of the back muscles in weightlessness. It was so called because it caused a test subject on Earth to hunch over and waddle like a penguin. It was tested on Salyut 1, and later, without the elasticated straps, was worn as a straightforward work suit.

Pion

An apparatus flown up to Salyut 6 to investigate crystallisation both in and out of solution, and the effect of convection currents in fluids under the influence of microgravity. The mixture to be studied was delivered in a clear disk-shaped container. It incorporated a lamp which illuminated the sample from one side while a camera recorded the transformation from the other side. The temperature and time were recorded on each frame.

Pion-M

An improved crystal growth apparatus on Salyut 7. It was used to study melting and crystallisation processes of various materials, and heat and matter transport in liquids to assist in the design of an apparatus to process large quantities of biological materials, particularly colloids. It was found that silica aerogel (a type of glass) in suspension formed small saucer-shaped structures, fluoroplastics formed tree-like structures, and glass pellets formed arbitrary but extremely robust lumps. It was transferred to Mir, and then used to study the thermocapillary process in microgravity.

Piramig

A French camera on Salyut 7. A highly sensitive camera which was to be used to study the upper atmosphere, the interplanetary medium and faint galaxies at visible light and near-infrared wavelengths. It was set up in the porthole in the forward transfer compartment normally used by the MKF-6M. It could be used only in orbital darkness. The compartment was isolated, with the hatches shut and the lights off (the cosmonauts could use only low-power torches) and with the station's thrusters off (to avoid ejecting glowing debris into the field of view). It used an extremely sensitive camera incorporating fibre optics and photomultipliers, and it had filters operating between 2,000 and 10,000 Angstroms, so could be used for ultraviolet, visible and near-infrared observations of the upper atmosphere, the interplanetary medium and a number of astrophysical sources. The Kaskad orientated the station to within 1 degree, then the camera was fine-tuned to achieve an accuracy of about 1 arcsec. It was difficult to keep steady for a five-minute exposure. The station was specifically orientated for such observations for up to six hours on some days, and during 20 sessions about 350 photographs were taken. These turned out to be very good, in particular those of high altitude cloud formations, so the resident cosmonauts were asked to use the hand-held PCN camera twice each week to photograph the horizon in the hope of recording more of these clouds (at an altitude of 80 kilometres) and other upper atmospheric phenomena (at altitudes up to their 350-kilometre orbit).

Pirin
A Bulgarian smelting experiment on Salyut 6. The Splav furnace was used to make 'foam metal' in microgravity. These combine low specific weight with softness and high mechanical strength. In this case, they produced foam aluminium. This was achieved by heating a capsule containing quartz ampoules of silumin, titanium hydride and silicon nitride for 10 minutes at 800°C. Foam steel would have all the strength of normal steel, but be as light as wood. In the future, such materials could be used in the construction of extremely large orbital complexes. Later, another experiment in this series was performed in the Kristall furnace to assess the stability and structure of zinc crystals grown using the diffusion process.

Plasma-1
A freezer on Salyut 7, used to store body fluid samples for return to Earth.

Pleven-87
An experiment on Mir which used fifteen psychological tests, many of which tested the locomotor functions and volition processes. (Its data was processed by the Zora computer.)

Plotnust
An ultrasonic device used on Salyut 4 to measure the density of bone calcium.

Pneumatik
An experiment on Salyut 6. It involved using the Pneumatik-1 apparatus while in the Tchibis suit to determine whether a blood distribution appropriate to gravity could be re-established in space, to limit the rush of blood to the cranial regions.

Pneumatik-1
A device to reduce the flow of blood to the upper torso.

Polarizatsiya
An experiment on Mir employing Bulgarian apparatus to make photometric observations of stars, galaxies and nebulae.

Polynom-2

An electrocardiograph flown on Salyut 3. It was a multifunction blood flow unit to investigate a range of circulatory conditions. It was to be used to measure blood flow before and after exercise to investigate the effect of weightlessness on the heart and brain, and could measure vascular tension. During the first few days in space as the human body adapted to weightlessness, blood gathered in the brain, and previous cosmonauts had reported this as an unpleasant feeling. These tests were intended to measure exactly how the flow of blood changed, in the hope that a way could be found to ameliorate the unpleasant symptoms.

Polynom-2M

A multipurpose electrocardio unit flown on Salyut 4 to monitor the performance of the heart. In addition to general cardiograms it could isolate the major blood vessels and monitor the various phases of the cardiographic cycle.

Polyus (Polar)

The payload of the first Energiya rocket was an engineering mock-up of a spacecraft which Nikolai Gerasimov, head of the Salyut Bureau, said was a multipurpose platform that was to be used to deliver 40 tonnes of cargo to a space station. Dr Vladimir Pallo, the platform's designer, said that it was essentially an enlarged version of the Mir core module, both longer and wider, incorporating a TKS tug for propulsion and attitude control; it was 38 metres long and had a mass of 80 tonnes. An alternative rôle would be as an industrial-scale microgravity factory for materials processing and biotechnology which would operate autonomously, but receive occasional visits by cosmonauts to service the equipment, restock the raw materials and take away the product.

Posa

A French apparatus on Salyut 7, comprising a set of motion sensors and muscle monitors for the Posture experiment.

Posture

A French experiment on Salyut 7 designed to investigate how the body maintained certain postures in microgravity. The subject donned the Posa sensors and used a foot restraint which defined the motion coordinates, and then, with both eyes closed, enacted a series of body movements to exercise sensory and locomotive functions so that the responses could be recorded for later analysis. A variety of sensory organs and muscle groups (different from those used when in motion) are used to maintain posture. The apparatus took considerably longer to set up than had been expected .

Potential

An experiment on Mir which investigated the interaction between the body's nervous and muscular systems.

Potok (Flow)

An experiment on Salyut 5 to evaluate the possibility of building cosmic capillary pumps for liquids which do not need electricity. It comprised two interconnected containers linked by a narrow channel. The capillary action and surface tension caused the fluid in the first container to flow into the second, and the cosmonauts recorded this process. It was an important experiment because the designers hoped that it would lead to a pump without moving or electronic components, in order to simplify the replenishment system of a future tanker spacecraft.

Pravda (Truth)
The newspaper in the Communist Party of the Soviet Union.

Priroda (Nature)
An expansion module of the Mir complex. It had a pressurised volume of 66 cubic metres.

Priroda-5
A multispectral Earth observation system on the Kristall module using two KFA-1000 cameras.

Prognoz
An experiment on Mir which assessed the operational performance of the cosmonauts. It involved using a video (on the Zora computer) to run a series of sensory-response tests. It used a series of mathematical puzzles and audiovisual cues to assess functional psychological state. If these tests were performed daily, degradations in performance could be tracked. It made use of the Pleven-87 apparatus.

Prognoz 1 (Forecast)
A class of automated research satellite (the first was launched in 1972) in highly elliptical orbits ranging between 950 and 200,000 kilometres at 65 degrees so that they spent most of their time high above the Soviet Union. It contained various detectors to study the emissions from the Sun, and their interaction with the Earth's magnetosphere.

Progress
A Soyuz-based spacecraft used as a resupply ferry for the Mir complex. It was essentially a Soyuz with the descent module replaced by a cylindrical compartment with tanks for fluids. Dry cargo was packed in containers bolted into racks within the orbital module. In addition to propellants, water, pressurised oxygen and nitrogen, it delivered food, scientific apparatus, tools, and replacement components for Mir's environmental and thermal regulation systems. Total capacity was 2,500 kilograms (1,000 kilograms of wet cargo, and 1,500 kilograms of dry cargo). It used the same unified ODU engine block as the second-generation Salyut. Each ferry was kept docked as long as possible to serve as storage, then loaded with rubbish just before its departure. Sometimes the docking assembly would be replaced by an experiment package to be activated while the spacecraft stood alongside the station.

Progress-M
An improved form of the Progress cargo/tanker ferry. It reintroduced solar panels. A Raduga Earth-return capsule could be inserted into the docking unit to return the results of experiments.

Proton
The Chelomei Bureau's medium rocket with a lifting capacity comparable to NASA's Saturn IB. It burned storable hypergolic propellants (unsymmetrical dimethyl hydrazine and nitrogen tetroxide) which ignited on contact. Its first launch in 1965 was a two-stage configuration which put the 12,200-kilogram

Proton high-energy physics satellite into orbit (the heaviest scientific satellite ever launched). With an upper stage, the rocket was later used to launch advanced planetary probes and send the Zond spacecraft on a circumlunar trajectory. In 1968, it put Proton 4 into orbit (at 18 tonnes, this satellite tested the main section of the pressure hull of the Almaz reconnaissance platform). It subsequently put the Salyut stations, the Mir base block, and their TKS-based expansion modules in low Earth orbit.

PRV

A precision radio-altimeter on the Priroda module operating at 2.25 centimetres.

PSY

A German experiment on Mir which used a portable computer to perform perception, speech and psychomotor tests.

Pulsar X-1

A spectrometer on Kvant 1. It incorporated four identical Phoswich detectors to study gamma-rays and hard X-rays at energies of 50–800 keV. It had a 3 degree square field of view.

Pulsar-2

A 1,000-kilogram X-ray telescope (similar to Pulsar X-1) which was to have been delivered by Buran and attached to Kristall's lateral APAS port.

Pulstrans

An Austrian experiment on Mir to analyse pulse transmission and heart frequency while the body was subjected to stress.

Puma

A viewfinder on Salyut 7. It had 15-times magnification.

QUELD

A NASA experiment on Mir. The Queen's University Experiment in Liquid Diffusion was supplied by Canada. It used an isothermal furnace that operated at temperatures up to 1,000°C, and was used to study boundary layer processes by measuring diffusion coefficients for semiconductors, binary-metal materials and glasses.

Raduga (Rainbow)

A small recoverable capsule ejected by a Progress-M immediately after the vehicle made its de-orbit manoeuvre. It could return 150 kilograms of compact cargo to Earth. Considering the experiment schedule, and the rate at which this was expected to produce results, it seemed likely that a capsule would be needed on every third or fourth resupply ferry. At 380 kilograms empty, however, it seriously diminished the cargo capacity of the vehicle which delivered it. It was delivered inside the orbital module, then inserted into the ferry's docking collar instead of the docking probe assembly. The truncated cone, 1.4 metres long with a bottle nose, projected in through the hatch into the cargo compartment (the capsule was 0.6 metres wide at it base and the hatch was 0.8 metres in diameter). After the ferry had withdrawn to make its de-orbit burn, tracking stations computed its trajectory in order to compute the optimum time to order it to eject the descent capsule, which had to be jettisoned at an altitude of about 120 kilometres. Obviously, in a change to the previous procedure, instead of de-orbiting over the Pacific, the capsule-returning ferry was de-orbited over Kazakhstan so that its capsule would descend in the normal recovery zone. A pressure sensor determined when the altitude had decreased to 15 kilometres and then commanded the parachute to release. When at 4 kilometres, it switched on the radio beacon. Following the break-up of the Soviet Union, recoveries were switched to a site near the Urals, on Russian territory. Most capsules were successfully recovered. It has the designation VBK (ballistic descent craft).

Raketa

A vacuum cleaner used on Salyut 4 to suck up the worst of the dust and debris which appeared as soon as a station entered weightlessness.

Rapana

A 26-kilogram, 5-metre long girder akin to Sofora, but using a different smart material. It was developed by the Institute of Electrical Welding, in Kiev. This Ferma-2 truss was erected on top of Kvant 1. It was a scaled-down test of a structure which was intended to be used to hold the parabolic dishes of a solar collector away from the body of the proposed Mir 2 orbital complex.

Reaktsia (Reaction)

An experiment on Salyut 5 to evaluate a technique being considered for assembly work in space. It used two chemical energy sources to smelt high-grade nickel and manganese solder to solder stainless steel pipes. The experiment was to reveal how the process of soldering differed in microgravity. On Earth, the solder does not spread evenly over the seam, but this should not occur in the absence of gravity. It was necessary to observe the formation and the crystallisation of the solder seam, and then to establish its mechanical characteristics. If it turned out that joints made in space were stronger than those made on Earth, soldering might well be able to play a part in the construction of structures in space. In the tests, the cosmonauts soldered two stainless steel 15-millimetre-diameter tubes with a wall thickness of 0.001 metres. It was later found that these seams were strong and airtight, and could withstand a pressure of 500 atmospheres. When the cosmonauts used a soldering iron to effect repairs to electrical equipment, they used a solder which did not include tin.

Rech (Speech)

An East German experiment on Salyut 6. This simply required reciting a series of specific phrases during each communication with the ground so that any variations in speech pattern could be correlated to state of health, mental attitude and general level of activity. These phrases, such as saying the number '226' in German, required elocutionary dexterity. Even a mundane experiment such as this could shed light on how various activities caused stress.

Reflotron

A biomedical kit on Mir, supplied by West Germany. It provided instant blood analysis, so that the changes in blood chemistry could be studied as they happened (rather than after analysis following return to Earth). Given a drop of blood, the analyser would automatically measure a wide range of parameters (including a report on haemoglobin, cholesterol, uric acid and glucose) and store the results for later transmission to Earth via the telemetry link.

Refraction
A Hungarian experiment conducted on Salyut 6, using the VPA-1M.

Rekomb
This was a biotechnology experiment on Mir. It cultivated hybrid micro-organisms having properties that would enable them to synthesise biologically active substances on Earth.

Relaks (Relax)
A psychological experiment on Salyut 6 carried out to identify stresses endured by cosmonauts whilst living and working in space on long-duration missions. It tested the influence of various recreational exercises on their psychological condition. It was part of a long-term programme to determine the most favourable conditions for living in space.

REM
A Radiation Environment Monitor supplied by the European Space Agency and placed outside Mir to measure charged-particle flux.

Reograf (Rheograph)
An apparatus on Salyut 4 which measured the distribution of blood by measuring blood flow in the head, torso and extremities.

Reporter
An East German experiment on Salyut 6. It involved conducting a methodical study of different types of film for use both for external photography and within the station. Jahn had two specially modified cameras (a Pentacon-6M and a Praktica EE-2) with a range of lenses for use in these studies.

Reservoir
An experiment on Mir which investigated the development of a system for filling and emptying a capillary tension reservoir.

Resistance
An experiment on Salyut 6 which used a detector to measure the loads imposed by drag from the upper atmosphere.

Resonance
An ongoing engineering experiment which measured the vibration modes of the complex in its various docked configurations, and with different apparatus operating, to identify any vibrations which might adversely affect the structure. In its simplest form, a cosmonaut strapped into the harness of the KTF and jumped up and down in time with a control signal. In another case they set up a number of spring-loaded masses and monitored their oscillations.

Resurs
A cassette outside Salyut 7, containing structural materials (16 plates of different alloys) whose physical properties were to be tested after prolonged exposure to the space environment.

Rezeda (Mignonette)
A spirometer on Salyut 4 which determined lung capacity. It was also on Salyut 5.

Glossary 387

Ritm
A plant growth apparatus that incorporated a radiation dosimeter. It was carried on Soyuz 16.

RMS-2
This respiratory measurement system was sent to Mir by the European Space Agency to study lung functions.

Rodnik (Spring)
An improved water supply system introduced on Salyut 7. Two 250-litre spherical tanks in the aft equipment section could be refilled directly by pumping from tanks in a Progress ferry (this was an improvement over the previous procedure, in which water was delivered in dozens of tiny spherical bottles, each of which contained 5 kilograms of water, and had to be transferred by hand). Water was fed to a squirt-nozzle by the table. Although the cosmonauts had hot water on demand for reconstituting dehydrated food packs, the heater could supply only 0.5 litres at a time. Each cosmonaut ingested about two litres of water per day (including that employed in reconstituted food). The Rodnik system was also used by the Mir base block, and expansion tanks were mounted externally on the Kvant 2 and Kristall modules.

Röntgen
The group of four X-ray instruments (TTM, HEXE, Sirene-2 and Pulsar X-1) located in the unpressurised compartment around the rear transfer compartment of Kvant 1.

ROK
A German experiment on Mir, developed by the Max Planck Institute to study the response to different orientations in weightlessness.

Rost
A plant growth apparatus on Salyut 7, used to grow oranges.

Rozhen
A Bulgarian apparatus on Mir. This was a sophisticated system incorporating a digitally processed electro-optical telescope (using the Parallax-Zagorka image intensifier and the Therma photometer). It was being evaluated for an autonomous space observatory to study deep sky sources. Its imagery was processed in orbit, then downlinked.

RS-17
An X-ray telescope on Salyut 7, mounted in the rear transfer tunnel. It was sensitive to high-energy photons in the 2,000–800,000 eV range, and was used to determine the temperature profile of the accretion disk in the Cygnus X-1 black hole system, and to study the Crab nebula and X-ray pulsars.

RSA
The Russian Space Agency

RSS
A hand-held spectrograph for Earth studies, tested on Soyuz 5 and Soyuz 7, and used on Soyuz 9. It could be used to measure the reflectance spectrum of the Earth's surface under direct insolation and so distinguish soils with different moisture content.

RSS-2
: A hand-held spectrograph on Salyut 3. It was designed in Leningrad to study the processes affecting the Earth's thermal environment, and to determine trends in climate changes which could be caused by the saturation of the atmosphere with aerosols such as industrial smoke, dust and chemical pollutants. These were important observations because it was believed that the disruption of the optical and radiation properties of the upper atmosphere would create the conditions necessary for climatic changes. It could be used either to observe the atmosphere at times of orbital sunrise or sunset by measuring the selective absorption of sunlight passing through the layers of the atmosphere, or to make observations of the Earth's surface and the oceans by measuring the reflectance spectrum under overhead solar illumination. It was used to measure the distribution of gaseous aerosol pollutants in the upper atmosphere and to measure the distribution of ozone (a thin layer in the upper atmosphere at an altitude of about 90 kilometres) to determine the extent to which it is contaminated by aerosol pollutants. Its output was recorded on film.

RSS-2M
: An improved form of the hand-held spectrometer, used on Salyut 5. As part of a long-term study of climatic changes, it was used to establish the presence of aerosols (particles such as smoke, dust and chemical pollutants) at different altitudes in the atmosphere to an accuracy of between 500 and 800 metres. It was also used to observe strong currents in the Atlantic. One study concentrated on the Volga region, measuring the distribution of water in the river, and the results correlated with simultaneous studies by ground teams and aircraft. The Soviet Ministry of Land Improvement and Water Conservation had plans to divert some of the outflow of rivers in the northern part of the USSR to areas further south, so these observations were intended to assess the extent to which the flow of southern rivers could be increased. The measurements of water capacity subsequently influenced planning for several hydroengineering projects in the Volga region.

RSS-3
: A spectrometer on Mir, sensitive in the 0.4–1.1 micrometre range. Its 600 lines per millimetre diffraction grating gave a resolution of 1 per cent. Although portable, it was usually mounted in a frame in a porthole and left to operate autonomously at sunrise and sunset, to study the atmosphere.

RT-4
: An X-ray telescope on Salyut 4. It was sensitive to X-rays with wavelengths in the range 44–60 Angstroms (that is, soft X-rays with energies under 1 keV). It used a telescope with a 0.2-metre diameter parabolic mirror to illuminate a sophisticated photon counting system to measure the intensity of a source. Its field of view was not wide enough for it to be used to scan the sky for new sources, but it could be used to make further studies of sources discovered by the Filin system.

RT-4M
: A mirror-based soft X-ray telescope on Salyut 7. It was an improved version of the apparatus tested on Salyut 4.

Ruchei (Brook)
An electrophoresis unit installed in Kvant 1, used to synthesise interferon and anti-influenza vaccines.
Ryabina
A spectrometer on Salyut 7, used to study gamma-rays and charged particles.
Ryabina-2
A telescopic spectrometer on Kvant 2.
Ryabina-4P
A telescopic spectrometer on the Spektr module.

SAFER
The Simplified Aid for EVA Rescue is a thruster pack attached to a NASA spacewalker's suit to effect a safe return in the event of a tether snapping.
Salyut 1
This orbital station comprised three compartments, arranged as a series of four cylinders with different diameters. The Soyuz ferry docked at the front, opposite the station's propulsive unit. As the cosmonauts left the tunnel they emerged into the transfer compartment. It was the narrowest section, barely 2 metres across, and contained some of the astrophysical apparatus and several control panels. At the other end of the transfer compartment was the hatch to the main work volume which extended right through the 3-metre diameter cylinder into the 4-metre diameter cylinder at the rear, behind which was the propulsive unit, which was the same as that in the Soyuz ferry's service module. It was 14.6 metres long (about 22 metres with a Soyuz docked). The main controls were on a number of panels, similar to those installed in the Soyuz ferry, in a console across the floor, just on the other side of this internal hatch. The rest of the station, which formed a long room rather than a cylinder, contained various experimental work stations. The length of the docked combination was about 30 metres and it had a mass of about 25 tonnes. The total habitable volume, with the ferry attached, was about 100 cubic metres. Its size was made evident during the first television transmissions, when the cosmonauts were shown performing somersaults for the camera. There were eight seats, each of which was positioned near a workstation so that the cosmonauts could strap themselves in while at work. In all, Salyut 1 incorporated 20 portholes, but only a few gave an unobstructed view, and the others were dedicated to various equipment. Two pairs of unfolding solar panels, like those used by Soyuz, were attached at the front on the new transfer section and at the rear on the engine unit. With a Soyuz docked, the total solar collector area was 36 square metres, which generated just 3.6 kW under full illumination. The fact that these could not be rotated to keep facing the Sun meant that whenever the batteries ran low, the station had to be reoriented to face the Sun so that they could be recharged, and this manoeuvre consumed propellant. The station was totally dependent on the batteries while it was in the Earth's shadow, so power had to be carefully managed.
Salyut 3, 5
This was the Almaz stripped of its Merkur capsule. It was 11.6 metres long (about 18 metres with a Soyuz docked). It had a two-element stepped-cylinder

configuration. The docking port was at the rear, set between the two engines. A small film-return capsule was accessible by a hatch within the wall of the transfer tunnel (this was a 400-kilogram, 0.85-metre wide drum-shaped spin-stabilised capsule with an ablative shield; it could return 120 kilograms of compact payload, which was far more than could be ferried back in a Soyuz descent module). The solar panels were stowed alongside the transfer tunnel and then unfolded. The primary instrument was a 6-metre focal length folded-optics telescopic camera for Earth observation with about 50-centimetres resolution. Since this viewed through the floor of the compartment, the station had to maintain given multi-axis rotation to keep the camera facing the Earth as the spacecraft progressed along its orbit. Because it could not hold a fixed orientation with respect to the Sun, its solar panels had to be able to rotate. To minimise the propellant used in adjusting the attitude of the station, an inertial system was installed which used electrically-driven magnetically-mounted flywheels whose gyroscopic action could finely control the orientation of the station. Salyut 3 introduced a water reclamation system that condensed water from the station's atmosphere. This proved to be very useful, as it could reclaim evaporated water to yield fresh water (1 litre per day per cosmonaut aboard). Recovered water was used for washing and food preparation. It had been an important experiment because recycling would help make a station self-sufficient and would reduce the amount of water which would have to be stored aboard at launch.

Salyut 4

This was the Salyut 1 configuration, but fitted with three large solar panels on the narrower part of the main compartment instead of four small fixed panels. The new panels were capable of tracking the Sun. Their combined area was 60 square metres and they generated 4 kW. It had a hatch in the side of the transfer compartment. (This compartment could be sealed off so that it could serve an airlock, although it isn't clear why this was deemed necessary, because the orbital module of a Soyuz has a hatch and could be used as an airlock in the event that the cosmonauts had to work outside.) It used the water reclamation system which condensed water from the station's atmosphere (introduced by Salyut 3). It was primarily an Earth observation platform.

Salyut 5B

A computer introduced by Mir to supersede the Argon 16.

Salyut 6, 7

This second-generation configuration combined the best features of the OPS and DOS designs. To accommodate a docking port at either end, the rear-mounted Soyuz propulsion unit was replaced by the OPS peripherally-mounted engine system. The clip-on solar panels were 5 metres long and 1.25 metres wide, and each pair added about 1.2 kW to the basic power level (about the same as did a docked Soyuz-T). They were primarily Earth-observation platforms.

SamoRöntgen

An experiment on Salyut 7 using an X-ray detector.

SAMS

A NASA experiment on Mir. The Space Acceleration Measurement System was

Glossary

to measure the microgravity environment. Its three-axis accelerometers continuously measured the vibration modes at the sites on Mir where NASA intended to install its experiments. It had already been used on shuttle flights that conducted microgravity research.

Saturn V

The NASA rocket which sent the Apollo spacecraft to the Moon.

SDRN (Satellite Data Relay Network)

The Luch relay satellite system.

Seeds

An experiment on Mir that simply exposed a bag of tomato seeds to radiation by storing them in the Kvant 2 airlock, which was a relatively unshielded compartment. They were planted after return to Earth, and examined for genetic irregularities.

Semyorka (Old Number Seven)

Korolev's affectionate name for the R-7 rocket which was developed as an intercontinental ballistic missile, but immediately adapted to launch spacecraft.

Sever

A side-looking camera on Mir. It had 'a fixed unit with a portable apparatus' that enabled it to take oblique images to either side of the ground track, to highlight surface relief.

SDI

The strategic defence initiative (SDI), announced by President Ronald Reagan on 23 March 1983, called for the development of a "peace shield" which would protect America from ballistic missiles. Basic research for weapons capable of incapacitating incoming strategic missiles was funded. Many space-based approaches were studied, including kinetic-kill interceptors, X-ray lasers and particle-beams, but no 'Star Wars' systems were deployed. The Department of Defense had (wrongly) said that Cosmos 1267, which docked with Salyut 7 in 1981, had converted that station into an "orbital battle station".

Sfera (Sphere)

A furnace on Salyut 5 to study the process of smelting and resolidifying metals by passing ingots of bismuth, lead, tin and cadmium through a furnace to sealed containers, in an effort to create perfect spheres from the liquid metal. It was designed to study the process of melting and hardening molten metals in microgravity. Small ingots were inserted by remote control, and after melting at $60°C$ the samples, each no bigger than the size of a match-head, were allowed to solidify in free fall in the hope that they would form perfect spheres. In fact, metal spheres cast under microgravity did *not* yield the smooth surfaces that had been expected.

ShK-1

The scientific airlock, consisting of two concentric spheres. The outer sphere had two hatches, one of which faced inwards, and the other outwards. The inner sphere, which could rotate inside the outer sphere, had a single hatch. The inner sphere was aligned with its hatch facing inwards so that an item could be inserted, and it was then rotated so that the outer hatch could be opened to expose the item to vacuum.

Silya-4
: An experiment on Salyut 4 which used a light nuclear isotope spectrometer to record the isotopic and chemical composition of cosmic rays. The apparatus would be refined for use on further flights, in order to establish a major new field of investigation.

Sirena
: A Polish materials-processing experiment on Salyut 6 in the Splav furnace. This used an ampoule of cadmium-mercury-telluride (CMT), prepared by scientists at the Warsaw Institute of Physics, which was to be heated to create a semiconductor which is one of the most sensitive detectors of infrared radiation yet discovered. The process involved fusing cadmium telluride with high density mercury telluride. In Earth's gravity, the process produced an imperfect stratified product. The first experiments, lasting 48 hours, yielded crystals with higher homogeneity than obtainable on Earth, with a purity of 50 per cent, compared with 15 per cent in the most effective terrestrial process. Although initially the process had a very low yield (fewer than 10 per cent of the crystals were suitable for use in industrial applications) subsequent fine tuning of the process substantially increased the yield. Other infrared-sensitive materials produced included tin-lead-tellurium, cadmium-mercury-selenium and lead-selenium-telluride. One run produced a CMT monocrystal with a mass of almost 50 grams.

Sirene-1 (Lilac)
: A French–Soviet X-ray spectrometer on Salyut 7. It was installed in the rear transfer compartment, and operated by remote control from a panel in the main compartment. The station was reorientated to aim it. Its data was recorded on magnetic tape.

Sirene-2
: A high-pressure gas scintillation proportional counter (GSPC) on Kvant 1, sensitive to X-rays in the 2–100 keV range with a 3 degree square field of view and a detector geometric area of 300 square centimetres. It was used to study emissions from extremely high-temperature rarefied cosmic gas sources. It was an improved form of the experiment flown on the EXOSAT observatory, and was supplied by the European Space Agency.

Skif
: A Belorussian experiment on Salyut 7. An improved form of the MSS-2M spectrometer, it was used to study the vertical structure of the atmosphere.

Skif-M
: This spectrometer on Mir sampled in five bands in the 0.4–1.1 micrometres range. Its output was stored on magnetic tape and processed by the Zora computer.

SKK-11
: An exposure cassette full of construction materials deployed on Kvant 2.

SKR-2M
: An X-ray spectrometer sensitive in the range 2–25 keV, installed aboard Salyut 7.

Smak (Taste)
: An experiment (also known as Vkus) on Salyut 6 devised by the Polish Military Institute of Aviation Medicine to investigate why the sense of taste changed in

space. (Some foods which were pleasant on Earth had turned out to be unpleasant in space, and others had been found to be much better when eaten in space.) By measuring the electrical stimulation of the taste buds they were able to correlate subjective reactions to taste with specific data.

SN1987A

A supernova discovered on 23 February 1987 in the Large Magellanic Cloud (LMC). As it was the nearest such event for several centuries, it was intensely observed, both by ground-based and satellite-borne telescopes, including the X-ray telescopes in Kvant 1.

Sofora

A girder developed by the Institute of Electrical Welding in Kiev. Unlike previous trusses deployed in space, it incorporated thermomechanical joints. Its tubular rods were connected by sleeve joints made from a 'smart' titanium-nickel alloy which contracted to re-establish a predefined shape when heated and, in doing so, extended and locked the truss. Once the first 1.5 by 1.5-metre square-section element had been affixed to a base plate mounted on Kvant 1, further elements were added. When the final (twentieth) element was in place, the girder stood 14 metres tall. The assembly process required 16 hours in all, but it demonstrated that orbital 'construction work' was feasible. There was a pivot about a third of the way up the girder, so that the upper part could be swung down to a point just above a ferry docked at the rear port. The VDU thruster block was eased out of a Progress ferry and mounted on top of the girder, which was then swung 11 degrees beyond vertical so that the thrusters would fire in the same plane as the roll-control thrusters set around the periphery of the rear of the base block.

Sokol (Falcon)

The lightweight (8-kilogram) pressure suit worn by cosmonauts for launch and re-entry; introduced with the Soyuz-T, superseding the pressure suit worn following the accident which killed the Soyuz 11 crew.

Solar constant

A measure of the energy radiated by the Sun which is received by the Earth: 1.35 kW per square metre.

Solar flare

A flux of ionised plasma ejected from the Sun by a magnetic disturbance on its surface (when seen on the limb of the solar disk, such a storm shows itself as a prominence arcing between sunspots). Although a solar storm typically lasts for only an hour, it spews forth a 'tongue' of plasma which expands out across the Solar System. This plasma is composed primarily of hydrogen and helium nuclei travelling at about half the speed of light, which gives them energies corresponding to up to 100 million electron volts. When these charged particles flood into the Earth's magnetosphere, they degrade global communications. It is an important component of cosmic radiation, and would pose a threat to Mir crews if it were to penetrate the magnetosphere to their low orbit, a risk only near the South Atlantic Anomaly.

Solar power

The nominal power available to orbital station elements at the time of launch, before the output of the solar transducers degraded through exposure to the space environment, before auxiliary panels were added to compensate, and counting only the output from the panels on the station itself (that is, discounting any contribution from docked ferries):

	m^2	kW
Salyut 1	24	2.4
Salyut 3, 5	–	5
Salyut 4	60	4
Salyut 6, 7	60	4
Cosmos 929, 1267, 1443, 1686	40	3
Cosmos 1870	86	10
Mir	76	9
Kvant 2	50	7
Kristall	72	8
Spektr	126	16

Solar wind
: The flux of ionised plasma emitted continuously by the Sun. It forms the background of cosmic radiation. It comprises primarily hydrogen and helium nuclei travelling at speeds of several hundred kilometres per second (slow compared with similar material ejected by a solar flare), corresponding to energies of around 1,000 electron volts. It poses no threat to a spacecraft in low Earth orbit because the Earth's magnetic field acts as a shield.

Son-K
: A Bulgarian experiment on Mir which gathered electrophysiological data (the electrical activity in the brain) while a cosmonaut slept, and recorded it on a long-duration cassette tape.

South Atlantic Anomaly
: A region above the South Atlantic where the innermost of the Earth's radiation belts, which forms a torus around equatorial latitudes at an altitude of 1,000–5,000 kilometres, dips towards the ionosphere, posing a threat to spacecraft.

Soyuz (Union)
: The Korolev Bureau's flagship spacecraft, designed to be a modular system which could be configured to suit a variety of roles. It comprised three modules: the service module contained the propulsion system, the descent module took the crew to orbit and back again, and the orbital module provided accommodation and equipment. The descent and orbital modules had an overall pressurised volume of 10 cubic metres. Soyuz 1 to Soyuz 11 provided a shirt-sleeve environment for a crew of three and the solar panels gave an endurance of several weeks. Following the depressurisation accident which killed the crew of Soyuz 11, a seat was deleted to accommodate an independent life-support system for the remaining crew of two, who wore Sokol pressure suits during launch, docking manoeuvres and re-entry. In this form (excepting Soyuz 13, Soyuz 16, Soyuz 19 and Soyuz 22; it took the series to Soyuz 40) the solar panels were replaced by batteries. This limited its endurance to two days, and it was used only as a

transport to Salyut stations. The Soyuz-T variant reintroduced the solar panels and the third seat, and it had the Argon 16 computer of Salyut 4 and the unified ODU engine of the second-generation Salyut. The solar panels had an area of 12 square metres and gave a peak of 1.2 kW. Soyuz-T could carry a crew of three and 150 kilograms of cargo to Mir, but return only 50 kilograms of cargo. The Soyuz-TM increased the to-orbit capacity by 250 kg and the return capacity to 120 kilograms. Although the to-orbit cargo could be stowed in racks in the orbital module, the return cargo had to be squeezed into the tiny descent module. Despite being 'old technology', the latest configuration is admirably suited to its solitary rôle of transporting crews to and from the Mir complex, and it is highly reliable.

Spektr (Spectrum)

An expansion module for the Mir complex. It had a pressurised volume of 62 cubic metres, and had four solar panels with a total area of 126 square metres (in effect a set of base block panels and a set of Kvant 2 panels) providing a total of 16 kW.

Spektr

An experiment on Salyut 4 which involved a variety of apparatus installed around the station to investigate the properties of the atmosphere at orbital altitudes. This expanded on studies made on previous flights. The various instruments measured the density, composition and temperature of flows of neutral gas and plasma encountered by the station as it followed its orbit. The hull of the station actually became charged as it passed through such plasma and this charge could interfere with some of the station's systems and experiments. These impacts also contributed to the drag which eroded the station's orbit. The results of these studies would influence the design of future spacecraft.

Spektr-15K

A portable (11-kilogram) multispectral spectrometer on Salyut 6, developed by the Bulgarian Academy of Sciences to study oceanic plankton, crop yield, and the propagation of atmospheric pollution from industrial sites. It could measure absorption of insolation at orbital sunrise and sunset, or measure the reflection of insolation from the surface. It sampled at 15 wavelengths in the visible and near-infrared range, between 4,500 and 8,500 Angstroms, and stored its data on magnetic tape.

Spektr-15M

A spectrometer on Salyut 7.

Spektr-256

A spectrometer on Mir developed by Bulgaria to study the atmosphere. It sampled 256 channels in the 0.4–0.8 micrometre range with a holographic grating and a CCD detector, and its data was processed by the Zora computer.

Spectrum

An experiment on Salyut 6. It used the Spektr-15K spectrometer to study the physical, chemical and biological characteristics of Cuba and the surrounding sea.

Spin-6000

A 256-channel X-ray and gamma ray spectrometer designed by the Radium Institute of Leningrad to study materials. It was used during a Mir spacewalk to assess the extent to which the structure radiated gamma rays as a result of its passage through the Earth's magnetic field.

Spiral

A cassette outside Salyut 7. It contained structural materials (springs, seals, threaded connectors, pipe fittings conditioned to expand and contract as their temperature varied, cables, and various metals under constant stress) which were to be tested following prolonged exposure to the extremes of temperature, levels of radiation, X-rays, ultraviolet light, cosmic rays and micrometeoroids of the space environment, to determine whether they were suitable for 'assembly work in space.'

Splav-1 (Alloy)

A 23-kilogram electric furnace on Salyut 6. It had a separate control panel. Physically, the furnace took the form of a disk attached to a narrow cylindrical stem. It was placed in the airlock so that it could radiate its excess heat to vacuum. It had hot, cool and thermal-gradient compartments. The 0.3 kW heating chamber operated at up to 1,100°C, and its electronic control system could maintain this temperature to within 5°C. It incorporated a set of molybdenum reflectors to focus heat onto a capsule 170 millimetres long and 21 millimetres in diameter, which contained three crystal ampoules. Upon being heated in the hot chamber, the ampoules would fuse. They were automatically transferred into a 'cooling chamber' which maintained a linear thermal gradient to ensure optimum conditions for the formation of a monocrystal. When the sample had cooled to 650°C, it would be transferred into a third chamber which was held at this temperature to facilitate three dimensional crystallisation. The crystallisation process, which would typically take several days, would be recorded by time-lapse photography. Samples were returned to Earth for further analysis. It was also used to study the process of diffusion in molten metal in the absence of gravity; capsules of copper-indium, aluminium-magnesium and indium-antimonide were heated, then the process of crystallisation observed. Such an experiment lasted 14 hours, during which time the station remained in gravity-gradient stabilised orientation, with its axis pointed at the ground, and the attitude control system was switched off so that it drifted freely (otherwise its tiny manoeuvres might interfere with the test). Despite this precaution, the cosmonauts reported that they could see small imperfections caused by vibrations as they had moved about within the complex. This prompted concern that the 'microgravity factory' of the future might have to be a free-flying module which, although it would be visited by cosmonauts from time to time to carry out maintenance, would function automatically. Other experiments saw the production of alloys to investigate the interaction between solid and liquid metals as a step towards understanding the processes of welding and soldering in the absence of gravity. In one novel experiment the station was set rotating to create a centrifugal force that would cause the sample to undergo directional solidification. This contrasted with the earlier practice of trying to hold the station stable so as to minimise disturbances.

Splav-2
  Used on Salyut 6. A similar unit was subsequently flown on Cosmos 1841 (a Vostok-based spacecraft devoted to microgravity research) in 1987.
Sprut-5
  A spectrometer deployed outside Kvant 2 to measure the flux of charged particles. It relayed its data to Earth with the telemetry stream.
Sputnik (Fellow Traveller)
  Any spacecraft.
SRVK
  The water reclamation system which condensed water vapour from the station's air. This water was used for food preparation. It could condense 1 litre of water vapour from the air per day per person aboard.
SSAS
  A NASA experiment on Mir. The Solid-Sorbent Air Sampler was to sample the air in the cabin for analysis of microbial lifeforms that might pose a long-term threat to its habitability.
STR
  Salyut 4 tested this sophisticated thermal regulation system. The outer surface was covered with 'screen vacuum heat insulation' composed of layers of synthetic film sprayed with aluminium that minimised heat loss. The station incorporated an intricate set of radiators which collected solar heat on the sunward side and radiated excess heat on the shaded side. These enabled the thermal regulation system to control heat transfer within a wide range of temperatures. Individual elements of the multiple-loop system, which comprised heating and cooling units, had three or more backups to provide a high degree of redundancy, and so safety. The computer commanded the entire system, correlating the heaters and coolers with the station's orientation with respect to the Sun and the Earth's shadow. Water could be discharged into space, so that the evaporation would rapidly reduce the thermal energy of the station, in an emergency. While a ferry was powered down docked to the station its thermal regulation systems were put under the station's control. An improved version of STR was used by Salyut 6, and again on Salyut 7.
Stratokinetika
  An experiment on Mir which studied the body's movement in the absence of gravity.
Strela (Arrow)
  The general name for the Mir base block's computer complex. It included an information system which monitored and reported on the status of the station's various systems, and provided on-line documentation (previously, limited information had been available in written form) that could be automatically updated by the ground controllers (this last feature superseded the Stroka teletype uplink).
Strela
  A pair of telescopic cranes installed on either side of the Mir base block. Although the 45-kilogram crane was only 2 metres long when packed, it could be extended

to a length of 14 metres. Operated by a pair of hand cranks, it could be raised above or lowered below the base block, and be rotated around the outside arc to reach as far back as Kvant 1. It could transfer a load of 750 kilograms between any two points on its side of the complex. It was 'parked' canted against Kvant 2, so that the cosmonauts could ascend it on their way back to the airlock, and then slip down it at the beginning of the next spacewalk to access the crank. The cranes were mounted on fixtures originally used to support the launch shroud.

Stroka
An uplink teleprinter to send lengthy communications to the crew. Introduced by Salyut 4, and used on all subsequent stations.

Strombus
This Ferma-3 girder was erected on top of Kvant 1. It was a four-segment, 6-metre long truss.

Struktura
A materials processing experiment on Salyut 6 which used the Pion apparatus to investigate heat exchange and mass transfer during crystal formation in an aqueous solution. A later experiment on Mir produced alloys of aluminium-copper-iron, aluminium-tungsten and aluminium-copper.

Sugar
An experiment on Salyut 6 to grow four monocrystals of sucrose in different solutions. They grew surprisingly quickly. It was the first time that organic crystals had been grown in space.

Superconductor
A French experiment on Mir. It employed the Krater furnace to investigate the crystallisation of a high-critical-temperature superconductor in microgravity.

Supercooling
An experiment on Salyut 7. A 3-millimetre-diameter sphere of an extremely pure alloy of silver and germanium was melted by first cooling it with liquid helium and then heating it with a laser. It was hoped that this would produce a much stronger material.

Superpocket
A French experiment on Mir to study the neurosensory system and the reconditioning of postural reflexes.

Support
A Cuban experiment on Salyut 6. It involved wearing a specially-designed adjustable shoe (dubbed the Cuban Boot) for six hours each day. This was designed to impose a load on the arch of the foot to simulate the forces the foot feels when standing, to determine whether this affected the ability of the vestibular system to adapt to the absence of gravity. This was based on the theory that on Earth the state of the muscles in the foot contribute to the sense of balance, so when this sense is denied in space it produces a vestibular reaction. This test would also enable doctors to investigate the recovery and readaptation of locomotive stability on returning to gravity, because cosmonauts had been observed to develop gait and postural peculiarities whilst recovering from their flight.

SUR
: A German experiment on Mir. It studied whether the body's circadian rhythm changed during the process of adaptation to the absence of gravity.

Svet (Light)
: A Bulgarian cultivator on the Kristall module for growing higher plants. It used a substrate infused with nutrient, and a high-intensity lamp. Radish and lettuce grew rapidly. It was adapted by NASA for its Greenhouse experiment.

Svet
: A spectrometer on the Spektr module.

Svet-1
: An apparatus on Salyut 7. Its mechanism provided an infinitely variable colouriser, for describing visual Earth observations. It replaced the booklet specifying 1,000 colours that had previously been used. To calibrate it, various dyes were released into the Black Sea.

Svetlana
: An 800-kilogram industrial-scale electrophoresis processing system in the unpressurised compartment of Kvant 1. It was an advanced, semi-automated form of Tavriya, named after Svetlana Savitskaya, who had tested the prototype. It was said to be able to process 100 kilograms of material per year, but this capability was never exploited. Its first trial was in separating micro-organisms to produce agricultural antibiotics to assist in stock rearing.

Svetobloc (Light Box)
: A cultivator on Salyut 7 which worked on the principle of a hothouse. Orchids and tomatoes were successfully grown in it.

Svetobloc-M
: A greenhouse in the Kristall module used to cultivate high-order crops such as lettuce and radishes.

Svezhest (Freshness)
: An experiment on Salyut 6 which was designed to ionise the air in the station and, it was hoped, make the station slightly more comfortable. The cosmonauts reported that they liked the fact that it made the station smell of a pine forest.

Synergies
: A French experiment on Mir which studied the function of the vestibular system in controlling the dynamic equilibrium and the stabilisation of references in body synergy during a series of complex movements.

Tamponazh
: An experiment on Salyut 7 designed to test how different sealants solidified. It involved setting up small pipes filled with prepared concrete solutions which were then left to set. It was sponsored by the oil and gas industry to see how pores formed in materials such as concrete (used to seal wells), because pores permit oil and gas to pass.

Tass
: The Soviet News Agency.

Taurus
: An X-ray detector on the Spektr module to measure emissions from the complex, generated by its passage through the Earth's magnetic field.

Tavriya
: An experimental processing system for biological materials, tested by Svetlana Savitskaya on Salyut 7 in 1984. It separated biologically active substances by passing an electric current through a fluid medium (that is, by electrophoresis). It was basically a column of biological compounds which separated into homogeneous fractions, each of which had the same characteristics. The main chamber of the apparatus was almost 1 metre long, and incorporated 230 needles positioned to draw material off from the different layers along its length. On its evaluation trial, human blood proteins (albumin and haemoglobin) were separated. The cosmonauts were able to see the process, which was filmed for later analysis. Once verified, it was used to process several samples. In one run, it was used to purify urokinase (an enzyme present in human urine). Later, for the first time, interferon was processed using the electrophoresis technique (the electric current acted regardless of molecular weight). The trial was extremely encouraging, as the apparatus proved to be hundreds of times more productive than a comparable unit on Earth, and its product was of 10 to 15 times greater purity. It was later used to purify albumen (protein solution) and produced enough pure protein from membranes of an influenza virus to satisfy the Pasteur Institute in Leningrad for many months. Even the small quantities of materials produced by these experiments represented significant results. Test production of vaccine refined from the early results was begun immediately. Later runs synthesised an anti-infection preparation produced by genetic engineering and an antibiotic for agricultural use which, when added to animal and poultry fodder, would increase the weight of the animals by up to 20 per cent. The electrophoresis experiments proved so successful that the design of an industrial-scale unit was begun. It was hoped that on later missions this would be able to yield extremely pure vaccines and other pharmaceutical products on a semi-commercial basis. Such apparatus was also carried on Vostok-based spacecraft. (In 1987 for example, the Kashtan apparatus on Cosmos 1841, a precursor for the Foton materials-processing satellites, synthesised alpha-1 thymosin by purifying thymus hormone, and produced interferon with which to treat viral and tumour diseases, for the Institute of Biomedical Technology in Moscow.)

Tchibis (Lapwing)
: A Lower-Body Negative Pressure suit comprising reinforced rubberised leggings with a seal at the waist, from which air could be extracted to create a pressure below the ambient (selectable down to 70 per cent ambient) to draw blood from the upper body. By measuring pulse and blood pressure at different suit pressures it is possible to determine cardiovascular capacity. It is worn for about an hour a day during the initial phase of adaptation and for half an hour once a week, then during the final ten days or so is once again worn on a daily basis. Although this has proven to be effective in ameliorating the unpleasant effects of adapting to weightlessness, it is insufficient preparation for the return to gravity, so is

supplemented in the final phase by exercises to increase cardiovascular capacity, drugs to increase stamina and saline solution for rehydration.

TDRS
NASA's geostationary Tracking and Data Relay System.

Teleassistance
A French experiment on Mir. It repeated the Orthostatisme experiment and evaluated how a link-up to an expert on the ground could provide a cosmonaut with technical backup during a complex task in orbit. One obvious difficulty was that without the Luch network, the limited communications with Mir seriously limited such interactive science.

TeleGeo-87
An Earth-observation programme organised by Intercosmos (such programmes were run each year during the summer), this time concentrating on Poland. The Mir cosmonauts participated.

TEPC
A NASA experiment on Mir. The Tissue-Equivalent Proportional Counter was a sophisticated radiation monitor that stored its data and downloaded it at regular intervals.

Terra-K
A commercial programme in which agricultural or industrial sites could request orbital investigations. Mir provided photography and spectrometry. Various high-resolution cameras were used to provide mapping imagery, and the MKF-6MA camera and the Spektr-256 spectrometer provided multispectral analysis of the growth of vegetation, assessed water purity, and tracked the spread of pollution.

TES
A German experiment on Mir that employed the Kristallisator furnace to measure the heat capacity of a supercooled metallic melt (in this case antimony, an alloy of silver and germanium, and two sapphire samples) and its variation with temperature. The study of the specific heat of supercooling fusions exploited the fact that in microgravity, contact between the material and the walls of the furnace can be eliminated. Determining the heat capacity enabled other thermophysical properties to be derived.

Therma
An impulse photometer used for the Rozhen experiment.

Tien-Shan-88
An Intercosmos study of Tadjikistan and Kirghizistan. The Mir cosmonauts contributed imagery.

Titus
A German-built six-zone tubular furnace capable of temperatures of up to 1,250°C, used by the European Space Agency to process semiconductors, alloys and glass.

TKS
The ferry for the Almaz platform. It comprised the FAB (the universal auxiliary block) and the VA (the Merkur descent module). The entire vehicle was 17.5 metres long, but much of this was taken up by the escape tower of the capsule.

To fit the Proton rocket, it had a 4.15-metre diameter base. TKS was an acronym for Transportnaya Korabl Snabscheniya-transporter and logistics spacecraft. It was tested as Cosmos 929 in 1977, and Cosmos 1267 docked with Salyut 6 in 1981, after that station had concluded its main programme, to test its ability to control the orientation of the joint complex. When operational, Cosmos 1443 delivered cargo to Salyut 7 in 1983, and Cosmos 1686 followed in 1985 with a cluster of scientific instruments. It was referred to by the Western press as a 'Heavy Cosmos' initially and later as a 'Star' module. It was basically a 3-metre diameter compartment around which the engines and propellant tankage were grouped. The Merkur capsule was accessed via a hatch in the end of the module, via a short tunnel that led to a hatch in the capsule's heatshield. At the other end the structure flared out to an adapter ring that mated to the Proton rocket. On the far side of the ring was a short conical cap which incorporated the docking unit. Overall, it had a mass of 20 tonnes (the main vehicle was 14 tonnes, including the 4,000 kilograms of miscellaneous cargo; the Merkur and associated propulsion unit was 6,000 kilograms). A stripped-down version (without a Merkur capsule) was the 'tug' that delivered Kvant 1 to Mir in 1987. The FAB was expanded (by adding integral pressurised compartments, as TKM) to serve as the basis for the modules used to expand the Mir complex between 1989 and 1996. All forms of TKS used a pair of 400-kilogram-thrust engines burning unsymmetrical dimethyl hydrazine and nitrogen tetroxide but their position and the capacity of the externally-mounted propellant tanks varied. The basic vehicle deployed a pair of solar panels (like those on the second-generation Salyuts) spanning 16 metres, with a total area of 40 square metres, and 3 kW capacity, and had a single cylindrical compartment with a typical internal volume of about 50 cubic metres.

TMS

A cassette outside Salyut 7. It contained thermomechanical compounds whose physical properties were to be tested after prolonged exposure to the space environment.

TON

A German experiment on Mir. Supplied by the University of Hamburg, it used a specially devised sensor to determine the interior pressure of the eye.

Tonus-2

A sophisticated electromechanical muscle stimulator used on Salyut 4.

Torsion

An experiment on Salyut 7 to investigate the strength of various materials when exposed to space.

TORU

A system which enabled a cosmonaut on Mir to fly an automated ferry by remote control. A pair of hand controllers for rotational and translational motions were installed on the main control panel, to perform the same function as did those on a Soyuz, and a screen presented the view from the camera in the ferry's docking camera with the same data-overlay. The cosmonaut flew the ferry just as if physically onboard. It was tested with Progress-M 15, then used to dock the errant Progress-M 24. The name is the Russian acronym for the 'remote-control flight system'.

Travers

An experiment on the Priroda module. This two-channel synthetic aperture radar radiated at 23 and 92 centimetres, and had medium (30-metre) resolution ground-imaging capability.

Trek

A cosmic ray detector supplied by scientists at the University of California. It was installed outside Kvant 2. The 1 square metre plate was a passive detector which contained layers of phosphate glass to record the passage of the super heavy nuclei component of the cosmic ray flux. It was retrieved after two years. It was only the second American experiment to be delivered to the Mir complex.

Tropex-74

An oceanic study of the Atlantic in which observations by Salyut 3 were correlated with data from the Meteor weather satellites and ship-based studies.

Tropics

An experiment on Salyut 6. It involved using the MKF-6M to assess Cuba's natural resources. It was actually Tropics-3, and continued earlier observations of other regions. As always, the orbital photography was coordinated with data gathered by airborne and ground teams. Unfortunately, although the weather in the Caribbean was excellent throughout the flight, the timing of the flight meant that there would be few suitable daylight passes, so the residents performed most of this work after the visitors had returned to Earth. The results showed that at both its eastern and western extremities the island of Cuba is criss-crossed by intersecting networks of faults.

Trud (Labour)

A newspaper in the Soviet Union.

TTM

The wide-angle COded-Mask Imaging Spectrometer (COMIS) on Kvant 1, sensitive to X-rays in the 2–20 keV range and with a 7.8-degree square field of view. It used a coded aperture mask to determine the location of X-ray sources. Supplied by the Netherlands Space Research Organisation, Utrecht, and Birmingham University, it was an improved form of apparatus flown on the Spacelab 2 shuttle mission. It developed an intermittent failure in late 1987, suffering from severe electron precipitation in the vicinity of the South Atlantic Anomaly, so the detector was replaced during a spacewalk.

UBD

This ultrasonic bone densitometer was supplied to Mir by the European Space Agency to measure microgravity-induced calcium loss from bones in the lower leg, as a means of assessing countermeasures.

Utrof-Pannoniya

A Hungarian Earth observation programme on Salyut 6, involved in using the MKF-6M camera and the Spektr-15K spectrometer. Their data were to be used to compile a geomorphological map of the Carpathian Basin and the Tisza

404 Glossary

River Basin, to assess the effect of the Kishkere reservoir on soil salination in the inland waterways linked to the Danube, and to assess the ecological state of Lake Balaton. The data were to be correlated with that simultaneously gathered by an Antonov-30 flying at 6,500 metres, an Antonov-2 flying at 2,000 metres, a low-level helicopter, and several ground teams. Of three faults which they identified in Hungary, one was found, within months, to have oil and gas reserves.

URI

The 30-kilogram Universal Manual Toolkit (URI) tested on Salyut 7 was a development of the Isparitel electron-beam apparatus. Developed by the Institute of Electrical Welding in Kiev, it incorporated a variety of tools designed to process metal in space, including a small hand-held non-contacting infrared thermometer to measure the temperature of the materials. Savitskaya first used an electron beam to cut 0.5-millimetre titanium and stainless steel plates. She reported this to be very easy and noted that she could actually see the beam. Then she joined two stainless steel and four titanium samples using tack-welding, and reported that she had produced three good seams. Next she soldered two metal plates together using solder composed of tin and lead, but reported that she did not think that the seam looked very good. Finally, she heated a small silver granule in a crucible (it took only 45 seconds) and then sprayed a silver coating onto an anodised black aluminium plate. She complained that it was more awkward than she had expected because it was hard to see the spray. Academician Paton, the toolkit's designer, reported that these simple tests demonstrated that in the future large space structures could be assembled by robotic systems equipped with such construction tools.

Vaporiser

An experiment on Salyut 6 using the Isparitel apparatus. Hundreds of disks (made of glass, carbon, titanium and various other metals) were sprayed with a variety of thin coatings (including gold, silver, copper, aluminium and various polymers). In one case a multiple-layer coating was formed in which one layer was sprayed whilst the airlock was exposed to direct sunlight, and the other in darkness.

Vazon

A plant cultivation experiment on Salyut 6. A similar experiment (ginseng, onion, and chlorella) was undertaken on Mir.

VDU

A 700-kilogram thruster block affixed to the Sofora girder above Kvant 1 to facilitate propellant-efficient roll-control. Rolling the complex consumed a lot of propellant using the thrusters on the periphery of the base block. An 85 per cent saving in propellant resulted from controlling the roll of the complex using these thrusters, because they were so much further from the axis of the complex. It was hoped that this would reduce the frequency with which cargo ferries had to be dispatched to replenish the base block's tanks. It was delivered in a specially-configured Progress ferry which carried the 2.2-metre-long oblong box in an unpressurised central section instead of the usual wet-cargo compartment. It was a completely self-contained unit, with its own propellant tanks. It required only a power and command umbilical to be connected, and this was run down the length of the girder and plugged into the Kvant module. Since no provision had been made for replenishment, it had been loaded with sufficient propellant for several years of use. Of the 700-kilogram mass, 400 kilograms was propellant gas. When empty, the unit was replaced.

Vektor

An experiment on Salyut 7. It used an Indian-built electrocardiograph to study the cardiovascular system.

Vibrogal

An experiment on Mir to study microscopic accelerations shaking the complex. It characterised the small dynamic loads resulting from the operation of the equipment. The data were collected near the Gallar furnace.

Vibroseismograph

An experiment on Mir to measure the microscopic accelerations acting on the complex.

Vika

An oxygen production unit included to satisfy a sudden demand, most frequently to repressurise the airlock on Mir following a spacewalk. When thermally decomposed at high temperature (1,000°C), sodium chlorate ($NaClO_3$) releases oxygen. An exothermic charge initiated the short-lived reaction, and generated a surge of oxygen production. A stock of cartridges was maintained, as the once-only cartridges had to be replaced after use.

Vinimal-92

A French experiment on Mir which studied sensor-motor performance in perception and orientation tests. It used the Echograph-II monitor screen to present cues, so that a cosmonaut could use a hand-operated key in response, to test visual acuity. When subsequently rerun, it used a miniature flight simulator to assess the effects of adaptation to absence of gravity.

Vita

A biotechnology experiment on Mir that used animal cells to cultivate protein compounds which were to be used later on Earth to produce pharmaceutical agents. It studied the growth dynamics of cells which produced luciferase, a biologically active albumen.

Voal

A materials experiment on Mir to produce an alloy of wolfram (a raw form of tungsten) and aluminium.

VOG

A German experiment on Mir that used video-oculography cameras to capture correlations between eye movement and the vestibular system.

Volkov

A spectrometer on the Spektr module.

Volna-2 (Wave)

Contained in Kvant 2, this 250-kilogram apparatus was designed to investigate fluid flows in capillaries, to assist in the design of propellant tanks. It involved setting up a special hydrotank, a control unit and photographic apparatus to record the results of the tests.

Vorotnik

An experiment on Salyut 6 to test an aid to adapting to microgravity. It involved wearing a collar to create an artificial load on the cervical vertebrae and limit the rotation of a cosmonaut's head, in the hope of eliminating vestibular disorientation.

Voskhod (Sunrise)

A version of the Vostok capsule adapted to accommodate several cosmonauts.

Vostok (East)

The 2-metre-diameter spherical spacecraft that carried Yuri Gagarin on his pioneering Earth orbit. It could control its attitude but it had no orbital manoeuvring capability. It remained attached to the upper stage of the rocket, which performed the de-orbit burn and was then jettisoned. The capsule was covered with ablative. It made a ballistic re-entry, and the occupant ejected once the parachute had opened, to make a separate landing.

Vozdukh

A 150-kilogram carbon dioxide scrubber in Kvant 1 which superseded the lithium hydroxide unit in the Mir base block. It took the complex another step towards a closed-cycle environment, eliminating the need to ferry up lithium hydroxide canisters in Progress ferries (once saturated a LiOH cartridge was discarded). This regenerative system used two desiccant and two regenerative molecular sieve beds of an absorbent similar to zeolite. The amount of gas that can be absorbed by the zeolite is directly proportional to the gas/absorbent contact area (so it operated by gas-capillary action). It is also a function of the pressure until the absorbent is saturated. However, it is inversely dependent on temperature, so the air has to be chilled first. Saturated zeolite is regenerated by exposing it to vacuum, which vents the CO_2. Fans drew air from the cabin and fed it through a silica-gel dryer that removed the water vapour. The heat released was removed by the base block's coolant loop. The dry but now freezing (–50°C) air was fed through the zeolite to remove the CO_2. The clean air was then heated to about 90°C (the exact temperature depending on the environmental control system settings) and fed into another silica-gel container where the hot dry air vaporised water in the silica-gel, simultaneously regenerating the silica-gel and rehumidifying and cooling the air for return to the cabin. Once the in-flow silica-gel unit was saturated with water, and the out-flow unit was dry, the two flows were switched. Similarly, there were two zeolite chambers. While one was in use, the other was leaking its gas to vacuum. Although it was a closed loop, about 0.25 kilogram of air was lost per day when the zeolite chambers were swapped over, but this was a negligible loss.

VPA-1

The visual polarisation analyser on Salyut 4, used to measure optical polarisation of the horizon as part of a study of the upper atmosphere.

VPA-1M

An improved form of VPA-1 flown on Salyut 6.

Vremya (Time)

An East German experiment on Salyut 6 which used the Rula timing device to measure the ability to react to stimuli, estimate specific periods of time and compare two intervals.

VSK-3, Vzor (Visor)
The 15-degree field of view optical periscope viewfinder of the Soyuz ferry.

VTL
The veloergometer, designed by the Lykachov Motor Works in Moscow, was a stationary bicycle which could be set to impose a load of up to 20 kilograms. In the series, VTL-1 was on Salyut 4, VTL-2 was on Salyut 5, and VTL-3 was on Salyut 6.

Vulkan
The welding apparatus tested on Soyuz 6.

Weightlessness
The common (and misleading) term for 'zero gravity', the effective absence of gravity associated with orbital motion; this too is incorrect, however, because microscopic vibrations caused by the routine operation of a spacecraft's systems induce accelerations indistinguishable from gravitation, so the seemingly 'weightless' state of everything aboard is more properly defined as microgravity.

Yakor (Anchor)
A series of work stations attached to the outside of a station to act as a foot restraint, and on which to set up apparatus for experiments during spacewalks.

Yantar (Amber)
An electron beam on Mir which extended earlier work with the Isparitel apparatus. It was set up in the scientific airlock and used to apply thin metallic coatings of alloys (such as silver-palladium and tungsten-aluminium) onto polymer film.

Yelena-F
A 22-kilogram gamma ray detector on Salyut 6. This measured the gamma-rays and charged particles in the near-Earth environment. It was designed by the Moscow Engineering and Physics Institute. It had a 30-degree field of view, and its detector was sensitive to energy in the 30–500 MeV energy range and drew only 10 W, so it could be left running for prolonged periods. It had a Cerenkov counter, eight scintillation counters and sixteen photomultipliers, and produced its results photographically. As expected, it recorded higher readings when the station passed the location of the South Atlantic Anomaly. It would be used to measure the strength, distribution and processes contributing to the gamma-ray background at orbital altitude. Observations were sometimes coordinated with similar detectors flown on radio-equipped high-altitude balloons. The results mapped high energy electrons flowing in the South Atlantic Anomaly, and demonstrated that the background flux was highly dependent on latitude; its level varied by a factor of 10 from the equator, where it was least, to the latitudes corresponding to the furthest extent of its 51-degree inclination orbital track. A similar unit was carried on Salyut 7.

Yevpatoria
The original flight control centre in the Crimea.

YMK
The 'Icarus' autonomous manoeuvring unit developed by the Zvezda Bureau for free-flying during spacewalks (YMK is an acronym for Yustroistvo Manevrirovania Kosmonautov, which translates as the cosmonaut manoeuvring unit). It was similar to the MMU built by NASA, but was somewhat larger and was covered with a thermal blanket. It had four T-shaped thruster packs, each of which had eight pressurised nitrogen thrusters (the use of a harmless propellant meant that there were was no corrosive exhaust to contaminate externally-mounted apparatus). A control panel was mounted on each armrest (left for translational, right for rotational control) along with a group of toggle switches and a joystick for specifying motions. It could be manoeuvred either manually, or in one of two semi-automatic modes, one for rapid response, the other to maximise propellant efficiency. The nitrogen bottles in the back-pack had to be replaced after use, and only a few had been supplied. It was not intended to be used on a regular basis, however. It was tested on Mir for subsequent use by spacewalking Buran cosmonauts. It had been designed to be compatible with the Orlan-DM spacesuit (without a suit capable of independent operation, it would be difficult to test the YMK properly); its frame fit snugly around the suit's backpack, and it was fastened by a waist belt. The docking port apertures were too narrow for a cosmonaut wearing it to pass through, so the test had to wait until Kvant 2, with its 1-metre wide airlock hatch, was available. It proved to be extremely stable, and responsive to commands. After the test, it was mounted on a frame outside so as not to clutter up the airlock.

Yusa
A radiometer on the Spektr module.

Zarya (Dawn)
A Hungarian experiment on Salyut 6 using the Spektr-15K spectrometer to study absorption lines at sunrise and sunset to determine the density and temperature of the air in the stratosphere and the troposphere.

Zentis
Because microscopic debris is knocked out of the skin of the station by particles of matter and molecules in the station's orbital path, an orbital station creates a cloud of debris around itself. After several weeks of exposure to this, sufficient dust had accumulated on the mirror of the OST-1 and tarnished its highly reflective coating. Taking advantage of the natural vacuum, Grechko used an automatic system to respray them. This was achieved by passing an electric current through a tungsten wire to melt a blob of aluminium, so that a vaporised spray of metal would settle on the mirror. Since it proved to be so easy to restore the mirror to its original reflective condition, this opened the way for the fitting of evermore sophisticated mirror-based systems on future stations.

Ziemia (Earth)
A Polish experiment on Salyut 6 to make Earth-resources observations of Poland by using the MKF-6M.

Znamya (New Light, or Banner)
: A space-mirror experiment. The 40-kilogram package was attached to the docking assembly of a Progress ferry which initiated a fast roll manoeuvre, so that the centrifugal force dragged out eight triangular petals which formed a 20-metre diameter reflector made of 5-micrometre-thick aluminium-coated Kevlar. The orientation of the ferry was chosen so that the mirror beamed sunlight towards the Earth, with the result that a spot of light 4,000 metres wide travelled along the ground track. This was to assess the feasibility of using orbital mirrors to illuminate extreme northerly regions which suffered extended darkness. The first experiment worked, but the second was spoiled when the mirror became tangled.

Zona-2
: A crucible-less furnace in the Kristall module to produce monocrystals of semiconductors.

Zona-3
: A crucible-less furnace located in the Kristall module.

Zond
: A circumlunar variant of the Soyuz spacecraft using only the descent module and a modified service module, launched by Proton. Several automated flights were made, but it never carried cosmonauts.

Zone
: A Cuban experiment on Salyut 6. It studied the dissolution of monocrystals of sucrose. Cuba had developed apparatus specifically to enable photographs to be taken of processes taking place in the Kristall furnace.

Zora
: A 3-kilogram portable computer delivered to Mir to analyse data in space.

Zvezda (Star)
: The manufacturer of the Orlan spacesuits and the YMK autonomous manoeuvring system.

Reading list

Although any study of space station development will ultimately rely upon the contemporary record of events published in *Spaceflight*, the *Journal of the British Interplanetary Society*, *Flight International*, *Aviation Week & Space Technology*, the excellent but short-lived *Spaceflight News*, and more recently the Internet, the following books offer useful snapshots of interpretation. They have been listed in order of publication to provide a sense of advancement.

Clarke, Arthur C., *The Exploration of Space*, Temple Press, 1951
von Braun, Werner, *Across the Space Frontier*, a series of articles, with artwork by Chesley Bonestell, published in *Collier's* Magazine, New York, 1952; reprinted and edited by Cornelius Ryan as *Across The Space Frontier*, Viking Press, 1952, and *Conquest of the Moon*, Viking Press, 1953.
Poole, Lynn, *Your Trip into Space*, Lutterworth Press, 1954.
Bates, D. R. (ed.), *Space Research and Exploration*, Eyre & Spottiswoode, 1957.
Burgess, Eric, *Satellites and Spaceflight*, Chapman & Hall, 1957.
Beard, R.B. and Rotherman, A.C., *Space Flight and Satellite Vehicles*, Newnes, 1957.
Burchett, Wilfred, and Purdy, Anthony, *Cosmonaut Yuri Gagarin: First Man in Space*, Panther, 1961.
Emme, Eugene, *A History of Space Flight*, Holt, Rinehart & Winston, 1965.
Shelton, William, *Soviet Space Exploration: the First Decade*, Arthur Barker, 1969.
Vladimirov, Leonid, *The Russian Space Bluff*, Tom Stacey, 1971.
Stoiko, Michael, *Soviet Rocketry: the First Decade of Achievement*, David & Charles, 1971.
Smolder, Peter, *Soviets in Space: the Story of the Salyut and the Soviet Approach to Present and Future Space Travel*, Lutterworth Press, 1973.
O'Neill, Gerard, *The Colonisation of Space*, in Physics Today, September 1974.
US Senate, *Soviet Space Programs: 1971-75*, Senate Committee on Aeronautical and Space Sciences, US Government Printing Office, 1976.
O'Neill, Gerard, *The High Frontier: Human Colonies in Space*, Jonathan Cape, 1977.
Heppenheimer, T.A., *Colonies in Space*, Warner Books, 1978.

Reading list

Cooper, Henry, *A House in Space: the First True Account of the Skylab Experience*, Panther, 1978.
Baker, David, *The History of Manned Space Flight*, Cavendish, 1981.
Furniss, Tim, *The Story of the Space Shuttle*, Hodder & Stoughton, 1982.
Shapland, David and Rycroft, Michael, *Spacelab: Research in Earth Orbit*, Cambridge University Press, 1984.
Furniss, Tim, *Spaceflight: the Records*, Guinness, 1985.
Smith, Melvyn, *Space Shuttle*, Foulis-Haynes, 1985.
US Senate, *Soviet Space Programs: 1976-80*, Senate Committee on Commerce, Science and Transportation, US Government Printing Office, 1985.
Furniss, Tim, *Manned Spaceflight Log*, Jane's, 1986.
Peebles, Curtis, *Guardians: Strategic Reconnaissance Satellites*, Ian Allan, 1987.
Bond, Peter, *Heroes in Space: from Gagarin to Challenger*, Blackwell, 1987.
Shkolenko, Yuri, *The Space Age*, Progress Publishing, Moscow, 1987.
Oberg, James and Oberg, Alcestis, *Pioneering Space: Living on the Next Frontier*, McGraw-Hill, 1987.
Lebedev, Valentin, *Diary of a Cosmonaut: 211 Days in Space*, Phytoresource Research, 1988.
Glushko, Valentin, *Soviet Cosmonautics: Questions and Answers*, Novosti, 1988.
Clark, Phillip, *The Soviet Manned Space Programme*, Salamander, 1988.
Harvey, Brian, *Race into Space: the Soviet Space Programme*, Ellis Horwood, 1988.
Gatland, Kenneth, *The Illustrated Encyclopedia of Space Technology*, Salamander, 1989.
Aldrin, Buzz, and McConnell, Malcolm, *Men from Earth*, Bantam, 1989.
Spangerburg, Ray, and Moser, Diane, *Space Exploration: Opening the Space Frontier*, Facts-on-File, 1989.
Spangerburg, Ray, and Moser, Diane, *Space Exploration: Living and Working in Space*, Facts-on-File, 1989.
Asimov, Isaac, *Beginnings: the Story of Origins of Mankind, the Earth, the Universe*, Berkley, 1989.
Hooper, Gordon, *The Soviet Cosmonaut Team*, GRH Publications, 1990.
Baker, James, *Planet Earth: the View from Space*, Harvard University Press, 1990.
Newkirk, Dennis, *Almanac of Soviet Manned Space Flight*, Gulf Publishing Company, 1990.
Rycroft, Michael (ed.), *The Cambridge Encyclopedia of Space*, Cambridge University Press, 1990.
Calder, Nigel, *Spaceship Earth*, Channel 4 Books, 1991.
Semenov, Yuri, Rymin, Valeri, Popov, Viktor, Pivnyuk, Vladimir, Gilberg, Lev, Novokshchenov, Nikolai, and Rebov, Mikhail, *Cosmonautics 1991*, Cosmos Books, 1992.
NASA, *Space Station Freedom Media Handbook*, NASA, 1992.
Sharman, Helen, and Priest, Christopher, *Seize the Moment: an autobiography of Britain's first Astronaut*, Victor Gollancz, 1993.
Bond, Peter, *Reaching for the Stars: the Illustrated History of Manned Spaceflight*, Cassell, 1993.

Matson, Wayne, *Cosmonautics: a Colorful History*, Cosmos Books, 1994.
Pivynuk, Vladimir, and Bockman, Mark, *Space Station Handbook: Mir User's Manual*, Cosmos Books, 1994.
Johnson, Nicholas, *The Soviet Reach for the Moon*, Cosmos Books, 1994.
US Senate, *US–Russian Coooperation in Space*, US Congressional Office of Technology Assessment, US Government Printing Office, 1995.
Jenkins, Dennis, *Space Shuttle: the History of Developing the National Space Transportation System*, Jenkins, 1996.
Schmidt, Stanley, and Zubrin, Robert (eds.), *Islands in the Sky: Bold new ideas for colonizing space*, Wiley, 1996; articles from Analog magazine.
Bizony, Piers, *Islands in the Sky: Building the International Space Station*, Aurum Press, 1996.
Harvey, Brian, *The New Russian Space Programme: From Competition to Collaboration*, Wiley-Praxis, 1996.
Harford, James, *Korolev*, Wiley, 1997.
Burrough, Bryan, *Dragonfly: NASA and the Crisis Aboard Mir*, Perennial, 2000.
Linenger, Jerry, *Off the Planet: Surviving Five Perilous Months Aboard the Space Station Mir*, McGraw-Hill, 2000.
Siddiqi, Asif, *Challenge to Apollo*, NASA, 2000.
Foale, Colin, *Waystation to the Stars*, Trafalgar Square, 2000.
Hall, Rex, and Shayler, David, *The Rocket Men: Vostok & Voskhod – the first Soviet manned spaceflights*, Springer-Praxis, 2001.
Linenger, Jerry, *Letters from MIR: An Astronaut's Letters to His Son*, McGraw-Hill, 2002.
Oberg, James, *Star-Crossed Orbits: Inside the U.S.-Russian Space Alliance*, McGraw-Hill, 2002.
Hall, Rex, and Shayler, David, *Soyuz – A Universal Spacecraft*, Springer-Praxis, 2003.
Zimmerman, Robert, *Leaving Earth*, Joseph Henry, 2003.

Index

Afanasayev, Viktor, 182, 202, 203, 205, 207, 227, 228, 278–281, 289, 290, 292, 295
Afghanistan, 108, 171, 172, 306, 309
Akers, Tom, 256, 257
Akiyama, Toehiro, 202, 203, 207, 289
Aksyonov, Vladimir, 46, 47, 90, 91, 101
Alexandrov, Alexander, 118, 119, 121, 122, 128, 138, 143, 160–166, 168, 289, 306, 307
Alexandrov, Alexander (Bulgarian), 168, 169, 289
Almaz space station, 9–12, 22, 26, 27, 29, 31, 32, 34, 51, 54, 102, 103, 163
  see Salyut 2, Salyut 3, Salyut 5
Anderson, Walt, 281
André-Deshays, Claudie, 255, 256, 281, 290
Andreyev, Boris, 88
Androgynous Peripheral Assembly System (APAS), 197, 220, 221, 227, 233, 234, 239, 240, 317
Apollo spacecraft (US), 298
  Apollo 8, 298
  Apollo 11, 11
  Apollo Soyuz Test Project (ASTP), 23, 26, 31, 41, 46, 197, 221, 298
arabidopsis plant, 78, 97, 107, 113, 116, 317
Argon 16 flight computer, 33, 88, 90, 95, 107, 108, 117, 142, 195, 298
Artsebarski, Anatoli, 206–211, 214, 289, 292
Artyukhin, Yuri, 22, 27, 28, 30, 32
Atkov, Oleg, 122, 125, 129, 138, 162, 307, 308
Atlantis, space shuttle, 226, 227, 234, 236, 238, 239, 242–244, 247–251, 256, 257, 258, 261, 264, 268, 271, 274
Aubakirov, Takhtar, 209, 211, 228, 289
Austria, 209
Avdeyev, Sergei, 217, 219, 220, 221, 245, 246, 247, 249, 250, 277, 278, 279, 280, 289, 290, 292, 293, 294, 295

Baikonur Cosmodrome, 3, 4, 228
Baker, Ellen, 240
Baker, Mike, 260, 261
Balandin, Alexander, 181, 190, 194–196, 199–201, 289, 292
Bania shower, 28, 97, 106, 192, 236
Baturin, Yuri, 277, 285, 290
Bean, Al, 308
Belgium, 238
Belitski, Boris, 129
Bella, Ivan, 278, 279, 290
Beregovoi, Georgi, 13, 65, 94
Berezovoi Anatoli, 49, 107, 110–115, 122, 138, 139, 200, 307, 308, 314
Big Bird satellites (US), 27
Blagonrovov, Anatoli, 15
Blagov, Viktor, 73, 114, 161, 163, 165, 168, 174, 182, 186, 187, 195, 196, 198, 222, 225, 272, 273, 275, 283
Blaha, John, 256–261, 290, 311
Boeing, 225, 226
Borman, Frank, 1, 5, 308
Brand, Vance, 41
Budarin, Nikolai, 240–242, 244, 245, 274, 275, 277, 283, 290, 294, 295

Bulgaria, 65, 81, 83, 168, 169, 173
Buran space shuttle, 126, 158, 164, 172, 183, 184, 192, 197, 204, 209, 239, 309, 317–318
Burnett, Mark, 283
Bushuyev Konstantin, 41
Bykovsky, Valeri, 1, 21, 46, 47, 73, 101, 307, 308
Byurakan Observatory, Crimea, 25, 127, 157

Cameron, Kenneth, 247, 248
Canada, 248, 253
cargo–tanker spacecraft, *see Progress*
Carr, Jerry, 308
Chelomei Bureau, 34, 102, 103, 116, 118, 159
Chelomei, Vladimir, 9, 13, 22, 23, 51, 297, 299, 300
Chilton, Kevin, 250
China, 250
Chrétien, Jean-Loup, 108–110, 138, 176–178, 190, 271, 289, 291, 314
Chukhrov Feodor, 15
Clark Phillip, 21, 31, 39
Clifford, Rich, 250, 294
Coca Cola, 209
Collins, Eileen, 233
Columbia, space shuttle, 318
Conrad, Pete, 308
Cooper, Gordon, 308
Cooperative Solar Array, 226, 249, 254, 256, 260, 282
Cosmos satellites, 43, 87
  Cosmos 110, 5
  Cosmos 496, 21
  Cosmos 557, 23, 32, 305
  Cosmos 613, 25
  Cosmos 573, 26
  Cosmos 869, 87
  Cosmos 881, 102
  Cosmos 882, 102
  Cosmos 929, 102, 103
  Cosmos 997, 102
  Cosmos 998, 102
  Cosmos 1001, 87
  Cosmos 1074, 87
  Cosmos 1100, 102
  Cosmos 1101, 102
  Cosmos 1267, 100, 103, 105, 116, 120
  Cosmos 1443, 116–120, 134, 137, 158, 302
  Cosmos 1669, 133, 137, 149, 323
  Cosmos 1686, 134, 136, 137, 141, 148, 149, 151, 154, 303
  Cosmos 1700, 148, 152
  Cosmos 1897, 156, 232
  Cosmos 2054, 190, 232
Crimean Astrophysical Observatory, 34, 41
Cuba, 65, 92
Culbertson, Frank, 257
Czechoslovakia, 65

Dauran, Mohammad, 172
Delta navigational system, 33, 51, 105, 107, 115, 116, 118, 119, 142
Demin, Lev, 30, 31
Dezhurov, Vladimir, 235, 236, 237, 240, 241, 244, 290, 293, 294
Discovery, space shuttle, 227, 233, 234, 240, 275, 276, 277, 284
Dobrokvashina, Yelena, 136
Dobrovolsky, Georgi, 15, 19–21, 308
Docking Module (DM), 244–251, 254, 259, 260, 263, 272, 280
Dunayev, Alexander, 172, 180
Dunbar, Bonnie, 227, 240, 242
Dyakonov, Robert, 68, 76
Dzhanibekov, Vladimir, 59, 60, 97, 101, 108, 110, 126–128, 130–134, 138, 139, 144, 149, 174, 190, 302, 306

Elektron apparatus, 157, 188, 261–265, 267
Endeavour, space shuttle, 274
Energiya Bureau, 159, 195, 203, 239, 247, 278, 280, 281, 283, 284
Energiya rocket, 159, 209, 299
Energomash factory, 225
European Space Agency, 210, 215, 220, 229, 235, 238, 242, 245, 249, 250, 261, 309
Ewald, Reinhold, 261, 262, 278, 290
Eyharts, Léopold, 269, 274, 290

Faris, Mohammed, 160–162, 289, 306
Farkas, Bertalan, 89, 90, 101
Feoktistov, Konstantin, 8, 15, 19, 43, 72, 78, 87, 88, 95, 309
Filipchenko, Anatoli, 13
fire, on Salyut 1, 18; on Salyut 7, 114; on Mir, 231, 261

Flade, Klaus-Dietrich, 215, 216, 224, 227, 289, 307
Foale, Michael, 263–265, 268, 269, 270, 290, 295
France, 60, 81, 108, 123, 129, 176, 217, 220, 223, 238, 245, 255, 269, 274, 278, 309
Freedom space station (US), 159, 210, 215, 220, 224
French Space Agency, 178

Gagarin, Yuri, 1, 13, 19, 182, 205, 308
Garriott, Owen, 308
Gazenko, Oleg, 67, 76
Gemini spacecraft (US), 1;Gemini 7, 1, 308
Gerasimov, Nikolai, 159
German Democratic Republic, 46, 65, 73, 151
German Space Agency, 220
Germany (unified), 215, 216, 225, 227, 236, 245, 248, 250, 261, 273
Gibson, Ed, 308
Gibson, Robert, 240, 241
Gidzenko, Yuri, 245, 246, 248–250, 284, 290, 294
Glazkov, Yuri, 48, 49
Globus positional indicator, 33
Glushko, Valentin, 42
Godwin, Linda, 250, 294
Goldin, Daniel, 240, 249, 270, 281, 282
Gorbachev, Mikhail, 209, 213
Gorbatko, Viktor, 48, 49, 91, 92, 101, 307
Gorshkov, Leonid, 195
gravity gradient, 28, 69, 73, 80, 95
Grechko, Georgi, 36–38, 44, 56, 57, 58, 60, 62, 64–69, 72, 73, 101, 102, 109, 133, 134, 138, 166, 306, 307, 308, 312
Grigoriev, Anatoli, 175
Grunsfeld, John, 261
Gubanov, Boris, 159
Gubarev Alexei, 36, 37, 38, 44, 65, 66, 67, 101
Gurovsky, Nikolai, 76
Gurragcha, Jugderdemidiyin, 97, 101
gyrodyne attitude control system, 29, 34, 51, 157, 158, 190, 196, 217, 222, 227, 231, 236, 245, 249, 250, 257, 260, 261, 263, 267, 280, 281, 283

Hadfield, Chris, 247, 248

Haigneré, Jean-Pierre, 223, 224, 278, 279, 280, 281, 289, 290, 295
Halsell, James, 247
Helms, Susan, 284
Hermaszewski, Miroslaw, 70, 71, 101
Heyerdahl, Thor, 64
Holloway, Tommy, 249
Hungary, 65, 83, 89

Igla rendezvous system, 52, 117, 144, 154, 156, 164, 196, 302
India, 123
Inspektor spacecraft, 273
integrated propulsion system (ODU), 51, 78, 80, 81, 150, 206, 298
Intercosmos Organisation, 55, 65, 70, 74, 83, 87, 96, 99, 108, 123, 151, 160, 163, 168, 171, 209, 251, 312
International Space Station, 226, 228, 242, 244, 245, 250, 251, 254, 258, 273, 275, 276, 278, 281, 283, 284, 319, 299, 318
Israel, 221
Italy, 282
Ivanchenkov, Alexander, 56, 68, 69, 70, 72–74, 76, 77, 101, 102, 108, 109, 110, 138, 307, 308
Ivanov, Georgi, 81, 82, 168
Ivanova, Yekaterina, 136

Jähn, Sigmund, 73, 74, 75, 101
Japan, 202, 250, 258, 278, 282
Jett, Brent, 260, 261

Kaleri, Alexander, 209, 215–218, 228, 255, 257, 260–262, 279, 281–283, 289, 290, 292, 294, 295, 306
Kaliningrad control centre, 23, 33, 41, 55–57, 66, 69, 73, 75, 76, 81, 83, 85, 86, 111, 115, 124, 132, 134, 174, 178, 198–200, 214, 223, 234, 240, 254, 263, 267, 274, 275, 280, 284, 302
Kaskad attitude control system, 33, 40, 51, 60, 65, 69, 106, 107
Kazakhstan, 4, 6, 202, 228, 285
Kazbek couch liner, 61, 241, 251, 256
Keldysh, Mstislav, 9
Kennedy, John, 9, 270
Kerimov, Kerim, 186

Kerwin, Joe, 308
Khrunichev factory, 12, 22, 23, 182, 198
Khrunov, Yevgeni, 12, 22, 56
Kizim, Leonid, 95, 101, 122–125, 127–130, 138, 143, 145, 147–151, 160, 162, 289, 308
Klimuk, Pyotr, 25, 26, 40, 41, 42, 44, 64, 65, 70, 71, 101
Komarov, Vladimir, 7, 8, 11, 19, 309
Kondakova, Yelena, 229, 230, 231, 235, 256, 264, 290
Konus docking assembly, 185, 190, 237, 238, 267, 268, 270, 271, 272
Koptev, Yuri, 213, 240, 259, 277, 281, 283
Korolev Bureau, 32, 34, 51, 102, 159, 299, 303, 309
Korolev, Sergei, v, 2, 9, 11, 297, 298, 299, 300
Korzun, Valeri, 255, 257, 260–262, 290, 294, 295, 306
Kovalyonok, Vladimir, 55, 68, 69, 70, 72, 73, 74, 76, 77, 96–99, 101, 102, 306–308
Kozesnik, Jaroslav, 65
Krikalev, Sergei, 176, 177, 180, 182, 183, 186, 206, 207–209, 211, 213, 214, 216, 227, 284, 289, 292, 306, 307, 309, 315
Kristall space station module, 196, 197, 200, 203–205, 207, 220, 221, 223, 224, 226, 227, 229, 232–234, 236, 237, 239, 240, 248, 250, 259, 260, 268, 272, 273, 280, 286, 287, 304, 305, 310, 317
KRT radio telescope, 84–86
Kubasov, Valeri, 16, 23, 41, 61, 89, 90, 101, 306, 307
Kurs rendezvous system, 144, 150, 153, 156, 169, 174, 184–187, 191, 196, 200, 204–208, 211, 215, 217, 220–222, 228, 229, 232, 260, 262, 274, 277, 282, 283
Kvant space station modules,
  Kvant 1, 154–158, 167, 168, 170, 183, 185, 186, 188, 190, 195, 196, 198, 203–205, 208, 211, 214, 216, 217, 219, 222, 223, 225, 229, 236, 247, 250, 253, 254, 262–265, 272, 282, 285, 303, 304, 305, 310
  Kvant 2, 186–190, 192, 194, 196, 197, 199, 200, 203, 204, 207, 214, 216, 217, 222, 226, 236–238, 244, 245, 248, 254, 259, 262, 263, 270, 271, 273, 280, 286, 303, 304, 305, 310
  Kvant 3, see Kristall
  Kvant 4, see Spektr
  Kvant 5, see Priroda
Lake Tengiz, 47
Laveikin, Alexander, 153–156, 158, 159–162, 164, 289, 291, 306, 307, 314, 315
Lawrence, Wendy, 270, 271
Lazarev, Vasili, 24, 38, 39, 40
Lazutkin, Alexander, 261, 263, 265, 269, 290
Lebedev, Valentin, 25, 26, 88, 89, 107, 110–115, 122, 138, 139, 306, 307, 314
Leonov, Alexei, 16, 23, 41, 48, 56, 61, 181, 190
Levchenko, Anatoli, 164–166, 289, 309
Linenger, Jerry, 260, 261, 263, 264, 270, 290, 295
Lisun, Mikhail, 49
Ljappa swingarm, 181, 187, 197, 237, 268, 303
Lousma, Jack, 308
Lovell, James, 1, 5, 308
Luch geostationary relay satellites, 57, 147, 148, 150, 152, 156, 190, 194, 214, 232, 314
Lucid, Shannon, 251, 253–258, 290, 312, 314
Lyakhov, Vladimir, 79–87, 101, 102, 118–122, 128, 138, 139, 171–174, 178, 228, 289, 308

Makarov, Oleg, 24, 38–40, 59, 95, 101
Malenchenko, Yuri, 228, 229, 231, 290, 293
Maleshko, Ganna, 196
Malyschev, Yuri, 90, 91, 101, 108, 123, 138
Manakov, Gennadi, 190, 200, 201, 203, 221–224, 228, 255, 289, 292, 293, 306
Manarov, Musa, 164–179, 202–207, 214, 216, 289, 291, 292, 308, 309, 311
Manber, Jeffrey, 281
Manned Orbiting Laboratory (US), 11, 27
McArthur, William, 248
Merbold, Ulf, 229–231, 246, 290
Meteor weather satellites, 18, 30, 315
Meteor-Priroda satellite, 119

Mir space station, 141–295, 299
  atmospheric regulation,
    see Elektron, Vika, Vozdukh
  build-up, 181, 302–305
  collisions (potential and actual), 155, 204,
      227, 228, 229, 240, 265, 266
  communications,
    see SDRN
  computer complex,
    see Argon 16, Salyut 5B
  de-orbited, 284, 288
  fee-paying visitors, 180, 202, 207, 218,
      221, 226, 229, 248, 259, 279, 310
  follow-on (Mir 2) proposal, 159, 214, 225,
      238
  legacy, 318, 319
  multiple docking adaptor,
    see Konus
  propulsion,
    see integrated propulsion system
        (ODU)
  rendezvous system,
    see Igla, Kurs
  suggestion that it be sold to NASA, 209
MirCorp, 281, 282, 283, 284
Mishin, Vasili, 12
Mitterrand, François, 176
MMU astronaut manoeuvring unit, 192
Mohmand, Abdul Ahad, 172–174, 178, 289,
    306
Molniya satellites, 29, 57
Mongolian People's Republic, 65, 97
Moscow Time, xxv, 311
Musabayev, Talget, 228–231, 274–277, 285,
    290, 293, 295

N-1 rocket, 11
National Aeronautics & Space
    Administration (NASA), 9, 18, 209,
    215, 220, 221, 224, 225, 228, 231,
    235, 239, 240, 242, 248, 250, 259,
    269, 270, 272, 277, 284, 309, 310,
    312, 318
Nikolayev, Andrian, 1, 2, 5, 9, 13, 16, 32, 308

Olesyuk, Boris, 198, 205
O'Neill, Gerard, 317
Onufrienko, Yuri, 249, 250, 253, 254, 256,
    290, 294

Orbiter Docking System (ODS), 238, 239,
    242, 247

Padalka, Gennadi, 277, 278, 279, 283, 290,
    295
Pallo, Vladimir, 159
Parazynski, Scott, 271, 295
Paton, Boris, 128
Patsayev, Viktor, 16, 19–21, 308
Pavlov, Vasili, 9
Payload Systems Inc., 189, 194, 213, 225, 227
Penguin suit, 37, 68, 76, 106
Pepsi Co., 254
Perseid meteor shower, 225
Petrov, Boris, 13, 19, 21, 22, 42, 65, 68
Plesetsk Cosmodrome, 213
Pogue, Bill, 308
Poland, 65, 70, 74
Poleshchuk, Alexander, 221–224, 289, 293
Poliakov, Valeri, 164, 171, 172, 174, 175,
    178–181, 183, 214, 227–232, 235,
    280, 289, 290, 307–310, 312, 315
Polyus spacecraft, 159, 209
Popov, Leonid, 88, 91–94, 99, 101, 113, 138
Popovich, Pavel, 22, 27, 28, 30, 32
Precourt, Charles, 240, 275, 276
Primakov, Yevgeni, 279
Priroda space station module, 198, 249–255,
    258, 259, 268, 274, 275, 280, 287,
    305

Progress cargo/tanker spacecraft, 32, 34, 43,
    51, 52, 62, 105, 106, 158, 299, 300,
    301
  model-M, 184
  Progress 1, 62, 100, 323
  Progress 2, 71, 73, 100, 323
  Progress 3, 73, 74, 100, 323
  Progress 4, 76, 78, 79, 100, 323
  Progress 5, 80, 81, 100, 323
  Progress 6, 83, 100, 323
  Progress 7, 84, 85, 87, 100, 323
  Progress 8, 88, 89, 100, 323
  Progress 9, 89, 100, 323
  Progress 10, 91, 100, 323
  Progress 11, 93–96, 100, 323
  Progress 12, 96, 97, 100, 323
  Progress 13, 107, 108, 137, 323
  Progress 14, 111–113, 137, 323

Progress cargo/tanker spacecraft, *continued*
  Progress 15, 114, 137, 323
  Progress 16, 116, 137, 323
  Progress 17, 120, 137, 323
  Progress 18, 121, 122, 137, 323
  Progress 19, 123, 137, 323
  Progress 20, 124, 125, 137, 323
  Progress 21, 125, 133, 137, 323
  Progress 22, 125, 126, 137, 323
  Progress 23, 129, 137, 323
  Progress 24, 133, 137, 323
  Progress 25, intended, 133
  Progress 25, actual, 147, 148, 285, 323
  Progress 26, 148, 285, 324
  Progress 27, 152, 153, 154, 285, 324
  Progress 28, 154, 285, 324
  Progress 29, 158, 285, 324
  Progress 30, 158, 160, 285, 324
  Progress 31, 162, 285, 324
  Progress 32, 162, 163, 285, 324
  Progress 33, 163, 164, 285, 324
  Progress 34, 166, 167, 285, 324
  Progress 35, 167, 168, 285, 324
  Progress 36, 168, 285, 324
  Progress 37, 171, 176, 286, 324
  Progress 38, 174, 176, 286, 324
  Progress 39, 179, 180, 286, 324
  Progress 40, 180, 182, 286, 324
  Progress 41, 182, 183, 286, 324
  Progress 42, 196, 286, 324
  Progress-M 1, 184, 185, 187, 286, 324
  Progress-M 2, 185, 189, 190, 194, 286, 324
  Progress-M 3, 195, 196, 286, 324
  Progress-M 4, 201, 286, 324
  Progress-M 5, 201, 286, 324
  Progress-M 6, 203, 204, 286, 324
  Progress-M 7, 204, 206, 286, 324
  Progress-M 8, 207, 208, 286, 324
  Progress-M 9, 209, 210, 286, 324
  Progress-M 10, 211, 213, 286, 324
  Progress-M 11, 213, 215, 286, 324
  Progress-M 12, 216, 217, 286, 324
  Progress-M 13, 217, 218, 286, 324
  Progress-M 14, 219–221, 286, 324
  Progress-M 15, 220, 221, 287, 324
  Progress-M 16, 222, 287, 324
  Progress-M 17, 222–225, 287, 324
  Progress-M 18, 222–224, 287, 324
  Progress-M 19, 224, 225, 287, 324
  Progress-M 20, 225–227, 287, 324
  Progress-M 21, 227, 228, 287, 324
  Progress-M 22, 228, 287, 324
  Progress-M 23, 228, 287, 324
  Progress-M 24, 228–230, 232, 235, 287, 324
  Progress-M 25, 231, 235, 287, 324
  Progress-M 26, 235, 287, 324
  Progress-M 27, 236, 237, 287, 324
  Progress-M 28, 245, 287, 324
  Progress-M 29, 246, 249, 287, 325
  Progress-M 30, 249, 287, 325
  Progress-M 31, 252, 255, 287, 325
  Progress-M 32, 255, 256, 287, 325
  Progress-M 33, 258, 260, 262, 288, 325
  Progress-M 34, 263, 265–267, 288, 325
  Progress-M 35, 267–269, 271, 288, 325
  Progress-M 36, 271–273, 288, 325
  Progress-M 37, 273, 274, 288, 325
  Progress-M 38, 274, 275, 288, 325
  Progress-M 39, 275, 277, 278, 288, 325
  Progress-M 40, 278, 279, 288, 325
  Progress-M 41, 279, 280, 288, 325
  Progress-M 42, 280, 281, 288, 325
  Progress-M 43, 283, 288, 325
  Progress-M1 1, 281, 282, 288, 325
  Progress-M1 2, 282, 283, 288, 325
  Progress-M1 3, 325
  Progress-M1 4, 325
  Progress-M1 5, 283, 284, 288, 325
Pronina, Irina, 117, 126
Proton rocket, 9, 11–13, 21, 23, 102, 116, 134, 141, 154, 187, 196, 297, 298, 299
Prunariu, Dumitru, 99, 101

R-7, *see Semyorka*
R-Bar rendezvous, 240
Raduga descent capsule, 163, 201, 204, 209, 213, 220, 221, 223, 225, 226, 228
Rapana girder, 225, 229, 254, 260, 274
Readdy, Bill, 256, 257
Reiter, Thomas, 245–250, 256, 290, 294, 314
Remek, Vladimir, 65–67, 101
Romanenko, Yuri, 56–58, 62, 64–69, 92, 93, 101, 102, 121, 153–155, 158, 160–166, 171, 172, 178, 179, 204, 289, 291, 306–308, 311, 314
Romania, 65
Ross, Jerry, 248

Rozhdestvensky, Valeri, 47
Rukavishnikov Nikolai, 14, 15, 81, 82, 123, 307
Russian Space Agency, 213, 216, 220, 224, 228, 240, 250, 254, 259, 277, 278
Ryumin, Valeri, 55, 79–81, 83–89, 91–94, 101, 102, 118, 151, 155, 173, 182, 229, 270, 275, 306, 308, 312

Sagdev, Roald, 120
Salyut 5B flight computer, 195, 196, 241, 242
Salyut Bureau, 159
Salyut space stations,
    Salyut 1, 13–20, 21, 24, 25, 27, 28, 33, 82, 300, 301, 305, 306
    'Salyut 2' intended, 21
    Salyut 2 actual, 22, 23, 26, 51, 305
    Salyut 3, 27–32, 34, 45, 51, 102, 300, 301, 305
    Salyut 4, 32–44, 45, 51, 52, 57, 62, 64, 65, 84, 88, 107, 259, 298, 300, 300, 301, 305
    Salyut 5, 44–49, 51, 102, 301, 305
    Salyut 6, 51–102, 105–108, 120, 128, 153, 166, 168, 298, 301, 302, 305–308, 312
    Salyut 7, 102, 105–139, 141, 143, 148–152, 154, 158, 160, 252, 299–302, 305–308, 310
Sarafanov, Gennadi, 30, 31
Satellite Data Relay Network (SDRN), *see Luch*
Savinykh, Viktor, 96–99, 101, 130–134, 138, 144, 168–170, 182, 289, 306
Savitskaya, Svetlana, 113, 114, 126–128, 136, 138, 139, 149, 156, 182, 309
Searfoss, Richard, 250
Semenov, Yuri, 194, 196, 278, 279, 283
Semyorka rocket, 2, 126, 297, 299
Sensenbrenner, James, 276
Serebrov, Alexander, 113, 117, 138, 153, 164, 181, 185–195, 224–227, 289, 291, 293
Sevastyanov, Vitali, 1, 2, 5, 9, 16, 22, 40–42, 44, 64, 65, 182, 202, 308
Severin, Gai, 192
Sharipov, Salizhan, 283
Sharma, Rakesh, 123, 138
Sharman, Helen, 206, 207, 289

Shatalov, Vladimir, 14, 15, 21, 31, 47, 49, 57, 67, 79, 106, 143, 161, 162, 183, 190
Shchukin, Alexander, 164, 309
Shepherd, William, 284
shower, *see Bania*
Shuttle–Mir programme, 226, 227, 233–276, 277
Skylab space station (US), 22, 23, 34, 42, 58, 65, 69, 238, 306, 308, 309, 312
Slayton, Deke, 41
Sovakia, 278
Smolders, Peter, 32
Sofora girder, 208, 213, 214, 219, 226, 254, 260, 274, 274, 280
Sokol pressure suit, 61, 95, 174, 241, 251, 256, 264
Solovyov, Anatoli, 168, 169, 190, 194, 195, 196, 199–201, 209, 217, 219–221, 228, 240–242, 244, 245, 268–274, 289–295
Solovyov, Vladimir, 122–130, 138, 139, 143, 145, 147–151, 160, 162, 168, 269, 270, 289, 308
South Atlantic Anomaly, 81, 93, 168, 183
Soyuz spacecraft, 4–8, 11
    as a lifeboat for NASA, 224, 318
    assessment, 297–299
    de-orbit procedure, 6
    landing procedure, 7, 8
    launch procedure, 2, 3, 4, 38, 39
    model-T, 26, 87, 105, 108, 114, 121, 122, 150, 298, 299, 302, 307
    model-TM, 144, 150, 299
    Soyuz 1, 7, 11, 321
    Soyuz 2, 321
    Soyuz 3, 13, 321
    Soyuz 4, 12, 14, 56, 321
    Soyuz 5, 12, 14, 56, 321
    Soyuz 6, 41, 321
    Soyuz 7, 13, 14, 16, 41, 321
    Soyuz 8, 14, 41, 321
    Soyuz 9, 1–9, 16, 19, 30, 308, 321
    Soyuz 10, 14, 19, 82, 321
    Soyuz 11, 15, 16, 19, 38, 88, 298, 300, 306, 307, 321
    Soyuz 12, 24, 26, 321
    Soyuz 13, 25, 298, 321
    Soyuz 14, 27, 28, 30, 32, 321
    Soyuz 15, 30, 31, 44, 300, 321

Soyuz spacecraft, *continued*
  Soyuz 17, 36, 38, 42, 44, 321
  'Soyuz 18' intended, 38, 39
  Soyuz 18 actual, 40, 42–44, 46, 321
  Soyuz 19, 41, 321
  Soyuz 20, 43, 44, 62, 84, 321
  Soyuz 21, 44, 46, 49, 321
  Soyuz 22, 46, 53, 298, 321
  Soyuz 23, 47, 48, 88, 321
  Soyuz 24, 48, 49, 321
  Soyuz 25, 55, 56, 61, 65, 68, 79, 100, 306, 321
  Soyuz 26, 56, 61, 68, 100, 321
  Soyuz 27, 58, 60, 61, 62, 67, 100, 321
  Soyuz 28, 65, 100, 321
  Soyuz 29, 68, 74, 75, 100, 321
  Soyuz 30, 70, 100, 321
  Soyuz 31, 73–75, 77, 78, 100, 322
  Soyuz 32, 79, 80, 82, 83, 100, 322
  Soyuz 33, 81, 83, 87, 99, 168, 307, 322
  Soyuz 34, 83–85, 87, 100, 200, 322
  Soyuz 35, 89, 90, 100, 322
  Soyuz 36, 89, 90, 92, 100, 322
  Soyuz 37, 91, 100, 322
  Soyuz 38, 92, 100, 322
  Soyuz 39, 97, 98, 100, 322
  Soyuz 40, 99, 100, 322
  Soyuz-T 1, 87, 100, 322
  Soyuz-T 2, 90, 95, 100, 322
  Soyuz-T 3, 95, 96, 100, 322
  Soyuz-T 4, 96, 98, 100, 306, 322
  Soyuz-T 5, 107, 109, 111, 114, 121, 137, 322
  Soyuz-T 6, 108, 109, 110, 130, 137, 322
  Soyuz-T 7, 113, 114, 117, 121, 126, 137, 322
  Soyuz-T 8, 116–118, 120, 130, 164, 302, 322
  Soyuz-T 9, 118, 120, 121, 137, 322
  'Soyuz-T 10' intended, 121, 164, 200
  Soyuz-T 10 actual, 122, 124, 137, 322
  Soyuz-T 11, 123, 124, 127, 129, 137, 322
  Soyuz-T 12, 126, 137, 322
  Soyuz-T 13, 130–132, 134, 136, 137, 322
  Soyuz-T 14, 133, 135, 137, 141, 176, 322
  Soyuz-T 15, 137, 143, 144, 148–151, 158, 285, 310, 322
  Soyuz-TM 1, 150, 285, 322
  Soyuz-TM 2, 155, 160, 162, 285, 322
  Soyuz-TM 3, 160–162, 165, 285, 322
  Soyuz-TM 4, 164, 166, 170, 285, 309, 322
  Soyuz-TM 5, 168–170, 173, 178, 285, 286, 322
  Soyuz-TM 6, 171, 174, 175, 286, 309, 322
  Soyuz-TM 7, 176, 179, 183, 286, 322
  Soyuz-TM 8, 182, 184–187, 189, 190, 195, 286, 322
  Soyuz-TM 9, 186, 188, 193–196, 198, 201, 286, 322
  Soyuz-TM 10, 200, 202, 203, 286, 322
  Soyuz-TM 11, 202, 204, 205–207, 286, 322
  Soyuz-TM 12, 206–208, 211, 286, 322
  Soyuz-TM 13, 209, 211, 214–216, 228, 286, 322
  Soyuz-TM 14, 209, 215, 218, 286, 322
  Soyuz-TM 15, 217, 221, 286, 322
  Soyuz-TM 16, 221, 223, 224, 287, 322
  Soyuz-TM 17, 223, 227, 229, 287, 322
  Soyuz-TM 18, 225, 227, 228, 287, 322
  Soyuz-TM 19, 228, 231, 287, 322
  Soyuz-TM 20, 229, 232, 287, 322
  Soyuz-TM 21, 235, 242, 287, 323
  Soyuz-TM 22, 245, 250, 287, 323
  Soyuz-TM 23, 249, 256, 287, 323
  Soyuz-TM 24, 255, 287, 323
  Soyuz-TM 25, 261, 264, 267, 323
  Soyuz-TM 26, 269, 271, 274, 323
  Soyuz-TM 27, 274, 277, 323
  Soyuz-TM 28, 277–279, 323
  Soyuz-TM 29, 279, 281, 323
  Soyuz-TM 30, 281, 282, 323
  Soyuz-TM 31, 284, 285, 323
  Soyuz-TM 32, 285, 323
Space Transportation System (STS),
  STS-60, 227
  STS-63, 233, 234
  STS-71, 238, 287
  STS-74, 247, 248, 287
  STS-75, 255
  STS-76, 250, 271, 287
  STS-79, 256, 257, 287
  STS-81, 260, 288
  STS-84, 264, 288
  STS-86, 270, 272, 288
  STS-89, 273, 274, 288
  STS-91, 275, 276, 288
  STS-102, 284
Spacehab, 250, 257, 260, 261, 264, 276

Spacelab, 239, 240
spacewalking, 56, 72, 102, 111, 112, 122, 124, 125, 127, 128, 139, 157, 159, 160, 167, 168, 170, 175, 177, 178, 190, 192, 199, 203, 204, 207, 208, 214, 219, 221, 222–226, 229, 231, 236–238, 244, 246, 249, 250, 251, 254, 260, 263, 268–274, 278–280, 291–295, 304, 314, 315
Spektr space station module, 198, 231, 235–238, 241, 244, 245, 247, 248, 251, 252, 259, 265–272, 274, 276, 279, 282, 287, 305
Sputnik satellite, 2, 76, 163, 297
Stafford, Tom, 41
Stankiavicus, Rimantas, 183, 309
Steklov, Vladimir, 281
Strekalov, Gennadi, 95, 101, 117, 121, 123, 138, 164, 190, 200, 201, 203, 235, 236, 237, 241, 244, 289, 290, 292–294
Strela crane, 203, 222, 223, 225, 229, 236, 237, 254
Strombus girder, 254, 260, 274
Sultanov, Ural, 164
Syria, 136, 158, 160, 161, 306, 310
Syromiatnikov, Vladimir, 203

Tamayo Méndez, Arnaldo, 92, 93, 101
Tchibis suit, 37, 38, 66, 76, 84, 109, 178, 179, 250
Thagard, Norman, 227, 235, 236, 238, 240–242, 244, 246, 250, 254, 255, 290, 312
Thomas, Andy, 271, 274–276, 290
Tito, Dennis, 283, 284, 285
Titov, Gherman, 1, 308
Titov, Vladimir, 117, 118, 121, 130, 153, 164, 166–168, 170–179, 214, 216, 227, 228, 233, 271, 289, 291, 295, 302, 308, 309
TKS spacecraft, 10–13, 22, 31, 102, 103, 116, 120, 154, 155, 158, 187, 299, 303
Tognini, Michel, 217–219, 223, 289
Tokyo Broadcasting System, 202
TORU remote piloting system, 222, 229, 251, 262, 265, 267, 269, 274
Tsibliev, Vasili, 224–227, 261–263, 265, 269, 289, 290, 293, 295

Tuan, Pham, 91, 92, 101

Usachev, Yuri, 227, 228, 249, 250, 253, 254, 256, 284, 290, 294

V-Bar rendezvous, 240
Vasyutin, Vladimir, 130, 133, 134, 138, 150, 160, 176, 306, 307
VDU thruster block, 219, 274, 275
Verne, Jules, 64
Viberti, Carlo, 282
Viehböck, Franz, 209–211, 289
Vietnam, 91
Vika oxygen generator, 157, 231, 261–264, 267
Viktorenko, Alexander, 143, 160–162, 181, 185–187, 189–192, 194, 195, 215–218, 228–231, 235, 289, 290–292
Vinogradov, Pavel, 255, 268, 269, 271–274, 283, 290, 295
Volk, Igor, 126, 138, 164, 183, 309
Volkov, Alexander, 130, 133, 134, 138, 176–178, 180–183, 186, 209, 211, 213, 214, 216, 228, 289, 291, 292, 309
Volkov, Vladislav, 16, 19–21, 308
Volynov, Boris, 44–46, 48, 49
Voskhod 1, 8, 15, 309
Voss, James, 284
Vostok spacecraft, 1, 2, 27, 297, 308
Vozdukh air purifier, 157, 247, 263, 264, 267, 272

Walz, Carl,
Wetherbee, Jim, 233, 234, 270
weightlessness, 5, 16, 309, 310–313
Weitz, Paul, 308
wheat plant, 107, 255, 260, 317
Wilcutt, Terry, 256
Wisoff, Jeff, 261
Wolf, David, 270–274, 290, 295

Yegorov, Anatoli, 64, 67
Yegorov Boris, 8, 309
Yeliseyev, Alexei, 12, 14, 15, 48, 56, 62
Yeltsin, Boris, 213, 275, 277
Yevpatoria, control centre, 4, 7, 13, 23, 41, 85

YMK cosmonaut manoeuvring unit, 184, 187, 190–192, 249, 317

Zalyotin, Sergei, 279, 281–283, 290, 295

Zholobov, Vitali, 44–46, 48, 49
Znamya experiment, 221, 222, 279
Zond spacecraft, 297
Zudov, Vyacheslav, 47, 88

ERAU-PRESCOTT LIBRARY

Printing: Mercedes-Druck, Berlin
Binding: Stein+Lehmann, Berlin